Future of Healthcare in Asia

Rong Xiang Ng · Weng Marc Lim ·
Huey Fen Cheong
Editors

Future of Healthcare in Asia

Advancing Equity, Ethics and Technology in Asian Healthcare

Springer

Editors
Rong Xiang Ng
Universiti Malaya
Kuala Lumpur, Malaysia

University of Glasgow
Glasgow, United Kingdom

Huey Fen Cheong
Universiti Malaya
Kuala Lumpur, Malaysia

Weng Marc Lim
Sunway Business School
Sunway University
Selangor, Malaysia

ISBN 978-3-032-03182-2 ISBN 978-3-032-03183-9 (eBook)
https://doi.org/10.1007/978-3-032-03183-9

This Springer imprint is published by the registered company Springer Nature Switzerland AG
The registered company address is: Gewerbestrasse 11, 6330 Cham, Switzerland

Titles on the *Futures of Asia*

Future of Business in Asia: Exploring Pathways to Tomorrow's Business Landscape
Future of Cities in Asia: Future Direction of Urban Planning, Technology and Environmental Sustainability
Future of Economy in Asia: Geopolitical Diplomacy, Technological Revolutions and Sustainable Pathways
Future of Education in Asia: Sustainability, Technology and Internationalisation
Future of Healthcare in Asia: Advancing Equity, Ethics and Technology in Asian Healthcare
Future of Media in Asia: Artificial Intelligence, Digital Technology and Media Practice
Future of Work in Asia: Transforming Work Models, Workforce Dynamics and Workplace Technology

Foreword

As the Dean of the Faculty of Medicine at Universiti Malaya, I am honoured to introduce "The Future of Healthcare in Asia", a groundbreaking contribution to the "Futures of Asia" series. This book arrives at a pivotal moment when healthcare systems across Asia and the world are undergoing profound transformation, especially in the wake of the COVID-19 pandemic.

Throughout my career and leadership role, I have witnessed firsthand the transformative power of research and collaboration in addressing complex healthcare challenges. This book exemplifies the kind of multidisciplinary, forward-looking scholarship that is essential for addressing the challenges of our time while reflecting our faculty's commitment to innovation, inclusivity, and global impact.

Asia, with its extraordinary diversity, rapid growth, and complex challenges, offers a unique lens through which to examine the future of health and wellbeing. The healthcare ecosystem spans advanced technological hubs and remote rural communities, blending traditional healing practices with cutting-edge innovations like personalised medicine. This complexity demands a nuanced understanding that goes beyond conventional healthcare paradigms, requiring innovative approaches that integrate digital tools and technologies.

Perhaps most importantly, this book maintains a steadfast focus on equity and inclusion. The contributors understand that technological advancement alone cannot address the profound healthcare challenges facing Asia. Social structures, political will, and ethical frameworks are equally critical in determining who benefits from innovation and how healthcare systems evolve to meet the needs of all people, particularly those most vulnerable.

I still remember sitting with a group of young medical students in Manila, listening as they described their dreams for healthcare in their communities. Their vision was not centred on the latest technological marvel but on creating systems where no one falls through the cracks, where traditional wisdom and modern science work in harmony, and where healing happens in context. Their perspective, pragmatic yet ambitious, rooted yet forward-looking, is the spirit that animates this book.

No project of this magnitude is possible without the dedication and collaboration of many individuals and institutions. My sincere thanks go to all the authors, not only from Malaysia but also from Singapore, the Philippines, Indonesia, Vietnam, China, Thailand, Australia, and beyond, for their dedication and expertise. Their contributions have enriched this volume with diverse perspectives and innovative ideas.

A special word of appreciation is due to Dr. Ng Rong Xiang, the Editor-in-Chief of this volume, as well as the series editors, Distinguished Prof. Weng Marc Lim and Dr. Cheong Huey Fen, who have provided invaluable support throughout the publication process. Together, they have skilfully navigated the various chapters into a cohesive narrative, ensuring that this book not only addresses current challenges but also inspires meaningful action toward healthier futures for all.

As you read this book, I encourage you to reflect on the possibilities it presents. The future of healthcare in Asia is not predetermined, it is shaped by the choices we make today. For policymakers, this book offers evidence-based insights and innovative frameworks to guide decisions that balance technological advancement with social equity. For practitioners, it provides practical strategies to address complex health challenges while respecting cultural contexts. And for researchers, it identifies critical gaps and emerging opportunities for advancing healthcare knowledge and solutions.

By embracing complexity, fostering collaboration, and prioritising equity, we can work collectively toward futures that offer not just longer lives but better health and wellbeing for Asia's diverse populations. This book serves as both a call to action for meaningful change, equipping stakeholders with the tools they need to navigate the evolving landscape of healthcare in Asia.

Finally, I am proud to note that this book aligns closely with the themes explored during the 2025 Alliance of Academic Health Centers International (AAHCI) Southeast Asia Regional Meeting hosted by our faculty in Universiti Malaya, under the theme 'Future-Proofing Healthcare in Southeast Asia.' Both the conference and this volume address critical challenges, such as equitable access, technological innovation, and sustainable health systems that are essential for building resilient healthcare futures across the region. Together, these initiatives reflect our Faculty's commitment to advancing healthcare innovation and equity in Asia.

On behalf of the Faculty of Medicine at Universiti Malaya, I am proud to present "The Future of Healthcare in Asia." May it serve as a valuable resource for academics, policymakers, practitioners, and anyone committed to building healthier societies.

Professor Dato' Dr. Yang Faridah Abdul Aziz
MBBS (UM), MRad (UM), MMedEd (UM), FAMM, DIMP
Dean, Academic Professor and Consultant Radiologist
Faculty of Medicine
Universiti Malaya
Kuala Lumpur, Malaysia

Acknowledgements

The editors thank all of those who contributed to the completion of this book and the *Futures of Asia* series that comprises seven volumes, including this one.

We extend our deepest appreciation to the chapter authors for their hard work and kind cooperation throughout the process. We are also grateful to the individuals and organisations that helped disseminate the calls for papers (CFPs) like the Regional Studies Association (RSA), various academic research groups on social media (particularly Facebook), and our respective institutions, including our faculties and international relations offices. Special thanks go to our assistant, Fatin Najihah, for her invaluable support in the early stages of this project.

We are also deeply grateful for the constructive feedback provided by the peer reviewers of this book, as well as the opportunities, guidance and support from Springer's commissioning editor, Doris Bleier.

Additionally, we also acknowledge and thank the following scholars for their insightful feedback on individual chapters, both published and unpublished in this volume:

Reviewed 2 Chapters

1. Diane Chong Woei Quan (Ministry of Health, Malaysia)
2. Khin Thandar Aung (International Islamic University Malaysia, Malaysia)
3. Maria Christina Bernadette C. Aguba (Batangas State University, The National Engineering University, Philippines)
4. Matome M. Sekhotha (University of Limpopo, South Africa)
5. Md Abdul Jalil (Shahjalal University of Science and Technology, Bangladesh)
6. Muhammad Nouman Akhtar (University of Genova, Italy)
7. Pragya Dev (Indian Institute of Technology Roorkee, India)
8. Patricia Pawa Pitil (Universiti Teknologi MARA Sarawak, Malaysia)
9. Sanghoon Park (Kangwon National University, South Korea)

10. Tze Yan Lee (Universiti Kuala Lumpur, Malaysia)
11. Yujia Qiu (University of Yale, United States of America)

Reviewed 1 Chapter

1. Catherine Labrooy (Sayfol International School, Malaysia)
2. Chung Han Yang (Malaysian AIDS Council, Malaysia)
3. Erwin L. Rimban (Cagayan State University, Philippines)
4. Kok Soon Lee (Ministry of Health, Malaysia)
5. Marcos da Silva Cruz (Federal University of Pará, Brazil)
6. Mounir Baala (National School of Mines of Rabat, Morocco)
7. Sajad Jamalpour (Universiti Malaya, Malaysia)
8. Samuel Musungwini (Midlands State University, Zimbabwe)
9. S M Nazmuz Sakib (Scholars Academic and Scientific Society, Bangladesh)
10. Wenjing Huang (Independent scholar, USA)

** We also thank peer reviewers who opted for anonymity.*

Again, we extend our deepest gratitude for the support, time, and expertise of these individuals. Their contributions were instrumental in publishing this book. In case we have unintentionally overlooked any individual or organisation that contributed to this project, we express our sincere appreciation and gratitude as well.

Contents

List of Figures

List of Tables

About the Editors

Rong Xiang Ng is an Infectious Diseases physician and medical lecturer at Universiti Malaya (Malaysia) and holds an affiliate lecturer position at the University of Glasgow (UK). He is engaged in understanding and shaping the future of healthcare in Asia. His clinical and research work focuses on optimising clinical pathways and exploring innovative solutions, including digital tools, to address infectious disease challenges across diverse populations. Dr. Ng's contributions to hospital treatment guidelines and research on Aritificial Intelligence (AI) in HIV care and infectious diseases, multidrug-resistant organisms, COVID-19, and telemedicine in sexual health directly reflect the key themes of technological innovation and equitable access explored in this introductory chapter. As a Fellow of the Asia Pacific AIDS & Co-infections Conference (APACC) and an affiliate of the Centre of Excellence for Research in Infectious Diseases and AIDS (CERIA, Universiti Malaya), Universiti Malaya), his active involvement in research and clinical trials bridges the crucial gap between current clinical practice and the forward-looking perspectives presented throughout this volume on the future of healthcare in Asia.

Weng Marc Lim is a Distinguished Professor and the Dean of Sunway Business School at Sunway University, an ASU-Cintana Alliance Global Partner Affiliate Faculty at Arizona State University in the United States of America, and an Adjunct Professor at Swinburne University of Technology's home campus in Melbourne, Australia, and international branch campus in Sarawak, Malaysia, where he served as Dean and Head of School as well as on the University Council. Professor Lim is the Record Holder for "Youngest Professor" in the Malaysia Book of Records and features in the Academy of Sciences Malaysia's Top Research Scientists Malaysia and JCI's Ten Outstanding Young Malaysians. With over 100 publications in prestigious "A*" and "A" ranked journals, Prof. Lim's research has been cited more than 45,000 times. As a thought leader, Prof. Lim has led high-level discussions at AppliedHE, Association to Advance Collegiate Schools of Business (AACSB), Association of MBAs (AMBA), Business Graduates Association (BGA), Quacquarelli

Symonds (QS), Times Higher Education (THE), United Nations Educational, Scientific and Cultural Organization (UNESCO) and Principles of Responsible Management Education (UNPRME), Nikkei Forum and World Economic Forum (WEF). Professor Lim holds a Bachelor of Business and Commerce with First Class Honors and a Doctor of Philosophy from Monash University, and multiple certificates post-Ph.D., including a Graduate Certificate in Learning and Teaching (Higher Education) with High Distinction from Swinburne University of Technology and a Certificate in School Management and Leadership from Harvard University.

Huey Fen Cheong is a Senior Lecturer at Universiti Malaya, Malaysia. Her interests have been multidisciplinary, mainly linguistics, sociology, education, marketing, and psychology. Driven by a cultural and postcolonial perspective, she collaborated with a group of established scholars to initiate this *Futures of Asia* series (which contains this book) and the earlier *COVID-19 in Malaysia* series. These initiatives aim to promote Global South research and publications, especially on significant events such as the COVID-19 pandemic, post-pandemic normality and transformations, and the rise of AI, which equally matter to non-Western regions and deserve equal scholarly attention (and acceptance) beyond Western-centric frameworks in academia.

Editors' Note

The clusters of future studies in each chapter are based on Fergnani (2019, 2020), who identified six (6) key research/thematic clusters from reviews of future studies since 1968: "corporate foresight; past and futures; humanity at the limen; environmental futures; post-normality and complexity, and technological trends". While the chapter topics present different aspects of healthcare in Asia, the clusters of future studies show different dimensions of futures and future studies.

References

Fergnani, A. (2019). Mapping futures studies scholarship from 1968 to present: A bibliometric review of thematic clusters, research trends, and research gaps. *Futures, 105*: 104–23.

Fergnani, A. (2020). Mapping fifty years of futures studies scholarship (1968–2017). In R. Slaughter & A. Hines (Eds.), *The Knowledge Base of Futures Studies 2020* (pp. 24–47). US: Association of Professional Futurists and Foresight International.

Chapter 1
Introduction: What the Future Holds for Healthcare in Asia

Rong Xiang Ng

Abstract This introductory chapter sets the stage for an exploration of the evolving landscape of healthcare in Asia, a region marked by immense diversity and dynamic change. It addresses the complexities inherent in defining "Asia", highlighting the heterogeneity that shapes healthcare needs and opportunities across the continent. As part of the "Futures of Asia" series, this chapter outlines the book's interdisciplinary approach, drawing upon conceptual frameworks, reviews, and empirical studies to examine the forces shaping the future of healthcare in this region. It introduces the book's structure, arranged into sections focusing on building healthy societies, advancing health and care, and leveraging health technology, including artificial intelligence. Furthermore, this chapter emphasises the cross-cutting themes of equity, innovation, and technology, particularly in the context of pandemic preparedness and the integration of artificial intelligence and telemedicine. By providing a roadmap of the diverse perspectives and methodologies employed, this introduction invites readers to delve into the pathways that will determine what the future holds for healthcare in Asia, ultimately aiming to stimulate critical dialogue and contribute to the development of more equitable, accessible, and effective healthcare systems.

Keywords Asia · Healthcare · Futures · Diversity · Equity · Technology · Artificial intelligence

1.1 Introduction

Asia stands at a critical juncture in its healthcare evolution in this era, reflecting an extraordinary diversity in healthcare systems (Frølich et al. 2010; Griffiths et al. 2005; Wilson 2019). It is a continent of unparalleled diversity. Home to nearly 4.7 billion people, representing approximately 60% of the global population, the

R. X. Ng (✉)
Department of Medicine, Faculty of Medicine, Universiti Malaya, Kuala Lumpur, Malaysia
e-mail: rxng@um.edu.my

University of Glasgow, Glasgow, UK

R. Xiang Ng et al. (eds.), *Future of Healthcare in Asia*,
https://doi.org/10.1007/978-3-032-03183-9_1

region encapsulates a vast spectrum of healthcare systems, each shaped by unique historical, cultural, and socioeconomic factors (United Nations 2023). This continent encompasses various epidemiological profiles, demographic transitions, technological adoption rates, and cultural approaches to health and healing (Ibeneme et al. 2017; Peabody 2000; Prinja et al. 2024; Raghavan 2024; World Health Organization 2021; Hudtohan and Zhang 2022). This diversity presents both unprecedented challenges and remarkable opportunities for healthcare across Asia.

The twenty-first century has brought rapid urbanisation and increasing interconnectedness, demanding innovative solutions for healthcare delivery and management (Chongsuvivatwong et al. 2011; Mohan and Dasgupta 2005; Tonne et al. 2021). The COVID-19 pandemic, a defining event of our time, has exposed vulnerabilities in existing healthcare systems while also accelerating the adoption of digital health solutions (Balasubramanian et al. 2023; Jovanović et al. 2020; Lee et al. 2024; Thottempudi et al. 2025). Simultaneously, the recent surge in artificial intelligence (AI) presents transformative possibilities for healthcare, from diagnostics and treatment to personalised medicine and public health surveillance (Abernethy et al. 2022; Ali et al. 2023; Balasubramanian et al. 2023; Santamato et al. 2024; Scott et al. 2020; World Health Organization 2021). However, we must not forget that equitable access to healthcare is imperative for the future of Asia. Disparities in access to care, affordability, and quality of services persist across the region, influenced by socioeconomic factors, geographic location, and cultural barriers (Al-Worafi 2024; Durairaj 2007; Kruk et al. 2018). Addressing these inequities is not only a matter of social justice but also a prerequisite for building resilient and sustainable healthcare systems.

This book, as part of the broader "Futures of Asia" series, examines these complex dynamics through three interconnected perspectives, as reflected in its distinct sections: Building Healthy Societies, Advancing Health and Care, and Leveraging HealthTech. These sections will explore how equity, ethical innovation, and technological advancement, especially in the context of pandemic preparedness and AI integration, will be crucial to the future of healthcare in Asia.

1.2 Defining "Asia"

One of the central challenges in studying healthcare in Asia is defining what "Asia" means. Asia encompasses more than 50 countries and various cultural traditions. The very term 'Asian' is fraught with controversy, encompassing a vast diversity that challenges any attempt at a singular definition. While official classifications and popular perceptions of 'Asian' vary significantly, as seen in the contrasting definitions used in the United States and New Zealand, among other countries, it is the acknowledgement of this heterogeneity that is paramount to this book (Lee and Ramakrishnan 2020; Rasanathan et al. 2006; United States Census Bureau 2024). From the technologically advanced urban centers of East Asia to the agrarian communities of South Asia, the region defies easy categorisation. Yet, it is precisely this diversity that makes Asia a crucial site for exploring the future of healthcare.

In this book, we adopt a broad yet inclusive definition of "Asia," encompassing East Asia, Southeast Asia, South Asia, and the Pacific Islands. While no single definition can fully capture the complexity of the region, our focus is on identifying shared challenges, opportunities, and innovations that transcend national boundaries. As Rasanathan et al. note, defining "Asia" or "Asian" in health research is fraught with challenges (Rasanathan et al. 2006). The region's vast diversity often makes generalisations problematic. Rather than attempting to rigidly define 'Asian,' we focus on the diverse areas, countries, and regions that constitute the geographical and cultural space we refer to as 'Asia.' By focusing on cross-cutting themes such as equity, innovation, and technology, we aim to provide insights that resonate across diverse Asian contexts while respecting local nuances. With the recognition of the diverse experiences and needs within these Asian communities, we can move towards developing healthcare strategies that are both nuanced and inclusive, ensuring a healthier future for all.

1.3 Key Concepts

Healthcare cannot be seen as merely delivery of medical services but we conceptualise healthcare as complex socio-technical systems embedded within particular cultural, economic, and political contexts in order to sustain it (Carayon et al. 2011; Khalil 2024; Kruk et al. 2018; Marjanovic et al. 2020). This expanded understanding recognises that healthcare encompasses prevention, treatment, rehabilitation, and palliation, extending beyond clinical settings to include community-based care, self-management, and broader public health initiatives, more so in the setting of chronic diseases as our population aged (Frølich et al. 2010; Griffiths et al. 2005; Wilson 2019).

This book is part of the "Futures of Asia" series, a pivotal intellectual nexus that draws upon a wide range of scholarly disciplines and research areas to explore the multifaceted, dynamic, and evolving futures across diverse Asian contexts. This series is dedicated to foster dialogues among academics and offering insights into the prospects of Asia, especially in terms of the six research clusters of future studies: "corporate foresight; past and futures; humanity at the limen; environmental futures; post-normality and complexity, and technological trends" (Fergnani 2019; Fergnani 2020; Fitzgerald and Davies 2022).

These clusters resonate throughout our exploration of Asian healthcare. For example, chapter on gut microbiota link 'past and futures' to inform current treatments. Chapters on HIV activism and Persons Who Use Drugs (PWUDs) highlight 'humanity at the limen,' addressing marginalised populations. Pandemic preparedness and ethical committees fall under 'post-normality and complexity,' focusing on adapting to uncertain situations. Finally, the HealthTech part directly engages with 'technological trends' in telemedicine and AI.

While future studies are usually future-oriented and forward-looking, the series echoes Dator's (Dator 2019) concept that future studies do not "predict" the real future, but rather study the ideas about the future: (1) forecast "alternative futures" and (2) "envision, invent, implement, continuously evaluate, revise, and re-envision" "preferred futures". Just as Dator noted that the Western images of the future may differ from the non-Western, we endeavour to provide a platform for non-Western research and voices, though they may be non-hegemonic in academia. We aspire to address the issues of marginalisation, especially towards the Global South. Nevertheless, we also welcome Western perspectives, voices, and research practices that contribute to the exploration of Asia's future trajectories.

Understanding the future of healthcare in Asia requires us to navigate this complex diversity. This book aims to provide insights into the forces shaping this future, focusing on cross-cutting themes that transcend national boundaries and address the unique challenges and opportunities within the region's diverse contexts.

1.4 Book Structure and Chapter Overview

This book is organised into three interconnected parts, each examining a critical dimension of the future of healthcare in Asia. It presents a diverse collection of chapters, incorporating conceptual frameworks, comprehensive reviews, and empirical studies to provide a multifaceted understanding of the future of Asian healthcare. This methodological diversity is crucial, as conceptual chapters offer innovative perspectives and theoretical foundations, review chapters synthesise existing knowledge and identify key trends, and empirical studies provide data-driven insights into real-world challenges and solutions. Together, they form a robust foundation for exploring the complex landscape of healthcare across Asia.

This introductory chapter sets the stage for our exploration of the future of healthcare in Asia. It begins by addressing the complexities of defining 'Asia' and 'Asian,' acknowledging the region's vast diversity and the challenges it presents for healthcare research and policy. We then outline the 'Futures of Asia' series, situating this book within a broader interdisciplinary framework focused on envisioning potential trajectories for the region. Finally, we provide an overview of the book's structure and the diverse methodologies employed, including conceptual analyses, comprehensive reviews, and empirical studies, which collectively contribute to a multifaceted understanding of the issues at hand.

The book is organised into three main parts, each reflecting one of its major themes, with chapters grouped under each part.

1.4.1 Part I: Building Healthy Societies: Equity and Inclusion for Social Health and Wellbeing

Dr. Martin Luther King, Jr. famously stated, "Of all the forms of inequality, injustice in health is the most shocking and the most inhuman ("King Berates Medical Care given Negroes" 1966)." Health care disparities are central human justice issues (Chin 2017). The first part of this book delves into the critical social determinants of health and the systemic approaches necessary to build equitable and inclusive healthcare systems. Chapters in this part explore the integration of traditional and complementary therapies with personalised medicine, addressing the unique healthcare needs of diverse populations across Asian countries. We examine the vital role of HIV activism in the Asia–Pacific region, highlighting the importance of community engagement and advocacy. Furthermore, this part offers a comparative analysis of pandemic preparedness and border management in Southeast Asia, drawing lessons from the experiences of Thailand, Cambodia, and Myanmar. Finally, we investigate the factors influencing health insurance for vulnerable populations, focusing on the case of farmers in Vietnam's Mekong Delta, and propose policy directions to enhance their access to care.

- **Chapter** 2: Integrating Complementary and Traditional Therapy with Personalised Medicine: The Future of Healthcare Across Asian Countries (conceptual chapter). This conceptual chapter explores integrating traditional therapies with personalised medicine to enhance Asian healthcare. It highlights the potential for holistic, patient-centered care by addressing cultural contexts and modern demands. The chapter advocates for scientific validation and policy frameworks, suggesting technology and AI can strengthen evidence-based practices for future healthcare systems.
- **Chapter** 3: It Takes a Village to Ruffle Feathers: The HIV Activism Agenda and Framework for the Asia–Pacific Region (conceptual chapter). The fight for equitable healthcare often requires passionate voices and unwavering advocacy (Chin 2017). This chapter addresses the critical issue of health inequities undermining HIV care and prevention across Asia. It proposes a "Four-fold HIV Activism Agenda and Framework" to combat these inequities, emphasising the need for community-driven initiatives and systemic change. By drawing on a structured review of literature and policy documents, the chapter highlights the urgency of addressing actionable drivers of health inequities. This framework is crucial for shaping the future of Asian healthcare by advocating for policies and practices that center communities most affected by HIV, improve individual well-being, and foster meaningful alliances. By emphasising the persistence of HIV activism, the chapter provides a roadmap for future health research, policy, and practice that prioritises equity and justice.
- **Chapter** 4: Navigating the Known Unknown: A Comparative Study of Pandemic Preparedness and Border Management in Southeast Asia (Thailand, Cambodia, Myanmar) (empirical chapter). The recent COVID-19 pandemic underscored the critical importance of being prepared for the unexpected (Lal et al. 2022; World

Health Organization 2023). This empirical chapter presents a comparative analysis of pandemic preparedness and response strategies in Thailand, Cambodia, and Myanmar, using a combination of ethnographic fieldwork, interviews, and document analysis. It examines the impacts of political, social, and economic factors on the effectiveness of cross-border management strategies. By highlighting the varied experiences of these countries, from Thailand's robust health surveillance to Myanmar's system collapse due to political instability, the chapter underscores the importance of regional interconnectedness and cooperation. This comparative study contributes to the future of Asian healthcare by providing crucial insights into effective pandemic management and the importance of political stability and regional cooperation in ensuring health security.

- **Chapter** 5: Enhancing Health Insurance for Farmers in Vietnam's Mekong Delta: Influential Factors and Policy Directions (empirical chapter). This empirical study focuses on the factors influencing Mekong Delta farmers' satisfaction with health insurance services, using a mixed-method approach including surveys, interviews, and focus group discussions. It identifies key factors impacting farmers' satisfaction, such as participation and payment procedures, hospital service quality, and staff support. By highlighting the importance of improving access to quality care and enhancing client satisfaction, the chapter recommends policy reforms to address the socio-economic and environmental challenges in the region. This research contributes to the future of Asian healthcare by advocating for the integration of technological innovations like telemedicine and AI diagnostic tools to enhance healthcare accessibility for rural populations. It provides valuable insights for policymakers and healthcare providers to improve health insurance services and social security systems, ensuring equitable access to care for vulnerable communities.

1.4.2 Part II: Health and Care: Advanced and Ethical Solutions for Physical and Mental Health

As Asia's healthcare landscape evolves, the pursuit of advanced solutions must be guided by a strong ethical compass. 'Health and Care,' our second part, navigates the advancements in physical and mental health. It focuses on the advancement of care models and the ethical considerations surrounding emerging medical practices. Chapters in this part address the critical issue of mental health service utilisation in Indonesia, laying the groundwork for future service enhancements. We explore the role of gut microbiota as a treatment target for mental illness in Asian settings and examine the self-management capacity of people who use drugs (PWUDs) with multimorbidity. This part also investigates the benefits of exergaming in promoting physical activity in urbanised East Asian cities and delves into the future of personalised medicine in Asia, highlighting trends, challenges, and collaborative pathways. Finally, we address the imperative for ethical stewardship in Asian healthcare, emphasising the role of clinical ethics committees in navigating a rapidly evolving landscape.

- **Chapter** 6: Assessment of Mental Health Service Utilisation Among Indonesian Adults: Basis for Enhancing Future Services in Indonesia (empirical chapter). Understanding the barriers that prevent individuals from accessing vital mental health support is the first step towards building more effective systems. This empirical study addresses the critical issue of low mental health service utilisation among Indonesian adults, a common challenge in low- and middle-income countries. Utilising an online survey, the research identifies significant factors influencing service usage. The findings revealed that only a small percentage of participants have accessed mental health services in the past three years, highlighting a significant gap in care. Based on these insights, the chapter recommends expanding insurance coverage and improving awareness of mental health services. This study provides a crucial foundation for enhancing future mental health services in Indonesia, contributing to the development of more equitable and accessible healthcare systems across Asia.
- **Chapter** 7: Future of Gut Microbiota as a Promising Treatment Target for Mental Illness in Asian Settings (review chapter). This review chapter explores the emerging field of gut microbiota as a potential treatment target for mental illness, a pressing concern in Asian countries with increasing mental health problems. The review highlights the intricate linkage between gut health and mental illness, emphasising the gut microbiota's role in immune system modulation. By synthesising recent publications, it focuses on microbiota-targeted precision treatments using probiotics, prebiotics, and postbiotics. The chapter aims to inform clinicians on alternative and proactive interventions, paving the way for innovative mental health strategies that can improve the future of healthcare in Asia.
- **Chapter** 8: Self-Management Capacity of People Who Use Drugs (PWUDs) with Multimorbidity: Insights from Asian Healthcare Perspectives and Future Trends (empirical chapter). This study investigates the self-management capabilities of Persons Who Use Drugs (PWUDs) with multimorbidity in the Philippines, focusing on a rehabilitation center. Utilising retrospective chart reviews and interviews, the research explores multimorbidity prevalence and self-management practices across seven domains. The study provides insights into chronic care strategies for PWUDs, highlighting the importance of integrated interventions and collaborative approaches. The findings contribute to future healthcare trends for PWUDs in Asia, offering a unique Asian country perspective that can inform more inclusive and effective healthcare models across the region, especially among the marginalised populations.
- **Chapter** 9: Exploring the Eco-Friendly and Health Benefits of Exergaming to Improve Physical Activity Levels in Urbanised East Asian Cities (empirical chapter). Changes in lifestyle and society have led to a significant decrease in the level of physical activity, worsened after COVID-19 pandemic (Dey et al. 2021; Al-Hajj et al. 2023; Lau et al. 2021). One of the major changes has been the growing popularity of leisure time sedentary activities brought about by new

media solutions such as television viewing, computer use, and video gaming (Kari 2014; Matthews et al. 2012). Finding engaging ways to promote physical activity in increasingly urbanised environments is important in this era and one potential solution is exergaming (Davis et al. 2024; Kari 2014). This case study addresses the growing issue of sedentarism in rapidly urbanised East Asian cities and its associated health problems. It explores the benefits and challenges of exergaming, a combination of exercise and gaming, as a tool to promote physical activity. The study utilises virtual reality-based sports exergaming with two sedentary urban Asians, measuring physiological improvements and collecting interview data. The findings demonstrate that home-based exergaming training can improve physical activity levels and cardiorespiratory fitness. The chapter discusses the potential of exergaming to overcome common barriers to exercise in urban Asian settings, while also acknowledging challenges like cost and hardware issues, offering a potential pathway to improve urban health and wellness in the future of Asian healthcare.

- **Chapter** 10: Future of Personalised Medicine in Asia: Trends, Challenges, and Collaborative Pathways (review chapter). This review chapter delves into the past, present, and future of personalised medicine in Asia, with a particular focus on the role of genomics. It examines the development of genomic technologies and their application in healthcare across the region, considering research, industry, governance, and ethical factors. The chapter categorises Asian countries into three tiers based on their genomic capabilities, projecting different future scenarios for each. It explores the potential for disparities in the implementation of genomic-driven personalised medicine, highlighting opportunities for global collaborations and data sharing. By providing recommendations for a multisectoral approach to the integration of genomics into healthcare, this review serves as a valuable resource for understanding the trends, challenges, and collaborative pathways in the future of personalised medicine in Asia, ultimately shaping the trajectory of advanced, genomic-informed healthcare technologies in the region.
- **Chapter** 11: Ethical Stewardship in Asian Healthcare: The Imperative for Clinical Ethics Committees in a Rapidly Evolving Landscape (conceptual chapter). This chapter addresses the increasing ethical complexities in Asian healthcare due to rapid technological advancements. It explores the role of Clinical Ethics Committees (CECs) in navigating these challenges, synthesising findings from literature and case studies. The chapter integrates principles of ethical stewardship, cultural relativism, and operational dynamics to provide insights into how CECs mediate between universal standards and localised practices. It offers a practical guide for establishing CECs within the Asian healthcare context, outlining actionable steps and operational guidelines. By focusing on Asia, this chapter serves as a timely resource for healthcare institutions, policymakers, and medical professionals striving to address ethical complexities while considering regional specificities, ensuring that the future of healthcare in Asia is built on a foundation of ethical integrity and patient-centered care.

1.4.3 Part III: HealthTech: Advances in Technology for Healthcare

The rapid development of technology offers new opportunities to transform healthcare delivery. In 'HealthTech,' the final part of this book, we explore the cutting-edge innovation and examine the transformative potential of health technology in shaping the future of healthcare across Asia. Chapters in this part explore the game-changing role of telemedicine in prison healthcare settings and investigate the feasibility of hospital-at-home models for delivering care. We delve into the use of telemedicine to address sensitive issues like sexual health, breaking taboos and improving access to care. Furthermore, we examine the legal implications of artificial intelligence in healthcare, focusing on the standard of care under the law of medical negligence in Malaysia and Singapore.

- **Chapter** 12: Telemedicine: The Game Changer in Future Prison Healthcare Settings (review chapter). Healthcare access should extend to all members of society, including those behind prison walls (Mathis and Schoenly 2008; Kamarulzaman et al. 2016; McLeod et al. 2020). This review explores the potential of telemedicine to address the significant healthcare challenges faced by incarcerated populations in Malaysia. The chapter highlights the ability of telemedicine to facilitate access to specialised medical care which is often limited in prison settings. However, it also acknowledges the challenges associated with implementation, such as patient privacy concerns, infrastructure requirements, and regulatory limitations. This review provides valuable insights into the opportunities and challenges of integrating telemedicine into prison healthcare, contributing to the development of more equitable and accessible healthcare solutions for marginalised populations in Asia, paving ways to be more inclusive for all, including incarcerated populations, leaving no one behind in the upcoming trend in healthcare in Asia.
- **Chapter** 13: Hospital Care Delivered at Home: The Future of Healthcare in Asia with Hospital-at-Home (review chapter). This review investigates Hospital-at-Home (HaH) as a potential strategy to address the growing healthcare demands in Asia, driven by aging demographics and the rise in non-communicable diseases. It synthesises evidence from scientific literature and policy documents to assess HaH's safety, effectiveness, and public acceptance compared to traditional hospital inpatient care. This review provides a comprehensive overview of HaH's role in shaping the future of healthcare delivery in Asia by offering innovative solutions to address the region's increasing healthcare needs.
- **Chapter** 14: Breaking Taboos on Sexual Health in Asia Through Telemedicine: A Conceptual Framework (conceptual chapter). Breaking down societal taboos is often necessary to address critical health needs (Dimitrov et al. 2022; Evans et al. 2000; Mohd. Tohit and Haque 2024). This conceptual chapter addresses the significant challenge of sexual health stigma in Asian societies, where issues like HIV

among men having sex with men (MSM), erectile dysfunction, and teenage pregnancy are often shrouded in taboo. It explores how telemedicine can bridge healthcare gaps in sexual medicine by offering privacy and convenience. The chapter proposes a conceptual framework, CAAT-Q (Cultural sensitivity, Anonymity, Accessibility, Trust-building, Quality assurance), to leverage telemedicine effectively in Southeast Asia, a region with particularly strong cultural sensitivities. By embedding these elements, telemedicine platforms can foster open discussions on sexual health and improve access to care, contributing to healthier communities. This framework offers a pathway to integrate culturally sensitive digital health solutions, advancing the future of sexual healthcare in Asia by breaking down taboos and promoting open accessible care.

- **Chapter** 15: Artificial Intelligence (AI) in Healthcare: Standard of Care Under the Law of Medical Negligence in Malaysia and Singapore (conceptual chapter). While AI is the new trend, we must not forget the legal aspects behind it, especially in the healthcare setting (Chikhaoui et al. 2022; Gerke et al. 2020; McGreevey et al. 2020; Schönberger 2019). There is a lack of governance guidance in preparedness for AI establishment in Asia (Alfiani and Santiago 2024; Tun et al. 2025). This conceptual chapter explores the legal and liability implications of the increasing integration of artificial intelligence (AI) in healthcare practices in Malaysia and Singapore. It examines how the standard of care may need to adapt to the use of AI, considering the potential for errors and harm to patients. The chapter analyses relevant legal frameworks, including court decisions and civil law acts, to establish the potential liabilities of medical practitioners and AI developers. This study emphasises the need for a balanced approach to AI regulation, ensuring patient safety without hindering technological advancements, which is crucial for the safe and ethical integration of AI into the future of healthcare in Asia.

1.5 Charting the Trajectory for Asian Healthcare

This book serves as a critical resource for anyone seeking to understand the evolving landscape of healthcare in Asia, offering a multidisciplinary perspective on the challenges and opportunities that lie ahead.

In this introductory chapter, we have laid the groundwork for understanding the multifaceted landscape of healthcare in Asia and the critical questions or issues surrounding its future. We navigate the complexities of defining 'Asia' itself, recognising the immense diversity that shapes healthcare needs and opportunities across the continent. As part of the 'Futures of Asia' series, this book embraces an interdisciplinary approach, drawing upon conceptual frameworks, comprehensive reviews, and empirical investigations to offer a rich and nuanced exploration.

Through the diverse perspectives presented in all the chapters, we aim to provide valuable insights for policymakers, healthcare professionals, researchers, and anyone invested in the health and well-being of Asia's diverse populations. Our goal is to stimulate critical dialogue and contribute to the development of more equitable, accessible, and effective healthcare systems that are responsive to the unique contexts within this dynamic region. We invite you to join us in exploring the pathways that will shape what the future holds for healthcare in Asia.

Understanding these emerging pathways is crucial not only for those working within Asian healthcare systems but for anyone interested in the diverse ways human societies organise to promote health and respond to illness in the twenty-first century.

Highlights

- Navigating Asia's diverse healthcare landscape is crucial to understanding its future.
- This chapter lays the groundwork by defining "Asia" and key concepts, highlighting the region's heterogeneity.
- It introduces the book's interdisciplinary approach and thematic focus on equity, innovation, and technology.
- By outlining the challenges and opportunities, this introduction provides the essential context for exploring the pathways that will shape the future of healthcare across this dynamic continent.

References

Abernethy A, Adams L, Barrett M, Bechtel C, Brennan P, Butte A, et al. The promise of digital health: then, now, and the future. NAM Perspect Discuss Pap, Natl Acad Med. 2022. https://doi.org/10.31478/202206e

Alfiani FRN, Santiago F. A comparative analysis of artificial intelligence regulatory law in Asia, Europe, and America. SHS Web Conf. 2024;204:07006. https://doi.org/10.1051/shsconf/202420407006.

Ali O, Abdelbaki W, Shrestha A, Elbasi E, Alryalat MAA, Dwivedi YK. A systematic literature review of artificial intelligence in the healthcare sector: Benefits, challenges, methodologies, and functionalities. J Innov Knowl. 2023;8(1). https://doi.org/10.1016/j.jik.2023.100333

Al-Hajj KM, Chia YH, Kamarudin S, Wan Muda AM, Lee YY. A scoping review of the changes in physical activity and sedentary behaviour and its health outcomes among adults in Asia during the COVID-19 outbreak. Malays J Med Health Sci. 2023;19(5):275–305. https://doi.org/10.47836/mjmhs19.5.36

Al-Worafi YM. Healthcare access and equitable access services in developing countries. In: Handbook of medical and health sciences in developing countries. Springer International Publishing; 2024. p. 1–24. https://doi.org/10.1007/978-3-030-74786-2_217-1

Balasubramanian S, Shukla V, Islam N, Upadhyay A, Duong L. Applying artificial intelligence in healthcare: lessons from the COVID-19 pandemic. Int J Prod Res. 2023. https://doi.org/10.1080/00207543.2023.2263102

Carayon P, Bass EJ, Bellandi T, Gurses AP, Hallbeck MS, Mollo V. Sociotechnical systems analysis in health care: a research agenda. IIE Trans Healthc Syst Eng. 2011;1(3):145–60. https://doi.org/10.1080/19488300.2011.619158.

Chikhaoui E, Alajmi A, Larabi-Marie-sainte S. Artificial intelligence applications in healthcare sector: ethical and legal challenges. Emerg Sci J. 2022;6(4):717–38. https://doi.org/10.28991/ESJ-2022-06-04-05

Chin MH. Movement advocacy, personal relationships, and ending health care disparities. J Natl Med Assoc. 2017;109(1):33–5. https://doi.org/10.1016/j.jnma.2016.11.003.

Chongsuvivatwong V, Phua KH, Yap MT, Pocock NS, Hashim JH, Chhem R, et al. Health and health-care systems in Southeast Asia: diversity and transitions. Lancet. 2011;377(9763):429–37. https://doi.org/10.1016/S0140.

Dator J. What futures studies is, and is not. In: Anticipation Science. Vol. 5. Springer Science and Business Media B.V.; 2019. p. 3–5. https://doi.org/10.1007/978-3-030-17387-6_1

Davis JC, Killen LG, Green JM, Waldman HS, Renfroe LG. Exergaming for physical activity: A systematic review. J Am Coll Health. 2024;72(7):2090–8. https://doi.org/10.1080/07448481.2022.2103377.

Dey KC, Zakrzewski-Fruer JK, Smith LR, Jones RL, Bailey DP. The prevalence of daily sedentary time in south asian adults: A systematic review. Int J Environ Res Public Health. 2021;18(17). https://doi.org/10.3390/ijerph18179275

Dimitrov R, Jelen A, L'Etang J. Taboos in health communication: Stigma, silence and voice. Public Relat Inquiry. 2022;11(1):3–35. https://doi.org/10.1177/2046147X211067002.

Durairaj V. Enhancing equity in access to health care in the Asia-Pacific region: remediable inequities. Report prepared for the UN Regional Thematic Working Group on Health. Geneva: World Health Organization; 2007 May. https://asiapacific.unfpa.org/en/publications/enhancing-equity-access-health-care-asia-pacific-region-remediable-inequities. Accessed 3 March 2025.

Evans RW, Avery PG, Pederson PV. Taboo topics: Cultural restraint on teaching social issues. Clearing House. 2000;73(5):295–302. https://doi.org/10.1080/00098650009600973.

Fergnani A. Mapping futures studies scholarship from 1968 to present: A bibliometric review of thematic clusters, research trends, and research gaps. Futures. 2019;105:104–23. https://doi.org/10.1016/j.futures.2018.09.007.

Fergnani, A. (2020). Mapping fifty years of futures studies scholarship (1968–2017). In R. Slaughter & A. Hines (Eds.), *The Knowledge Base of Futures Studies 2020* (pp. 24–47). US: Association of Professional Futurists and Foresight International.

Fitzgerald LM, Davies AR. Creating fairer futures for sustainability transitions. Geogr Compass. 2022;16(10). https://doi.org/10.1111/gec3.12662

Frølich A, Høst D, Schnor H, Nørgaard A, Ravn-Jensen C, Borg E, Hendriksen C. Integration of healthcare rehabilitation in chronic conditions. Int J Integr Care. 2010;10: e033. https://doi.org/10.5334/ijic.507.

Gerke S, Minssen T, Cohen G. Ethical and legal challenges of artificial intelligence-driven healthcare. In: Artificial intelligence in healthcare. Elsevier; 2020. p. 295–336. https://doi.org/10.1016/B978-0-12-818438-7.00012-5

Griffiths S, Jewell T, Donnelly P. Public health in practice: The three domains of public health. Public Health. 2005;119(10):907–13. https://doi.org/10.1016/j.puhe.2005.01.010.

Hudtohan ET, Zhang D. Traditional medicine and healthcare in the 21st century. Int J Manag Educ Hum Dev. 2022;2(02):362–5. Available from: https://ijmehd.com/index.php/ijmehd/article/view/1. Accessed 13 Apr 2025.

Ibeneme S, Eni G, Ezuma A, Fortwengel G. Roads to health in developing countries: understanding the intersection of culture and healing. In Current therapeutic research—clinical and experimental, 2017, Vol. 86, pp. 13–18. Excerpta Medica Inc. https://doi.org/10.1016/j.curtheres.2017.03.001

Jovanović A, Klimek P, Renn O, Schneider R, Øien K, Brown J, et al. Assessing resilience of healthcare infrastructure exposed to COVID-19: emerging risks, resilience indicators, interdependencies and international standards. Environ Syst Decis. 2020;40(2):252–86. https://doi.org/10.1007/s10669-020-09779-8.

Kamarulzaman A, Reid SE, Schwitters A, Wiessing L, El-Bassel N, Dolan K, et al. Prevention of transmission of HIV, hepatitis B virus, hepatitis C virus, and tuberculosis in prisoners. Lancet. 2016;388(10049):1115–26. https://doi.org/10.1016/S0140-6736(16)30769-3.

Kari T. Can exergaming promote physical fitness and physical activity? A systematic review of systematic reviews. Int J Gaming Comput-Mediat Simul. 2014;6(4):59–77. https://doi.org/10.4018/ijgcms.2014100105.

Khalil FG. Socio-technical platforms for care transformation: An integrative synthesis and conceptualization. Technol Soc. 2024;77. https://doi.org/10.1016/j.techsoc.2024.102532.

King berates medical care given Negroes. Oshkosh Daily Northwestern. 1966 Mar 26. https://www.newspapers.com/article/the-oshkosh-northwestern-oshkosh-daily-n/12049661. Accessed 30 Mar 2025.

Kruk ME, Gage AD, Arsenault C, Jordan K, Leslie HH, Roder-DeWan S, et al. High-quality health systems in the sustainable development goals era: time for a revolution. Lancet Glob Health. 2018;6(11):e1196–252. https://doi.org/10.1016/S2214-109X(18)30386-3.

Lal A, Abdalla SM, Chattu VK, Erondu NA, Lee TL, Singh S, et al. Pandemic preparedness and response: exploring the role of universal health coverage within the global health security architecture. Lancet Glob Health. 2022;10(11):e1675–83. https://doi.org/10.1016/S2214-109X(22)00341-2.

Lau JH, Nair A, Abdin E, Kumarasan R, Wang P, Devi F, et al. Prevalence and patterns of physical activity, sedentary behaviour, and their association with health-related quality of life within a multi-ethnic Asian population. BMC Public Health. 2021;21(1). https://doi.org/10.1186/s12889-021-11902-6

Lee J, Park S, Jang SN, Reyes KA, Garcia F, Canila C, et al. Differential impacts of health systems and sociocultural environment on vulnerable populations during the COVID-19 pandemic: lessons from four Asia-Pacific countries. BMC Public Health. 2024;24(1). https://doi.org/10.1186/s12889-024-18949-1

Lee J, Ramakrishnan K. Who counts as Asian. Ethn Racial Stud. 2020;43(10):1733–56. https://doi.org/10.1080/01419870.2019.1671600.

Marjanovic S, Altenhofer M, Hocking L, Chataway J, Ling T. Innovating for improved healthcare: Sociotechnical and innovation systems perspectives and lessons from the NHS. Sci Public Policy. 2020;47(2):283–97. https://doi.org/10.1093/scipol/scaa005.

Mathis H, Schoenly L. Healthcare behind bars: what you need to know. Nurse Pract. 2008;33(5):34–41. https://doi.org/10.1097/01.npr.0000317487.35167.c7.

Matthews CE, George SM, Moore SC, Bowles HR, Blair A, Park Y, et al. Amount of time spent in sedentary behaviours and cause-specific mortality in US adults. Am J Clin Nutr. 2012;95(2):437–45. https://doi.org/10.3945/ajcn.111.019620.

McGreevey JD, Hanson CW, Koppel R. Clinical, legal, and ethical aspects of artificial intelligence-assisted conversational agents in health care. JAMA—J Am Med Assoc. 2020;324(6):552–3. https://doi.org/10.1001/jama.2020.2724

McLeod KE, Martin RE, Butler A, Young JT, Southalan L, Borschmann R, et al. Global prison health care governance and health equity: A critical lack of evidence. Am J Public Health. 2020;110(3):303–8. https://doi.org/10.2105/AJPH.2019.305465.

Mohan R, Dasgupta S. The 21st century: Asia becomes urban. Econ Polit Wkly. 2005;40(2):213–23. https://doi.org/10.2307/4416078.

Mohd Tohit NF, Haque M. Forbidden conversations: A comprehensive exploration of taboos in sexual and reproductive health. Cureus. 2024. https://doi.org/10.7759/cureus.66723

Peabody JW. Policy and health: implications for development in Asia. Cambridge University Press; 2000.

Prinja S, Purohit N, Kaur N, Rajapaksa L, Sarker M, Zaidi R, et al. The state of primary health care in south Asia. Lancet Glob Health. 2024;12(10):e1693–705. https://doi.org/10.1016/S2214-109X(24)00119-0.

Raghavan A. Asia's digital health innovations: The role of cross-border health data sharing. 2024. https://www.asiahouse.org/research-analysis/aarthi-raghavan-asias-digital-health-innovations-the-role-of-cross-border-health-data-sharing. Accessed 7 Apr 2025.

Rasanathan K, Craig D, Perkins R. The novel use of "Asian" as an ethnic category in the New Zealand health sector. Ethn Health. 2006;11(3):211–27. https://doi.org/10.1080/13557850600565525.

Santamato V, Tricase C, Faccilongo N, Iacoviello M, Marengo A. Exploring the impact of artificial intelligence on healthcare management: A combined systematic review and machine-learning approach. Appl Sci. 2024;14(22). https://doi.org/10.3390/app142210144

Schönberger D. Artificial intelligence in healthcare: A critical analysis of the legal and ethical implications. Int J Law Inf Technol. 2019;27(2):171–203. https://doi.org/10.1093/ijlit/eaz004.

Scott BK, Miller GT, Fonda SJ, Yeaw RE, Gaudaen JC, Pavliscsak HH, et al. Advanced digital health technologies for COVID-19 and future emergencies. Telemed E-Health. 2020;26(10):1226–33. https://doi.org/10.1089/tmj.2020.0140.

Thottempudi P, Konduru RM, Valiveti HB, Kuraparthi S, Kumar V. Digital health resilience: IoT solutions in pandemic response and future healthcare scenarios. Discov Sustain. 2025;6(1). https://doi.org/10.1007/s43621-025-00886-7

Tonne C, Adair L, Adlakha D, Anguelovski I, Belesova K, Berger M, et al. Defining pathways to healthy sustainable urban development. Environ Int. 2021;146: 106236. https://doi.org/10.1016/j.envint.2020.106236.

Tun HM, Naing L, Malik OA, Rahman HA. Navigating ASEAN region Artificial Intelligence (AI) governance readiness in healthcare. Health Policy Technol. 2025;14(2). https://doi.org/10.1016/j.hlpt.2025.100981

United Nations, Economic and Social Commission for Asia and the Pacific (ESCAP). Asia-Pacific population and development report 2023 (ST/ESCAP/3112). 2023.

United States Census Bureau. About the topic of race. 2024. https://www.census.gov/topics/population/race/about.html. Accessed 10 Apr 2025.

Wilson C. Improved access to health care and self-management of long-term and disabling conditions including end-of-life care: a corpus of published work incorporating evidence appraisals and critical approaches to public health [dissertation]. Lancaster: Lancaster University; 2019. https://eprints.lancs.ac.uk/id/eprint/136798/1/2019wilsonphd.pdf. Accessed 7 April 2025

World Health Organization. Global Strategy on Digital Health 2020–2025. World Health Organization; 2021. https://www.who.int/publications/i/item/9789240020924. Accessed 6 April 2025.

World Health Organization. Managing epidemics: key facts about major deadly diseases. World Health Organization; 2023. https://www.who.int/publications/i/item/9789240083196. Accessed 4 April 2025.

Rong Xiang Ng is an Infectious Diseases physician and medical lecturer at Universiti Malaya (Malaysia) and holds an affiliate lecturer position at the University of Glasgow (UK). He is engaged in understanding and shaping the future of healthcare in Asia. His clinical and research work focuses on optimising clinical pathways and exploring innovative solutions, including digital tools, to address infectious disease challenges across diverse populations. Dr. Ng's contributions to hospital treatment guidelines and research on artificial intelligence (AI) in HIV care and infectious diseases, multidrug-resistant organisms, COVID-19, and telemedicine in sexual health directly reflect the key themes of technological innovation and equitable access explored in this introductory chapter. As a Fellow of the Asia Pacific AIDS & Co-infections Conference (APACC) and an affiliate of the Centre of Excellence for Research in Infectious Diseases and AIDS (CERIA, Universiti Malaya), his active involvement in research and clinical trials bridges the crucial gap between current clinical practice and the forward-looking perspectives presented throughout this volume on the future of healthcare in Asia.

Part I
Building Healthy Societies: Equity and Inclusion for Social Health and Wellbeing

Chapter 2
Integrating Complementary and Traditional Therapy with Personalised Medicine: The Future of Healthcare Across Asian Countries

Clusters of Future Studies: Past & Futures; Humanity at the Limen

Ping Lei Chui and Nant Thin Thin Hmwe

Abstract The integration of complementary and traditional therapy (CTT) with personalised medicine is emerging as a transformative approach to healthcare across Asian countries. This conceptual article examines how this integration is shaping the future of healthcare in the region, highlighting its potential to address diverse health challenges, enhance patient-centred care, and drive medical innovation. This discussion explores the cultural contexts and definitions of CTT, emphasising its deep-rooted practices across Asia. It highlights integration as a strategy to align with spiritual beliefs and increase acceptance while addressing modern healthcare demands. However, challenges such as regulatory frameworks, safety, efficacy, standardisation, and economic disparities affecting access and adoption are acknowledged. The article underscores the importance of rigorous scientific validation, comprehensive policies, and improved infrastructure to support the integration of CTT with personalised medicine. Future directions include leveraging technology, AI, and big data to strengthen evidence-based practices, enhance credibility, and ensure quality. Advancing research, fostering interdisciplinary collaboration, and addressing safety concerns are central to achieving holistic care. Integrating CTT with personalised medicine offers the potential to transform healthcare delivery by improving patient outcomes and promoting holistic well-being. Collaboration among medical practitioners, complementary therapists, and traditional healers is essential for creating a more inclusive, effective, and patient-centred healthcare system.

P. L. Chui (✉) · N. T. T. Hmwe
Department of Nursing Science, Faculty of Medicine, Universiti Malaya, Kuala Lumpur, Malaysia
e-mail: chuipinglei@um.edu.my

N. T. T. Hmwe
e-mail: nantthinthin.h@um.edu.my

R. Xiang Ng et al. (eds.), *Future of Healthcare in Asia*,
https://doi.org/10.1007/978-3-032-03183-9_2

Keywords Complementary therapy · Traditional medicine · Personalised medicine · Integrative healthcare · Asian healthcare · Future of medicine

2.1 Introduction

The integration of complementary and traditional therapy (CTT) with personalised medicine is emerging as a transformative approach in healthcare across Asian countries. The integration of CTT with personalised medicine would play a crucial role in addressing contemporary healthcare needs and the future of healthcare practice. A broader understanding of CTT, the potential integration of CTT in the Asian context and its implications are areas to be discussed.

This chapter explores the broad definitions and cultural contexts of complementary, alternative, and traditional medicine, highlighting the diverse practices and their deep cultural roots in Asia. The subsequent sections discuss the evolving role of personalised medicine in integrating CTT within Asian healthcare systems. It highlights the challenges and future directions necessary to fully realise their potential.

2.1.1 Definitions and Cultural Context of Complementary and Traditional Therapy (CTT)

The definition of complementary and alternative medicine (CAM) is broad. It generally refers to a variety of approaches that have not been previously considered a part of Western medicine. At the core, complementary medicine refers to therapies used together with conventional medicine, while alternative medicine refers to therapies used to substitute conventional medicine and is therefore generally not recommended (National Centre for Complementary and Integrative Health (NCCIH) 2021). As CAM is often culturally dependent, the terms 'complementary medicine' or 'alternative medicine' are used interchangeably with traditional medicine in some countries (WHO 2019). A more comprehensive and evolving term that could be used is complementary and integrative therapies (CIT), which combines evidence-based complementary therapies with conventional methods. Traditional medicine encompasses a rich history, comprising the collective wisdom, expertise, and methodologies rooted in the theories, beliefs, and experiences unique to various cultures that differ in evidence base from modern Western medicine. These practices, whether scientifically validated or not, play a role in maintaining health and addressing physical and mental illnesses through prevention, diagnosis, improvement, or treatment, as recognised by the World Health Organization (WHO).

Certain aspects, such as herbal remedies or acupuncture techniques, have been researched in clinical trials and observational studies, offering evidence of their effectiveness for specific health conditions. However, traditional medicine may not always align with the rigorous standards of evidence-based medicine in Western healthcare systems. It often lacks extensive clinical trials and systematic reviews. Nevertheless, cultural and historical factors may influence the acceptance and use of traditional medicine within communities, regardless of scientific evidence.

In Asian countries, CTT is shaped by diverse cultural practices and contemporary healthcare needs. The adoption and implementation of personalised medicine alongside complementary traditional therapy are intrinsic to the region's healthcare landscape. Religious and spiritual beliefs are fundamental to health and healing practices in many Asian societies. Therefore, integrating personalised medicine with CTT may need to align with these spiritual practices and rituals to achieve broader acceptance. In China, TCM is deeply rooted in culture, guided by principles such as yin and yang, qi, and the five elements guiding healthcare practices. These cultural beliefs influence the perception and acceptance of personalised medicine within the context of TCM (Wang and Zhang 2017). Traditional healers, such as Ayurvedic doctors and herbalists, are highly trusted and influential within communities in India and Indonesia. Their endorsement or integration of personalised medicine can significantly impact patient acceptance. Verma et al. (2024) highlight the importance of trust in traditional healers when integrating personalised medicine with Ayurveda in India. In Japan and South Korea, healthcare decisions often involve family members and community elders. The acceptance of personalised medicine alongside CTT may be influenced by collective decision-making processes. Economic disparities and the availability of healthcare resources can significantly influence the adoption of personalised medicine (Chong et al. 2018). The presence of CTT practitioners and facilities in the healthcare infrastructure is vital for integrating personalised medicine effectively. The affordability and accessibility of CTT compared to conventional medical treatments strongly influence patient preferences and adoption rates in many Asian countries. Additionally, regions with well-established CTT infrastructures are better equipped to incorporate personalised approaches more readily (Park and Canaway 2019). To address this, regulatory frameworks for integrating personalised medicine with CTT are indispensable for ensuring safety and efficacy, maintaining quality control and public health protection, fostering innovation and research, and upholding ethical considerations.

In countries such as China and India, robust regulatory bodies oversee CTT, ensuring high standards of safety, efficacy, and quality. In healthcare systems where CTT is deeply embedded, regulatory frameworks are designed with integration strategies in mind. These strategies encompass accreditation processes for CTT practitioners, the standardisation of practices, and the implementation of monitoring mechanisms to ensure safety and effectiveness. Despite ongoing efforts, challenges remain in harmonising diverse CTT due to variations in terminology, diagnostic methods, and treatment approaches, complicating the standardisation of regulations. This chapter discusses the evolving role of personalised medicine in integrating CTT within Asian healthcare systems. It highlights the challenges and future directions necessary to fully realise their potential.

2.1.2 Prevalence and Types of Complementary and Traditional Therapy (CTT) Across Asian Countries

The prevalence of CTT use across studies has to be interpreted with caution as there are differences in what constitutes CTT. Therefore, it's essential to consider the specific therapies and practices involved. The prevalence of using CTT could be influenced by many factors including unmet patient needs, the higher costs of conventional treatments, potential complications associated with such treatments, and the public's perception that traditional remedies offer greater safety and cost-effectiveness (Ashrafizadeh and Rassouli 2023). Convenience also plays a crucial role in the adoption of CTT (Tangkiatkumjai et al. 2020). CTT is often favored for its accessibility without a prescription, self-administration potential at home, and reduced need for frequent healthcare facility visits. There are many types of CTT, which include natural products; herbs, vitamins and minerals, and dietary supplements, mind and body practices; yoga, tai chi, meditation, and hypnotherapy, traditional Chinese medicine (TCM); acupuncture, moxibustion, auriculotherapy; other practices of traditional healers, homeopathy, and naturopathy (NCCIH 2018). In some literature, self-care practices such as exercise, music, prayers, and home remedies (honey, lemon, olive oil, etc.) are considered CTT (Bauer and Rayner 2012). Similarly, the adoption of CTT varies in different cultures and countries; for example, acupuncture is a traditional medical practice in Chinese culture, however, it has been used as a complementary therapy in Western and non-Chinese populations (Bauer and Rayner 2012; Johnson et al. 2018).

Complementary therapy encompasses various health-related practices outside mainstream medical care, while traditional medicine involves plant, animal, and mineral-based medicines, along with spiritual therapies and manual techniques. Generally, these can be classified into two subgroups: mind and body practices or natural products. Mind and body practices encompass a broad spectrum of techniques and therapies aimed at fostering holistic well-being by integrating the mind, body, and spirit. These practices include meditation, yoga, Tai Chi and Qigong, biofeedback, and hypnotherapy. However, there are potential risks associated with these practices. Psychological risks may arise, leading to psychological distress, while physical strain can result in muscle or joint injuries if performed incorrectly or excessively. Moreover, the lack of standardisation in training or regulation can lead to variations in quality and safety among practitioners. It's important to note that while mind and body practices are beneficial for stress reduction and relaxation, they may not be effective for managing all health conditions or suitable as standalone treatments for severe medical conditions.

Natural products encompass a diverse array of substances sourced from plants, animals, and minerals, serving therapeutic purposes. These include herbal medicine and dietary supplements. It can manifest in various forms, including fresh or dried preparation, ingestion as pill or powders, administration as tinctures, or topical application on skin using gels, ointments, or lotions. While natural products may offer therapeutic benefits, individuals should exercise with caution and be vigilant about potential risks associated with their use. Concerns regarding natural products include lack of regulation, potential drug interactions, allergic reactions, and contamination.

Traditional medicine refers to a broad range of healing practices, knowledge, and beliefs that have been passed down through generations within various cultures. These practices are often based on centuries-old traditions and are deeply rooted in cultural and historical contexts. One of the key characteristics of traditional medicine is its holistic approach, which considers the interconnectedness of the body, mind, and spirit in maintaining health and well-being. Traditional healers often take into account not only physical symptoms but also emotional, mental, and spiritual aspects when diagnosing and treating individuals.

The common use of CTT in healthcare in Asian countries reflects a deep-rooted cultural heritage and a growing recognition of the benefits these therapies can offer in promoting holistic well-being and improving health outcomes. Many people continue to rely on CTT for a wide range of health conditions, including chronic diseases, pain management, mental health issues, and preventive care. These therapies are often viewed as holistic approaches that consider not just the physical symptoms but also the emotional, mental, and spiritual aspects of health.

The traditional medicine practices in East Asia, including Traditional Chinese Medicine, Japanese Kampo-Medicine, and Korean Medicine, have evolved within their cultural contexts. Originating from a common heritage, these traditional medicine practices encompass herbal prescriptions, acupuncture, moxibustion, massage, cupping therapy, physical therapy, Taiji, Qigong, and food therapy (Zinboonyahgoon et al. 2023). Chinese Tibetan medicine and Mongolian medicine have their own unique characteristics. These therapies are often viewed as holistic approaches that consider not just the physical symptoms but also the emotional, mental, and spiritual aspects of health. South Asian traditional medicine encompasses diverse practices like Ayurveda, yoga, meditation, mindfulness, naturopathy, homeopathy, Sowa Rigpa, Siddha, and Unani, along with region-specific herbal medicines. These treatments, administered by various practitioners, including yoga therapists and naturopaths, require further research before integration into patient care for pain management and healthcare in general.

Traditional medicine in Southeast Asia represents a fusion of influences from neighbouring Asian regions, gradually evolving to develop distinct philosophies. These practices, shaped by cultural exchanges and historical developments, are cherished and passed down through generations. For example, Thai traditional medicine, which incorporates herbal remedies and massage, promotes holistic health. On the other hand, the use of cannabinoids and cannabis-based medicines to alleviate chronic pain has gained attention. However, their integration requires caution due to potential long-term harm and abuse risks (Zinboonyahgoon et al. 2021). In Malaysia, a rich diversity of traditional medicine practices is observed, including traditional Malay, Chinese, and Indian medicines, Islamic medical practices, homeopathy, as well as chiropractic and osteopathy treatments. This diversity is influenced by the country's ethnic makeup, contributing to a multifaceted approach to healthcare and wellness. It is imperative to enhance the safety and quality of traditional medicine through the implementation of specialised legislation and the adoption of a systematic approach to evaluate the suitability of administrative policies and measures. (Park et al. 2022). Traditional Indonesian herbal medicine, known as Jamu, has been practiced for centuries in Indonesian society and remains highly popular for maintaining health and treating diseases (Pradipta et al. 2023). This popularity stems from the belief that Jamu is safer than chemical drugs, contributing to its continued use and cultural significance in Indonesian healthcare practices. The extensive history and longstanding usage highlight significant potential for its ongoing development and refinement (Pradipta et al. 2023).

In Central Asia, traditional healing practices heavily rely on various plant-based remedies that have been used for generations (Sitpayeva et al. 2021; Pawera et al. 2017). These medicinal plants play a crucial role in the region's traditional medicine. Traditional healers and patients in Mongolia commonly utilise products of animal origin as part of their therapeutic approaches, particularly in addressing health issues like cancer (Oyunchimeg et al. 2017). This diversity of natural remedies underscores the rich heritage and continued relevance of traditional healing methods in Central Asia. Traditional healing practices such as herbal healing, cupping, cauterising, orthopedic therapy, Qur'anic healing, and massage have a long heritage in Arab culture in Western Asia (Moe et al. 2014). Expertise in these therapies, especially Qur'an therapy, is often passed down through inheritance, although some skills, such as memorising Qur'an verses, are acquired. Divination was commonly used for diagnosis, and religious rituals combined with herbal remedies were typical treatments (Moe et al. 2014).

It's indeed a positive development that governments in several Asian countries are recognizing the value of integrating CTT into mainstream healthcare systems. These policies not only support the training and certification of practitioners but also regulate herbal medicines and enhance the inclusion of these therapies in public healthcare services. This recognition contributes significantly to the widespread acceptance and accessibility of CTT in the region, aligning with the shift towards personalised medicine and holistic approaches in healthcare.

2.2 Methodology

This article adopts a conceptual approach, focusing on the exploration and development of ideas, and concepts, rather than gathering empirical data. It aims to advance understanding by synthesising existing literature, and philosophical perspectives to address specific issues or questions (MacInnis 2011). Drawing on interdisciplinary sources, the article relies on logical reasoning and conceptual analysis to develop coherent arguments. Unlike systematic literature reviews, which follow predefined protocols (Kraus et al. 2022), and empirical studies, which test hypotheses through data (Rana et al. 2020, 2022), this article does not rely on data collection. Instead, it enriches the conceptual landscape by offering new perspectives on the integration of CTT with personalised medicine in shaping the future of healthcare across Asian countries, while also highlighting areas for further exploration.

2.3 Key Concepts and Issues

In many countries across Asia and worldwide, there is a high demand for CTT services to support chronic disease prevention and management, symptom relief, and the healthcare needs of older populations (World Health Organization 2018). The increasing prevalence of chronic conditions, along with an aging population, are key drivers of this demand, underscoring the need to integrate CTT into mainstream healthcare practice.

Facilitating the safe and effective utilisation of these therapies can significantly improve health outcomes across the general population and promote healthy aging (Chung et al. 2021; Park and Canaway 2019). The recognition by governments in several Asian countries of the value of integrating CTT into mainstream healthcare systems marks a pivotal and positive development. These policies strengthen both the training and certification of practitioners, regulate herbal medicines and support the integration of these therapies into public healthcare services. Such recognition notably contributes to the widespread acceptance and accessibility of CTT in the region, harmonising with the shift towards personalised medicine and holistic healthcare approaches. Integrating scientifically proven effective and low-risk CTT into mainstream healthcare practice holds the potential to enhance the efficiency of health services and improve healthcare consumers' satisfaction.

The integration of traditional Chinese medicine into the mainstream healthcare system has been well-established in Taiwan. The general hospitals are incorporated with the Department of Chinese Medicine, providing both Western medicine and Chinese medicine services to the public (Ministry of Health and Welfare Taiwan 2023). To enhance the quality of Chinese medicine service, modern Chinese medicine physicians are trained to receive Western and Chinese medical knowledge. Universities and colleges with nursing departments are encouraged to open a Chinese medicine nursing program and elective course for Chinese medicine nursing. This scenario is a good example of integrating traditional medicine into the healthcare system, along with service and education programs.

An integrative healthcare approach is aligned with holistic and person-centered healthcare (Hall et al. 2018). The World Health Organisation is supporting the appropriate integration of CTT in the healthcare system (WHO 2018; WHO 2013). The WHO (2018) stated that the global burden of chronic diseases is the main reason for strengthening collaboration between conventional healthcare and complementary and traditional medicine. For the successful integration of CTT into mainstream healthcare practice, the policymakers need to facilitate communication and collaboration amongst complementary and traditional medicine practitioners, conventional clinicians, regulatory bodies, healthcare financing bodies and the public (Chung et al. 2021). Based on the strategic directions and actions provided in the WHO Traditional Medicine Strategy 2014–2023 (WHO 2013), the following actions could be considered for integration of CTT into the mainstream healthcare practice:

- Identify healthcare consumers' needs and preferences for the use of CTT.
- Determine the benefits, safety and cost-effectiveness of CTT based on the best available evidence.
- Determine the resources and training requirements and provide the required training and resources to healthcare providers.
- Develop practice guidelines for CTT in conventional healthcare.
- Support the conducting of research to investigate the effectiveness of CTT.

2.4 Discussion of Implications

Reflecting the action points suggested, the implications of integrating CTT in the mainstream healthcare system are discussed in the sub-sections: Healthcare consumers' needs and preferences, benefits, safety, and cost-effectiveness of CTT, resources and training for healthcare providers, regulatory system and practice guidelines, and research on CTT. Lastly, the integration challenges and issues are discussed with highlights on ethical concerns.

2.4.1 Healthcare Consumers' Needs and Preferences for the Use of Complementary and Traditional Therapy (CTT)

Effective integration of CTT in healthcare requires identification and analysis of healthcare consumers' preferences on the use of CTT. Based on their needs and preferences, strategies for integration could be considered. In fact, the utilisation of complementary and traditional medicine is driven by a broad spectrum of motivations, shaped by individuals' preferences, beliefs, and personal experiences. Integrating these therapies alongside conventional medicine highlights the significance of delivering patient-centred care that honours individual choices and values in healthcare decision-making. People seek out complementary and traditional medicine for a multitude of reasons, frequently integrating them with conventional treatments to amplify their efficacy. These therapeutic modalities provide comprehensive approaches to health, acknowledging and attending to the interconnectedness of physical, mental, emotional, and spiritual aspects of well-being. The allure of natural remedies, deemed gentler and safer, draws many to these practices. With deep roots in cultural and spiritual traditions, they empower individuals to actively engage in their own healthcare journey. Additionally, positive firsthand experiences and challenges accessing conventional care contribute to their adoption. By placing a strong emphasis on preventive care, these therapies advocate for holistic wellness, striving to maintain harmony and equilibrium within the body.

2.4.2 Benefits, Safety, and Cost-Effectiveness of Complementary and Traditional Therapy (CTT)

Understanding the needs and preferences of healthcare consumers, the subsequent action for effective integration of CTT is analysis and evaluation of the benefits, safety and cost-effectiveness of CTT. It's essential to approach traditional medicine with critical thinking and caution, recognising that while some traditional practices may have merit, others may lack scientific validation or could pose risks if used inappropriately. Integrating traditional medicine into modern healthcare systems often involves a careful evaluation of its safety, efficacy, and compatibility with evidence-based approaches to healthcare.

Research evidence on the health benefits of CTT is emerging. Acupuncture, tachi, yoga, aromatherapy, art therapy, music therapy, mindfulness and meditation-based interventions are shown to be beneficial in promoting the health and well-being of people with chronic conditions. Acupuncture and related therapies are found to be effective in reducing pain, insomnia, and fatigue (He et al. 2020; Lau et al. 2016). Mind and body interventions such as meditation, tai chi, qigong, and yoga practices are shown to improve sleep, cognitive function, and mental well-being and moderately reduce pain (Garland et al. 2020; Vasudev et al. 2019; Wang et al. 2019). Aromatherapy, music and art therapy are among the recommended therapeutic approaches to promoting physical and psychological well-being (Kievisiene et al. 2020; Maujean et al. 2014).

With regard to economic evaluation, there is a potential cost-effectiveness and cost-saving of complementary and integrative medicine (Herman et al. 2012; Ventegodt 2019). Amongst different types of CTT, mind and body interventions are suggested to be low-risk and cost-effective in promoting health outcomes. Based on the best available evidence, low-risk CTT practices could be integrated into routine healthcare practice provided health benefits outweigh the potential risks.

2.4.3 Resources and Training Requirements for Healthcare Providers

Considering healthcare professional perspectives, the majority have favorable views on complementary and traditional medicine, however, they are concerned about the safety of certain practices (Chang and Chang 2015; Johannessen and Garvik 2016). In addition to safety concerns, the lack of knowledge about complementary and traditional medicine among healthcare providers has been a major challenge in supporting the effective integration of complementary and traditional medicine in conventional healthcare (Anheyer et al. 2018; Chang and Chang 2015; Johannessen and Garvik 2016). There is a need to have educational programs for healthcare providers about commonly used CTT so that they can communicate the risks and benefits, and support people to make informed choices (Chang and Chang 2015). Providing educational talks or seminars about complementary and traditional medicine, in continuous medical/nursing education sessions, would be appropriate to expand healthcare providers' knowledge.

2.4.4 Regulatory Systems and Practice Guidelines for Complementary and Traditional Therapy (CTT)

The development of regulatory systems and practice guidelines is vital for the effective integration and appropriate practice of CTT. Regulations and guidelines serve as vehicles for CTT practitioners, adopting consistent standards and ensuring the health and safety of healthcare consumers. In the Asian context, traditional medicine has been used for a wide range of health conditions, and the majority of the countries are integrated into the national healthcare system (WHO 2013). In China, Japan, Korea, and Vietnam, regulatory systems for traditional medicine have been well established, resulting in improved healthcare services in these countries (Park and Canaway 2019). Ensuring quality and safety in health services, the integration of CTT could improve healthcare services through regulation of the products, practitioners, and services (Park and Canaway 2019). Strengthening regulatory mechanisms and the development of practice guidelines based on the best available evidence could enhance the efficacy and safety of complementary and traditional practices.

2.4.5 Research on the Effectiveness of Complementary and Traditional Therapy (CTT)

For strengthening practice guidelines and effective integration of CTTs, more research projects are required to investigate the efficacy and safety of a specific type of therapy on a specific health condition or outcome. Based on valid research findings, evidence-based practice guidelines could be developed. However, in reality, a lack of valid research data and insufficient funding support for research in CTT have been reported as major challenges in many countries (WHO 2019). Priorities for complementary and traditional medicine research should be considered to strengthen evidence-based practice and the effective integration of complementary and traditional medicine in mainstream healthcare settings.

The Asian Complementary and Traditional Research Network could be founded as a potential initiative that fosters knowledge sharing and joint research across Asian countries. Countries with inadequate research facilities and expertise should be given support to promote knowledge-based practice and evidence-driven integration of traditional medicine in their healthcare systems. Organising annual Complementary and Traditional Research Meetings could facilitate collaboration among member countries and communication on emerging issues and discussion on problem-solving strategies.

2.4.6 Integration Challenges and Issues Associated with Complementary and Traditional Therapies (CTT)

The challenges surrounding CTT are multifaceted and require attention from various stakeholders. The common issues highlighted in this section included insufficient research evidence, lack of regulatory framework, quality control and standard protocols for CTT practice and products, and exploitation and misleading the use of CTT products. A key issue is the lack of robust scientific evidence supporting the efficacy and safety of these therapies, which makes it challenging for healthcare providers and patients to make informed decisions. Also, a lack of standard treatment protocols and evidence-based guidelines is a significant challenge of CTT. Despite research efforts to assess the safety and efficacy of CTT, inconclusive findings suggest that the quality of evidence still needs improvement. This highlights the need for more rigorous research methodologies, such as well-designed randomized controlled trials and high-quality systematic reviews, to generate reliable evidence that can guide clinical practice and policy decisions related to CTT. This evidence not only ensures patient safety but also enhances the credibility and acceptance of these therapies among healthcare providers and patients.

Integrating complementary therapies with conventional medicine faces hurdles due to differences in treatment approaches and regulatory frameworks. Ensuring quality control, particularly with herbal supplements and alternative treatments, is a major concern that requires focused attention. In many developing countries, the absence of effective regulations and quality standards for herbal products and supplements allows unregistered and non-standard items to enter the market, posing potential risks. Quality control of herbal products encompasses the entire process from harvest and processing to shelf life and content, ensuring the postharvest value of plant- and animal-based products is maintained. Safety risks and potential interactions with conventional medications highlight the importance of thorough risk assessment and monitoring. Furthermore, there is an urgent need for increased education and awareness among healthcare providers, patients, and the public regarding the benefits, limitations, and potential risks of CTT. Continuing education opportunities and certification programs also ensure ongoing competency and quality assurance among practitioners.

Other challenges include exploiting the popularity of natural remedies for profit, marketing herbal products as anticancer remedies despite a lack of supporting evidence and leveraging religious beliefs to attract customers (Ashrafizadeh and Rassouli 2023). Cultural beliefs, ethical considerations, and varying acceptance levels further complicate the use and perception of these therapies, emphasising the necessity for comprehensive solutions to effectively address these challenges.

All the above issues concern with ethical use of CTT and the ethically appropriate integration of CTT in mainstream healthcare. While there is insufficient evidence on the safety of certain therapies, healthcare providers are in a dilemma to recommend the use of the patient's preferred type of therapy. For the appropriate integration of CTT in a personalised healthcare approach, a clear understanding of the indications, therapeutic effects, contraindications and potential risks of specific therapy is imperative to prevent serious complications. Ultimately, all parties involved in the integrative healthcare approach: conventional healthcare personnel, CTT providers and practitioners, suppliers and promoters of CTT products must be enforced to comply with ethical conduct.

2.5 Future Directions

The advancement of Complementary and Traditional Therapy (CTT) in healthcare depends on the combined efforts of healthcare professionals, researchers, policymakers, and stakeholders. Innovations in big data, genomics, artificial intelligence (AI), and telemedicine are transforming personalised medicine, enabling tailored treatments that boost the effectiveness of traditional therapies. AI and machine learning enhance real-time health assessments, while wearable technology and personalised wellness plans promote proactive health management. For safe and ethical integration, standardized policies, global collaboration, and public awareness are essential. This holistic approach has the potential to transform healthcare by enhancing patient outcomes and broadening access to culturally sensitive, evidence-based CTT.

Advancement in CTT in healthcare necessitates collaborative efforts from healthcare professionals, researchers, policymakers, and stakeholders. These endeavours should emphasise the adaptation to the advancing technological landscape to pave the way for the future of CTT, the promotion of evidence-based practices, the assurance of safety and quality standards, and the improvement of education and awareness. Technological innovations drive both progress and accessibility in healthcare. The integration of the evolving technological landscape with personalised medicine, which tailors healthcare to individual characteristics, has the potential to revolutionise the incorporation of complementary and traditional medicine. The advancements in big data, genomics, AI, and telemedicine might impact this field. Big data can monitor health trends and outcomes, enabling population-specific health strategies that incorporate traditional medicine. Integrating large databases with genomic data and traditional medicine outcomes can promote evidence-based practices and identify effective treatments. Machine learning algorithms can identify the most effective traditional treatments tailored to individual health profiles. AI can enhance diagnosis and treatment plans by analysing complex data from both traditional medicine practices and modern diagnostics.

Personalised medicine integrates traditional diagnostic methods like pulse diagnosis in Ayurveda or tongue diagnosis in TCM with conventional diagnostic methods for health assessments, providing comprehensive health insights and guiding treatment strategies. In the upcoming decade, facilities are expected to embrace portable, non-invasive diagnostic devices with AI features. This device will analyse a mix of conventional medicine data and traditional diagnostic information. This progress may lead to personalised health recommendations in real-time, enabling early and precise health assessments that promote proactive and preventive healthcare approaches. In the realm of preventive medicine, there's a potential trend toward personalised wellness plans that combine modern health screenings with traditional wellness practices. For example, during a comprehensive health risk assessment, individuals undergo genetic testing and TCM pulse diagnosis, leading to a tailored wellness plan. This plan includes nutritional guidance, personalised herbal supplements, regular physical activity, and mindfulness practices. Looking ahead, personalised wellness centres may emerge, offering integrated health assessments and tailored interventions. These centres could leverage advanced AI algorithms to continuously refine and improve wellness plans based on ongoing health data and emerging research. In community health, a holistic approach aims to provide comprehensive, culturally sensitive, and personalised care to individual needs and preferences. Technological advancements like AI and big data analytics will play a crucial role in enabling real-time customisation and continuous improvement of health interventions. It is anticipated that in the future, health service providers may offer both conventional screenings and traditional health consultations. These programs will integrate community-specific traditional practices, enhancing engagement and effectiveness while respecting cultural contexts.

Personalised medicine combines mind–body interventions like mindfulness, yoga, meditation and cognitive-behavioral techniques with traditional healing practices, which has the potential to enhance well-being and address specific health concerns. Additionally, tailored herbal and nutritional therapies, considering genetic, metabolic, and lifestyle factors, align with individual health needs alongside conventional medicine, and could optimize overall healthcare service. In cancer treatment, patients may receive a comprehensive treatment plan that combines precision oncology techniques and personalised herbal medicine. This approach is further enhanced by integrating mind–body therapies into the treatment regimen. By combining these modalities, healthcare providers aim to achieve better treatment outcomes for cancer patients, addressing not only the physical aspects of the disease but also considering the patient's emotional and mental well-being. In the coming years, cancer care is poised to witness the emergence of highly integrated multidisciplinary teams. These teams will likely include oncologists, geneticists, herbalists, and mind–body therapists, among other specialists. In the management of chronic diseases, such as hypertension, personalised medicine integrates conventional tools like a blood pressure monitor with an advanced mobile application for tailored dietary guidance and stress-reduction practices such as Tai Chi.

The incorporation of wearable devices, such as smartwatches and biosensors, allows individuals to monitor physiological parameters, providing real-time feedback to manage stress and anxiety effectively. In the future, sophisticated AI-driven platforms are poised to revolutionise chronic disease management. These platforms will seamlessly integrate continuous health data with personalised lifestyle and medication adjustments, ensuring comprehensive care round-the-clock. By combining conventional medicine with complementary practices, these systems aim to optimise disease control and enhance patients' quality of life.

Given the widespread use of complementary traditional therapy, it is crucial to seamlessly integrate these therapies into mainstream healthcare systems to unlock their full potential. This integration will indeed facilitate better access to evidence-based CTT and ensure that they are effectively incorporated into treatment plans. In the future, a mutual understanding of the role of these therapies among conventional medical practitioners, complementary therapy providers, and traditional healers is essential. This shared understanding will foster collaboration resulting in better patient outcomes. The interdisciplinary approach promotes a holistic understanding of health and disease, allowing diverse therapeutic modalities to address complex healthcare needs comprehensively. This integration requires policies and regulations that focus on recognition, accessibility, affordability, and reimbursement, thereby shaping the landscape of personalised medicine in conjunction with complementary traditional therapy. Implementing standardised guidelines and fostering international collaboration are crucial to ensuring consistent quality and promoting knowledge exchange in evidence-based CTT, which in turn aids in promoting recognition and accessibility. Consequently, establishing integrative care centres that combine conventional medicine with complementary traditional therapies can offer a holistic approach to healthcare. These centres provide services such as acupuncture, herbal medicine, and mindfulness practices alongside conventional treatments.

Governments can enhance healthcare infrastructure, integrate AI and big data analytics, and expand telemedicine services to improve access to evidence-based CTT. Providing specialised training for healthcare professionals and conducting public awareness campaigns, such as health fairs and outreach programs, further support accessibility. Integrating complementary traditional medicine into telemedicine platforms could reach a broader population through virtual consultations, online resources, and remote monitoring. Affordable policies may include offering subsidies and mandating insurance coverage for complementary traditional therapy consultations, aligning with established guidelines for safe and ethical practice. This integration expands patient treatment options, leading to improved outcomes and enhancing overall healthcare effectiveness.

2.6 Conclusion

The integration of Complementary and Traditional Therapy (CTT) with personalised medicine holds significant promise for transforming healthcare across Asian countries. This comprehensive approach acknowledges the cultural, spiritual, and holistic dimensions of health that are deeply rooted in the region's diverse practices. While challenges remain, such as the need for rigorous scientific validation, standardisation, and effective regulatory frameworks, the potential benefits are substantial. Embracing technological advancements like AI and big data, fostering interdisciplinary collaboration, and promoting evidence-based practices are essential steps forward.

The future of healthcare in Asia depends on the seamless incorporation of CTT into mainstream medical systems, ensuring accessibility, affordability, and cultural relevance. This integration requires policies that support training, certification, and public awareness, as well as the development of integrative care centres. By fostering a mutual understanding among conventional medical practitioners, complementary therapy providers, and traditional healers, healthcare systems can deliver more holistic and patient-centred care.

The synthesis of CTT and personalised medicine can lead to better health outcomes, enhanced patient satisfaction, and more effective healthcare services. By addressing current challenges and leveraging future opportunities, Asian countries can set a global example for a truly integrated and comprehensive healthcare model. This transformation will ultimately improve the well-being of individuals and communities, paving the way for a healthier future.

Highlights

- Appropriate integration of complementary and traditional therapy (CTT) into mainstream healthcare systems has the potential to significantly improve health outcomes in the future.
- Establishing regulatory systems and practice guidelines for CTT are vital for ensuring patients' safety.
- Integrating traditional way of healing in modern healthcare practice, personalised healthcare approach could provide comprehensive health screening and treatment plans.

References

Anheyer D, Kern C, Dobos G, Cramer H. "I think you can achieve quite a lot if all of the staff stands behind it"-A qualitative study about the experience, knowledge and application of complementary therapies and integrative medicine in pediatrics. Complement Ther Med. 2018;41:186–91. https://doi.org/10.1016/j.ctim.2018.09.025.

Ashrafizadeh H, Rassouli M. Traditional, complementary, and alternative medicine in cancer care: Challenges and opportunities. Asia Pac J Oncol Nurs. 2023;11(1). https://doi.org/10.1016/j.apjon.2023.100321

Bauer M, Rayner JA. Use of complementary and alternative medicine in residential aged care. J Altern Complement Med. 2012;18(11):989–93. https://doi.org/10.1089/acm.2011.0582.

Chang HY, Chang HL. A review of nurses' knowledge, attitudes, and ability to communicate the risks and benefits of complementary and alternative medicine. J Clin Nurs. 2015;24(11–12):1466–78. https://doi.org/10.1111/jocn.12790.

Chong HY, Allotey PA, Chaiyakunapruk N. Current landscape of personalised medicine adoption and implementation in Southeast Asia. BMC Med Genomics. 2018;11(94). https://doi.org/10.1186/s12920-018-0420-4

Chung VCH, Ho LTF, Leung TH, Wong CHL. Designing delivery models of traditional and complementary medicine services: a review of international experiences. Br Med Bull. 2021;137(1):70–81. https://doi.org/10.1093/bmb/ldaa046.

Chung VCH, Wong CHL, Zhong CCW, Tjioe YY, Leung TH, Griffiths SM. Traditional and complementary medicine for promoting healthy ageing in WHO Western Pacific Region: Policy implications from utilisation patterns and current evidence. Integr Med Res. 2021;10(1). https://doi.org/10.1016/j.imr.2020.100469

Garland EL, Brintz CE, Hanley AW, Roseen EJ, Atchley RM, Gaylord SA, Faurot KR, Yaffe J, Fiander M, Keefe FJ. Mind-body therapies for opioid-treated pain: A systematic review and meta-analysis. JAMA Intern Med. 2020;180(1):91–105. https://doi.org/10.1001/jamainternmed.2019.4917.

Hall H, Brosnan C, Cant R, Collins M, Leach M. Nurses' attitudes and behavior towards patients' use of complementary therapies: A mixed methods study. J Adv Nurs. 2018;74(7):1649–58. https://doi.org/10.1111/jan.13554.

He Y, Guo X, May BH, Zhang AL, Liu Y, Lu C, Mao JJ, Xue CC, Zhang H. Clinical evidence for association of acupuncture and acupressure with improved cancer pain: A systematic review and meta-analysis. JAMA Oncol. 2020;6(2):271–8. https://doi.org/10.1001/jamaoncol.2019.5233.

Herman PM, Poindexter BL, Witt CM, Eisenberg DM. Are complementary therapies and integrative care cost-effective? A systematic review of economic evaluations. BMJ Open. 2012;2(5):e001046. https://doi.org/10.1136/bmjopen-2012-001046.

Johannessen B, Garvik G. Experiences with the use of complementary and alternative medicine in nursing homes: A focus group study. Complement Ther Clin Pract. 2016;23:136–40. https://doi.org/10.1016/j.ctcp.2015.04.003.

Johnson PJ, Jou J, Rockwood TH, Upchurch DM. Perceived benefits of using complementary and alternative medicine by race/ethnicity among midlife and older adults in the United States. J Aging Health. 2019;31(8):1376–97. https://doi.org/10.1177/0898264318780023.

Kievisiene J, Jautakyte R, Rauckiene-Michaelsson A, Fatkulina N, Agostinis-Sobrinho C. Corrigendum to "The effect of art therapy and music therapy on breast cancer patients: what we know and what we need to find out-a systematic review". Evid Based Complement Alternat Med. 2021;12. https://doi.org/10.1155/2021/9870102

Kraus S, Breier M, Lim WM, et al. Literature reviews as independent studies: guidelines for academic practice. Rev Manag Sci. 2022;16:2577–95. https://doi.org/10.1007/s11846-022-00588-8.

Lau CHY, Wu X, Chung VCH, Liu X, Hui EP, Cramer H, Lauche R, Wong SYS, Lau AYL, Sit RWS, Ziea ETC, Ng BFL, Wu JCY. Acupuncture and related therapies for symptom management in palliative cancer care: Systematic review and meta-analysis. Medicine (Baltimore). 2016;95(9): e2901. https://doi.org/10.1097/MD.0000000000002901.

MacInnis DJ. A framework for conceptual contributions in marketing. J Mark. 2011;75(4):136–54. https://doi.org/10.1509/jmkg.75.4.136.

Maujean A, Pepping CA, Kendall E. A systematic review of randomized controlled studies of art therapy. Art Therapy 2014;31(1):37–44. https://doi.org/10.1080/07421656.2014.873696#.U3y0Eygq_7A

Moe T, Al Obaidly FA, Al Khoder R, Schmid D. Folk medicine and integrative healing in West Asia: Analysis of historical and modern practice and perceptions. 2014.https://doi.org/10.5339/avi.2014.1

Ministry of Health and Welfare Taiwan. (2023). The Development of Traditional Chinese Medicine in Taiwan. https://www.mohw.gov.tw/cp-3700-57710-2.html

Oyunchimeg B, Hwang JH, Ahmed M, Choi S, Han D. Complementary and alternative medicine use among patients with cancer in Mongolia: a National hospital survey. BMC Complement Altern Med. 2017;17(1):58. https://doi.org/10.1186/s12906-017-1576-8.

Park JE, Yi J, Kwon O. Twenty years of traditional and complementary medicine regulation and its impact in Malaysia: Achievements and policy lessons. BMC Health Serv Res. 2022;22(1):102. https://doi.org/10.1186/s12913-022-07497-2.

Park YL, Canaway R. Integrating traditional and complementary medicine with national healthcare systems for universal health coverage in Asia and the western pacific. Health Systems & Reform. 2019;5(1):24–31. https://doi.org/10.1080/23288604.2018.1539058.

Pawera L, Łuczaj Ł, Pieroni A, Polesny Z. Traditional plant knowledge in the white carpathians: ethnobotany of wild food plants and crop wild relatives in the Czech Republic. Hum Ecol. 2017;45:655–71. https://doi.org/10.1007/s10745-017-9938-x.

Pradipta IS, Aprilio K, Febriyanti RM, Ningsih YF, Pratama MAA, Indradi RB, Gatera VA, Alfian SD, Iskandarsyah A, Abdulah R. Traditional medicine users in a treated chronic disease population: A cross-sectional study in Indonesia. BMC Complement Med Ther. 2023;23(1):120. https://doi.org/10.1186/s12906-023-03947-4.

Rana S, Raut SK, Prashar S, Hamid ABA. Promoting through consumer nostalgia: A conceptual framework and future research agenda. J Promot Manag. 2020;27(2):211–49. https://doi.org/10.1080/10496491.2020.1829773.

Rana S, Raut SK, Prashar S, Quttainah MA. The transversal of nostalgia from psychology to marketing: What does it portend for future research? Int J Organ Anal. 2022;30:899–932. https://doi.org/10.1108/ijoa-03-2020-2097.

Sitpayeva G, Yerekeyeva S, Grudzinskaya L, Gemejieva N, Anarbekova G, Saikenov B. Bringing medicinal plants of the native flora of the northern Tien Shan into cultivation. J Water Land Dev. 2021;50:207–19. https://www.jwld.pl/files/2021-03-JLWD-23-Sitpayeva.pdf.

Tangkiatkumjai M, Boardman H, Walker DM. Potential factors that influence usage of complementary and alternative medicine worldwide: a systematic review. BMC Complement Med Ther. 2020;20(1):363. https://doi.org/10.1186/s12906-020-03157-2.

Ventegodt S. Comparative analysis of the world's ten major medical systems based on data from evidence-based medicine (preferably meta-analyses): Which types of medicine are effective for a majority of clinical conditions, safe, affordable, and sustainable? J Alternative Med Res 2019;11(2). https://novapublishers.com/shop/volume-11-issue-2-2019-journal-of-alternative-medicine-research/

Vasudev A, Torres-Platas SG, Kerfoot K, Potes A, Therriault J, Gifuni A, Segal M, Looper KJ, Nair V, Lavretsky H, Rej S. Mind-body interventions in late-life mental illnesses and cognitive disorders: A narrative review. Am J Geriatr Psychiatry. 2019;27(5):536–47. https://doi.org/10.1016/j.jagp.2018.10.020.

Verma SK, Pandey M, Sharma A, Singh D. Exploring Ayurveda: principles and their application in modern medicine. Bull Natl Res Cent. 2024;48:77. https://doi.org/10.1186/s42269-024-01231-0.

Wang X, Li P, Pan C, Dai L, Wu Y, Deng Y. The effect of mind-body therapies on Insomnia: A systematic review and meta-analysis. Evid Based Complement Alternat Med. 2019 Feb 13. https://doi.org/10.1155/2019/9359807

Wang WJ, Zhang T. Integration of traditional Chinese medicine and Western medicine in the era of precision medicine. J Integr Med. 2017;15(1):1–7. https://doi.org/10.1016/S2095-4964(17)60314-5.

World Health Organization. (2018). Traditional and complementary medicine in primary health care. https://apps.who.int/iris/bitstream/handle/10665/326299/WHO-HIS-SDS-2018.37-eng.pdf

World Health Organization. (2019). WHO global report on traditional and complementary medicine 2019. World Health Organization. https://www.who.int/traditional-complementary-integrative-medicine/WhoGlobalReportOnTraditionalAndComplementaryMedicine2019.pdf

World Health Organization (WHO). (2013). WHO traditional medicine strategy 2014–2023 https://apps.who.int/iris/bitstream/handle/10665/92455/9789241506090_eng.pdf;jsessionid=B42B0FC78051F2AD453BBEFDC664DB7A?sequence=1

Zinboonyahgoon N, Sharma S, Song X (2023). Traditional medicine practice across Asia, examples of non-western approaches. International Study For the Pain. https://www.iasp-pain.org/resources/fact-sheets/traditional-medicine-practice-across-asia-examples-of-non-western-approaches-2/

Zinboonyahgoon N, Srisuma S, Limsawart W, Rice ASC, Suthisisang C. Medicinal cannabis in Thailand: 1-year experience after legalization. Pain. 2021;162(Suppl 1):S105–9. https://doi.org/10.1097/j.pain.0000000000001936.

Ping Lei Chui is a Associate Professor in the Department of Nursing Science at the Faculty of Medicine, Universiti Malaya, Malaysia. She actively teaches both undergraduate and postgraduate students, contributes to curriculum development and review, and coordinates the Master of Nursing Science program. An accomplished researcher with numerous publications in international refereed journals, her expertise spans cancer care (including health promotion, symptom management, and patient education), complementary and integrative therapies (such as oral cryotherapy, mindful breathing, mind-body practices, and kangaroo mother care), various nursing issues (encompassing health behavior, professional quality of life, workplace violence, and information literacy), as well as stoma and wound care, and both adult and pediatric nursing.

Nant Thin Thin Hmwe is a Senior Lecturer in the Department of Nursing Science, Faculty of Medicine, Universiti Malaya. She graduated with a Bachelor of Nursing Science from the University of Nursing Yangon, Myanmar. She obtained Master of Nursing Science from Universiti Malaya, and a PhD in Nursing from the University of Newcastle, Australia. She is teaching undergraduate and postgraduate nursing programmes. Her research supervision involves undergraduate, Master's by coursework and PhD (Nursing) projects. Her research interests are aged care, renal and chronic diseases, sleep and wellbeing, traditional and complementary medicine, nursing education, and mixed-method research.

Chapter 3
It Takes a Village to Ruffle Feathers: The HIV Activism Agenda and Framework for the Asia–Pacific Region

Clusters of Future Studies: Past and Futures; Humanity at the Limen

Norman Chong, Valerie A. Earnshaw, and Iskandar Azwa

Abstract Health inequities are undermining life-saving HIV care, prevention, and harm reduction services for people with and affected by HIV. HIV activism has been a driver for change in improving the HIV response, from securing access to life-saving HIV treatment and challenging socio-cultural structures to demanding greater and meaningful involvement of communities. However, many initiatives and actions to progress the HIV response face a myriad of structural challenges and resistance. Harnessing the against-the-grain persistence of HIV activism, our conceptual article proposes a direction for HIV activism in Asia using a structured review of scholarly literature and policy documents to identify the successes and gaps in the HIV response, extrapolating the cost of inaction and indifference to the future of the HIV response, and providing recommendations for future health research, policy, and practice and forming meaningful alliances. We introduced a Four-fold HIV Activism Agenda and HIV Activism Framework for Asia to improve the HIV response across Asian countries. We emphasise the urgency to address the immediately actionable drivers of health inequities in the HIV response in the region, especially in centring communities most affected by HIV, and improving individual well-being. It necessitates addressing health inequities more systematically and collaboratively with a scaled approach.

N. Chong (✉)
Kirby Institute, University of New South Wales, Sydney, Australia
e-mail: nchong@kirby.unsw.edu.au

V. A. Earnshaw
Department of Human Development and Family Sciences, University of Delaware, Newark, DE, USA
e-mail: earnshaw@udel.edu

I. Azwa
Center of Excellence for Research in Infectious Diseases and AIDS (CERIA), Department of Medicine, Faculty of Medicine, Universiti Malaya, Kuala Lumpur, Malaysia
e-mail: iskandarshah@ummc.edu.my

R. Xiang Ng et al. (eds.), *Future of Healthcare in Asia*,
https://doi.org/10.1007/978-3-032-03183-9_3

Keywords HIV · Activism · Stigma · People with HIV · Health determinants · Empowerment

3.1 Introduction

> 'I started my work in the HIV clinic when I was pregnant in 2009. A frequently asked question then from colleagues and even friends was, "Why did you choose to work in an HIV clinic while pregnant? Aren't you worried about transmitting HIV?" My answer was simple yet profound: I understand well the modes of HIV transmission and feel confident that I can provide care without compromising my health or that of my coming baby.' Chan Shan, HIV specialty nurse in the International AIDS Society Me and My Healthcare Provider Campaign (2024)

Historically, HIV activism has been pivotal in advancing HIV prevention and care as well as dismantling HIV-related stigma. Like Chan Shan, many HIV activists in Asia, including healthcare champions and community leaders, often go out of their way to debunk HIV myths or confront jokes, statements, or innuendoes laced with prejudice, stereotypes, and intentions to discriminate against communities affected by HIV. As an integral component of the HIV response, HIV activisms are direct actions aimed at challenging established norms, and demanding democratic involvement to promote health, socioeconomic, and political justice to communities most affected by HIV. Preliminary findings support the benefits of HIV activism on well-being, active coping with stigma, social integration, and life satisfaction (Earnshaw et al. 2016a, b). However, in the face of substantial challenges, HIV activism also comes at the expense of the activist's mental health (Earnshaw et al. 2016a, b; Gilster2012, Owens et al. 2021) and potential legal implications (Bong 2020; Muhammed and Amuda 2018; Lim et al. 2020).

HIV activism intersects with activism related to ethnicities, gender, sexuality, generation, (dis-)abilities, and other social statuses (Klar and Kasser 2009; Bowleg 2021). It includes a broad spectrum of direct and indirect actions aimed at becoming an agitating force for societal change, advancing restorative justice (Freudenberg et al. 2011; Laverack 2013), and ensuring the well-being of communities affected by HIV who are historically othered and experience marginalisation (Earnshaw et al. 2016a, b). We draw the conceptualisation of activism arising from feminist scholars amidst the COVID-19 pandemic. Anderson and Eskandari (2023) posit a multidimensional continuum of activism that spans spatiotemporal, scalar, and typological dimensions. Activism arises through four dimensions: temporal (evolving before, during, and post-conflict), spatial (occurring in diverse arenas such as domestic spheres, streets, or communal spaces), scalar (addressing insecurity of varying magnitudes, be it local, regional, or global) and typology (challenging structural, cultural or symbolic oppression). These actions can range from lobbying for legislative change, initiating an advocacy group, product boycotts, and peaceful civil demonstrations or strikes to launching large-scale campaigns (Corning and Myers 2002; Laverack 2013). Empowerment and activism share a commonality of the sentiment of tension, shared grievance, struggle, and liberation that are inherent in the process of achieving hegemony (Ozduzen et al. 2023).

Over the past five years, the concurrent AIDS and COVID-19 pandemics, exacerbated by economic and humanitarian crises, have increasingly jeopardised the global HIV response (UNAIDS 2022). These crises have notably undermined life-saving health services for people living with and affected by HIV, including screening, linkage to care, and harm reduction services. It has also impacted the 95-95-95 UNAIDS goals, i.e., 95% of individuals with HIV to be aware of their status, 95% of PWH on continuous antiretroviral therapy, and 95% of those on treatment to achieve viral suppression. As of 2023, the Asia–Pacific region stands at 78–67-65 (UNAIDS 2022).

In Asia, where progress has stagnated amidst exhausted resources, current trends demonstrate inadequate efforts to reduce widening health inequities (UNAIDS 2022). This stagnation may have emerged as a result of prioritising profits over people (Dyer 2020), pathologising marginalised communities (Earnshaw et al. 2014; Earnshaw et al. 2016a, b; Jin et al. 2014; Ni et al. 2023; Tee et al. 2019) instead of repealing punitive laws, and undermining harm reduction services instead of promoting inclusive, community-centred service delivery (Nkengasong and Ratevosian 2023).

3.2 Methodology

The epistemological assumptions of this chapter are based on a social constructionist or interpretivist paradigm. Our positionalities are united by a shared commitment to addressing stigma and advancing health equity through research and community engagement. Our diverse backgrounds, ranging from clinical practice and research to activism, inform our approach to understanding and addressing the multifaceted challenges in the HIV response. We bring a wealth of experience, interdisciplinary perspectives, and a deep commitment to improving the lives of those affected by HIV, guided by principles of equity, community empowerment, and evidence-based practice. Our work and collaborations in the Asia–Pacific region, while considerably wide, are focused mainly on the Southeast Asia region. Thus, the propositions in this chapter stem (and are limited to) from our knowledge and expertise in geographical areas that are most familiar to us.

This book chapter is a conceptual article. A conceptual article distinguishes itself from empirical articles by not presenting or relying on data analysis (Rana 2020) and from systematic reviews as it follows a structured, bias-accountable, rigourous review protocol (Kraus et al. 2022; Tricco et al.2018). As a recognised form of scholarly work, conceptual articles synthesise existing literature based on comprehensive knowledge, expertise, and experience. Conceptual articles highlight key insights, irregularities, gaps, and gaps, offering a synthesis of a consolidated perspective of current research as well as outlining an agenda for future research, practice, and policy (Gilson and Goldberg 2015).

In developing the operational framework for this conceptual article, we followed recommendations from Jaakkola (2020) in framing a conceptual article. Our approach incorporates all four conceptual designs, as we (1) synthesised and integrated current successes and gaps in the HIV response in Asia (i.e., theory synthesis); (2) incorporated current understanding using a theoretical perspective to introduce a Four-fold HIV Activism Agenda and HIV Activism Framework for Asia (i.e. theory adaptation); (3) offered tangible applications of the HIV activism agenda on various socio-ecological levels and outcome measures (i.e. typology); and (4) presented recommendations driven by a theoretical framework (i.e. model). This integrated approach enables us to make a theoretical contribution by reshaping and expanding HIV activism in Asia.

3.3 Key Concepts and Issues

This article maintains its conceptual focus on setting up time-sensitive actions across socio-ecological levels to improve the collective HIV response in the Asia–Pacific region. In setting up the agenda for HIV activism for Asia, we reviewed the successes and gaps Asia has yet to achieve in the HIV response over time. Secondly, we leveraged current trends and evidence in Asia to illustrate the cost of continued inaction and indifference to the HIV response. Lastly, we offer tangible recommendations for future research, practice, and policy grounded in the socio-ecological model of health to bolster access to HIV care, improve individual well-being, and cultivate stigma-resistant communities. By employing the socio-ecological model of health, this chapter provides a crosscutting perspective that shifts from individual-based activism to collectivism, which includes structural activism and allyship.

3.3.1 HIV Activism: Lessons from History

HIV activism has a rich history of advancing treatment and playing a vital role in dismantling HIV-related stigma. Addressing the stigma contributing to the marginalisation and disenfranchisement of affected populations is a key aspect of this activism. Parker and Aggleton (2003) also emphasised that stigma is intrinsically fuelled by power and dominance. Community-driven solidarity and resistance have been essential in the HIV response to drive change. Additionally, there is a ballooning discussion on shifting stigma research from deficit-focused to empowerment-based. Deficit-based stigma research emphasises the detriments of stigma as leverage to advocate for social justice and equity. Exploring perspectives centred on resilience, empowerment, and resistance may offer more effective strategies for driving change (Logie et al. 2021, Stangl et al. 2019; Earnshaw et al. 2013).

The AIDS Coalition to Unleash Power (ACT UP) in the United States of America (USA) and France has significantly influenced the landscape of HIV activism, emphasising the importance of focusing on the lives of people living with and affected by HIV and organising efforts to identify and problem-solve. As demonstrated by ACT UP, collective action can lead to various positive outcomes, including greater personal empowerment (Zimmerman and Rappaport 1988), reforming immigration laws, advancing safer sex education, and enhancing medication approval procedures. ACT UP provided a powerful means for individuals and groups to challenge discrimination against people with HIV by government agencies, social service providers, medical institutions, the media, and the public (Brashers et al. 2002). HIV activists primarily drive change through intra- and inter-group communication, including networking, active recruitment meetings, mass media, engagements with allies, and counter-movement groups.

The HIV counter-movement groups in Asia evolved differently from the USA and France, mainly because of the influence of strong British colonisation. Many laws with British colonial roots, which are often punitive in nature, are still in place in many Asian countries (e.g., Malaysia, India, Pakistan, Brunei, and Indonesia). This has shaped social and cultural norms that are difficult to disentangle even with the radical changes in the UK legal landscape over the years. These laws include criminalisation of drug use, sexual and gender diverse individuals, and sex work, including HIV status, in various contexts.

3.3.2 *Cost of Continued Inaction and Indifference*

Should the gaps in the HIV response remain unimproved, healthcare systems in Asia may be forced to bear the cost of inaction and indifference. The following disparities serve as projections for the price of stagnancy and entry points for greater responsiveness. Evidence supporting these projections stems from Asia-based studies and beyond that can be applied to diverse contexts.

3.3.2.1 Structural Stigma

Structural stigma results from societal conditions, norms, and policies that limit the mobilisation of resources and opportunities in favour of the stigmatised (Hatzenbuehler 2016). Eighteen countries in the Asia–Pacific continue to criminalise consensual adult same-sex relations, freedom of gender expression, and marriage equality, including the death penalty. Anti-LGBT laws (particularly Sect. 377 of the Penal Code prohibiting ‘unnatural acts’) rooted in the British colonial legacy still exist in many Commonwealth nations such as Pakistan, Malaysia, Brunei, and Sri Lanka. Countries in Asia with punitive LGBT legislation have the highest HIV burden in the world—with HIV risk acquisition, late HIV diagnoses, late presentation to care, and AIDS-related mortality comparatively higher in countries that criminalise same-sex activity than countries that do not (The Lancet 2022). Rates of HIV screening, HIV status awareness, and ART initiation were significantly lower in countries with severe anti-LGBT laws relative to countries with the least stringent (Kavanagh et al. 2021). Studies evaluating this disparity demonstrated that HIV-related health-seeking behaviours were significantly lower with the introduction of the legislation (Schwartz et al. 2015). Fear of disclosure, including perceived and anticipated stigma, reportedly mediated the relationship between structural stigma and engagement in the HIV care cascade and HIV prevention. Mathematical modelling comparisons of the influence of stigma on testing and treatment cascade indicate that, in the absence of addressing stigma, only 88-88-89 of the 95-95-95 targets are likely to be achieved (John 2022).

3.3.2.2 Inadequate Endorsement from National and Local Governing Bodies on Community-Centred HIV Response.

National and local governing bodies must focus on community-centred HIV response, i.e., a HIV response that is by and for the people living with and affected by HIV. A top-down HIV response that seeks one-size-fits-all solutions is not only a denial of people affected by HIV but also perpetuates an ideology of oppression. With the toolbox and the prospect of innovative modalities for HIV treatment and prevention, the importance of wide-ranging choices and more products arises because individuals have diverse needs, priorities, developmental processes, and social contexts. Community-centredness does not only consider needs and aspirations across key populations but also age, gender, geographical areas, and subpopulations, in line with the Global AIDS Strategy targets to end AIDS by 2030, appreciating the 95-95-95 within all subpopulations and sociodemographic characteristics. We foresee several potential costs of not accounting for subpopulations and sociodemographic diversity:

(1) The (re)production of microlevel inadequacies can culminate into macro-level disparities, consistent with the frameworks of social production of disease (Krieger 1994, Krieger 2001, Link and Phelan 1995, Link and Phelan 1996) and the political economy of health (Raphael and Bryant 2006; Baer 1982);
(2) Macrolevel successes may overlook microlevel inadequacies—potentially leading to 'good enough' nonchalance.
(3) Top-down approaches can be economically cost-ineffective and limit projects to their pilot stages without emphasising scaling-up or service integration.

3.3.2.3 Widening of HIV Funding Gap

Another obstacle to maintaining equitable public health expenditure for epidemic control is micro and macro-fiscal restrictions, including the accrued debt at the national level due to the redistribution of resources arising from the COVID-19 pandemic. Community-led initiatives are a cornerstone in meeting the aspirations and needs of the individuals they serve. The UNAIDS 30-60-80 targets made measuring how well a response centres communities tangible (Anam et al. 2023), i.e., 30% of testing and treatment services to be provided by community-led organisations, 60% of programs aimed at supporting societal enablers to be delivered by community-led organisations, and 80% of service delivery for HIV prevention programs targeting key populations and women to be administered by community, key population, and women-led organizations. Partly attributable to the economy's volatility and regulatory environments that do not make it flexible for the community to receive the trust and empowerment to create continuity in projects, the inadequate capacity and supportive monitoring and evaluation structure to assess implementation impacts potential influx of funding.

The actions of Donald Trump's administration endanger the health, well-being, and lives of key populations across the Asia–Pacific region. It commenced with the withdrawal of the United States of America from the World Health Organization (WHO) and the Paris Agreements. It proceeded with the halting of the United States Agency for International Development (USAID) and a 90-day suspension of United States foreign aid, particularly towards the President's Emergency Plan for AIDS Relief (PEPFAR) supported by Elon Musk, who characterised USAID as 'evil' and a form of 'money laundering'. Even with a waiver for "life-saving humanitarian programs," services exempt from the radical and abrupt funding cessation remain uncertain, especially those focused on HIV prevention and key populations. Censorship directives on US Government websites and scientific journals, including the use of the terms 'activism', 'gender', 'transgender', 'LGBT', 'diversity and inclusion', and 'equity', are threatening the health research community (Lancet 2025). Many community-based organisations in the Asia–Pacific region have highlighted the impacts of these changes on the HIV response in their local communities, including closures of health services provided at the district and provincial, abrupt unemployment, and disruption of supply chains (APCOM 2025). Should this continue to escalate, it may undo the decades of progress achieved thus far with our response to major public health threats, including HIV, tuberculosis, hepatitis C, and malaria.

While efforts to transition from the traditional external donor-dependent funding models to domestically led models are commendable, continued inadequate funding structures may influence the resilience and sustainability of initiatives that we know are effective. Social contracting, a legally binding agreement between a governmental body and a non-governmental entity (e.g., community-based organisation), could be a strategy to mitigate this gap. Under this arrangement, the government commits to remunerating non-governmental entities at mutually determined costs. However, significant commitments were accompanied by slow implementation, wherein only 30% of 59 countries have established social contracting. Limited funding allocations impede the scalability, effectiveness, and sustainability of the wide variation of services as well as slow or partial adoption of innovative modalities, such as long-acting PrEP (Bekker et al. 2024) and Dapivirine vaginal ring (Bunge et al. 2024). These exciting innovations not only add to the basket of choices for HIV prevention strategies but empower end users especially women in ownership over their health.

3.3.3 The Successes and Gaps in HIV Response in the Asia Pacific Region

It is pertinent to understand the commendable strides and areas of improvement in the HIV response in the Asia Pacific region. This subsection aims to highlight several countries in the Asia–Pacific region because of their noteworthy successes in response to HIV alongside challenges they have and yet to overcome. We hope to inspire more discussions on how other countries in the region can adopt these successes into various contexts in the region as well as to invite solutions to the challenges.

The HIV epidemiology in several countries (e.g., Malaysia, China, Indonesia, Thailand, Myanmar) in the Asia–Pacific region has unique features relative to other countries in the region and globally in that the earlier HIV incidence affected predominantly people who inject drugs (PWID) compared to different modes of transmission (Ministry of Health, 2022). In Malaysia, much of the HIV activism earlier in the epidemic focused on providing harm reduction services such as the needle-syringe exchange programme and opioid agonist therapy (Reid et al. 2007). This resulted in an endorsement from the Ministry of Health to incorporate harm reduction as part of the national healthcare services, a substantial reduction in HIV prevalence among PWID, and a compelling return on investment (Herlianna Naning et al. 2013). Recent strides are the abolition of the mandatory death penalty bill (Revision of the Sentence of Death and Imprisonment for Natural Life, 2023) and a preliminary move towards decriminalisation of drug use and possession. In 2003, Malaysia, the first nation to authorise compulsory licensing for three antiretroviral medications used in the treatment of HIV/AIDS (i.e., Didanosine, Zidovudine, and a fixed-dose combination of Lamivudine and Zidovudine) reduced the cost of these medications by 81%. It doubled treatment capacity (Ling 2006). Compulsory licensing has now been extended to other infectious diseases (e.g., sofosbuvir-based Hepatitis C treatment regimen) with an expansion for voluntary licensing agreements (Azzeri et al. 2023). Currently, with over 95% of new cases attributable to sexual transmission (Ministry of Health, 2023), HIV activists in Malaysia face a myriad of challenges. This is particularly the case because individuals most affected by HIV are communities who experience pervasive stigmatisation, i.e., gay, bisexual, and other men who have sex with men (gbMSM), transgender women (TGW), and female sex workers (FSW). Efforts to scale up HIV prevention with HIV pre-exposure prophylaxis (PrEP) through free-of-charge PrEP dispensing at public health clinics face strong opposition from clinicians and faith-based groups with the rationale of promoting "homosexual lifestyles" (Carvalho 2022, DOC2US 2023).

Thailand's HIV response is one that many Southeast Asian countries revere. Already achieving 90-90-97 in the treatment cascade in 2022, Thailand is on their way to achieving the 95-95-95 UNAIDS goals (Lertpiriyasuwat et al. 2024). Thailand was also the first in the region to eliminate perinatal HIV transmission (Wong et al. 2019) and has reduced AIDS-related deaths by 65% since 2010 (Siraprapasiri et al. 2016). Recently, Thailand was the first in South East Asia to recognise same-sex marriages legally (Jonathan Head 2024). Thailand's success can be attributable to the rollout of key population-led healthcare delivery of HIV prevention and care. This demedicalised and decentralised model empowers certified lay providers who are members of key populations to provide services for their communities. Importantly, this model embraces the essence of "nothing about us, without us" (Barmania 2023; Pengnonyang et al. 2022). The client-focused, cost-effective model makes healthcare more accessible because clients no longer need to receive care at healthcare facilities. It reduces barriers to entry as it destigmatises healthcare services and includes the provision of same-day PrEP and more recently ART provision, which otherwise is largely accessible only at a healthcare facility or pharmacy in other countries. However, a policy change that rescinded authorisation of lay providers to provide PrEP and disallowed PrEP medication to be kept at key-population-led clinics, impacted community-led PrEP services. 80% of PrEP users in Thailand received services from key-population-led clinics and new PrEP clients have plummeted by 21% after the policy change (Phanuphak et al. 2024; Versteegh et al. 2022). HIV activists in Thailand are now lobbying for a full endorsement to provide PrEP and same-day ART as well as store medications across all key-population-led clinics in all localities because preliminary trends suggest that the policy change would set the ending of AIDS in Thailand off course.

While lessons learned from the COVID-19 pandemic unveiled inadequacies in many countries' healthcare systems, some countries adapted at warp speed to the circumstances. Cambodia and Viêt Nam surfaced with commendable strides in HIV prevention. Cambodia attained the 90-90-90 goals in 2017. Cambodia recently moved not to conduct pilot projects for PrEP but to concentrate on larger scale-ups and leverage community-led service delivery models with tested efficacy (UNAIDS 2022). COVID-19 lockdowns disrupted access to methadone in Viêt Nam; thus, the Ministry of Health, UNAIDS, and UN agencies implemented a multiprovincial take-home methadone programme that showed a promising retention rate of 88% over six months (Vi et al. 2023; Nguyen et al. 2022). Similarly, a low threshold service delivery model for opioid agonist therapy with flexible prescribing indications has also become commonplace in many countries in Asia because of its low barrier to entry. However, Viêt Nam warrants an expanded section in their HIV response by embracing innovative models of HIV self-testing, differentiated PrEP service delivery models, and multi-stakeholder 'Undetectable = Untransmittable' (U = U) national campaigns.

3.3.4 The Four-Fold HIV Activism Agenda and HIV Activism Framework for Asia

Disparities in the HIV response in Asia underscore the need for specific, timely, and scalable attention and actions. Understanding the synergistic effects of up- and downstream determinants of health can better position healthcare systems to be more

Fig. 3.1 Four-fold HIV activism agenda to reduce HIV disparities in Asia

appropriately responsive. We harness the lessons learned from over four decades of responding to set up a four-fold HIV activism agenda to reduce HIV disparities in Asia (see Fig. 3.1) alongside suggestions that serve as modifiable building blocks that can be integrated into implementation strategies to prioritise and make contextually appropriate adaptation to improve fit and effectiveness.

Building upon the successes and deficiencies observed in the HIV response alongside forecasts detailing the repercussions of inaction, the first fold in the HIV activism agenda concentrates on tackling root causes of inequities that are immediately actionable. We recommend addressing the substantiated three costs of indifference to propel progress in the HIV response with a greater sociopolitical will, urgency, and reimagination. Next, we strongly recommend centring affected groups at the focal point of the HIV response. Placing communities at the forefront cultivates collectivism, instilling a sense of belonging and ownership of programs to achieve optimal HIV outcomes. This approach is vital for addressing the social and structural barriers that contribute to heightened risks of HIV transmission. The second fold must be accompanied by the radical requirement of the third fold: the recognition of human rights and fundamental liberties. The livelihoods of key populations and PWH that the arms of domination have shaped must be nourished by life-affirming humanisation beyond merely meeting basic life sustenance such as healthcare, food, shelter, and transport. Pushing the envelope on individual liberties, such as freedom of expression, right to love, security, and access to protective legal structures, cannot be timelier than now. Lastly, an effective HIV response must involve forming alliances and working in meaningful collaborations with affected communities. These partnerships should extend to policy development, programme design, implementation, training initiatives, and other impactful endeavours. Moreover, these partnerships must be genuine and intentional rather than superficial and tokenistic. The individuals should also be girded with the necessary support to thrive, including building capacity, economic empowerment, and other resilience-bolstering resources.

In achieving the four-pronged HIV activism agenda, we recommend two guiding principles alongside supportive frameworks. The first principle focuses on the multitude of actors who would contribute and benefit from the four domains, suggesting that addressing HIV disparities requires multilevel involvement of various stakeholders through population groups instead of only individuals. An approach to understanding the multilevel involvement is to employ the socio-ecological model of health (Bronfenbrenner 1977), wherein the well-being of an individual is influenced by interrelated environmental systems that range from immediate surrounding radius (e.g., family and friends) to societal-level structures (e.g., community and policies).

The second principle is to employ multiple strategies in tandem to address inequities in the HIV response. The Model for Adaptation Design and Impact and Adaptation (MADI) and Framework for Reporting Adaptations and Modifications to Evidence-based Implementation Strategies (FRAME-IS) are valuable frameworks to guide the adaptation of strategies into diverse contexts (Kirk et al. 2020; Wiltsey Stirman et al. 2019). We recommend the frameworks because multilevel strategies and adaptations, if not guided by an operational framework, can be done unsystematically or unilaterally without community consensus, leading to interventions that do not meet the needs of the end users. Adaptations must also be time-bound and include fidelity measures to ensure programs are not overly modified and their essence is watered down.

Figure 3.2 (see the next page) operationalises the four-pronged HIV activism agenda into a framework for strategies and opportunities at various socio-ecological levels, health and social outcomes, and the overarching impact measures. The framework positions affected populations at its core and coalesces communities and institutions' involvement with the ultimate goal of holistic well-being.

3.3.5 Ethical Consideration for HIV Activism in Asia

Acknowledging potential challenges stemming from socio-cultural and political disparities across countries in Asia, including personal safety concerns and legal ramifications, we urge careful consideration of the cost-to-harm ratio. The strength of HIV activism comes from assets such as strong leadership, evidence-based positions, good media relations, a network of strategic alliances with other groups, the ability to use multiple strategies, well-defined organisational structures, and sufficient, autonomous financial resources. HIV activists must be adaptive and willing to adjust their tactics. While activism is commonly perceived as radical, this is not universally true.

Putting oneself in a precarious position can be risky and negatively influence public opinion. Such publicity can dissuade individuals from supporting or joining activist groups (Brashers et al. 2002). Despite these risks, some activists are willing to take such chances to raise awareness for their cause. Therefore, the exercise of equipoise is imperative in formulating an agenda for social transformation.

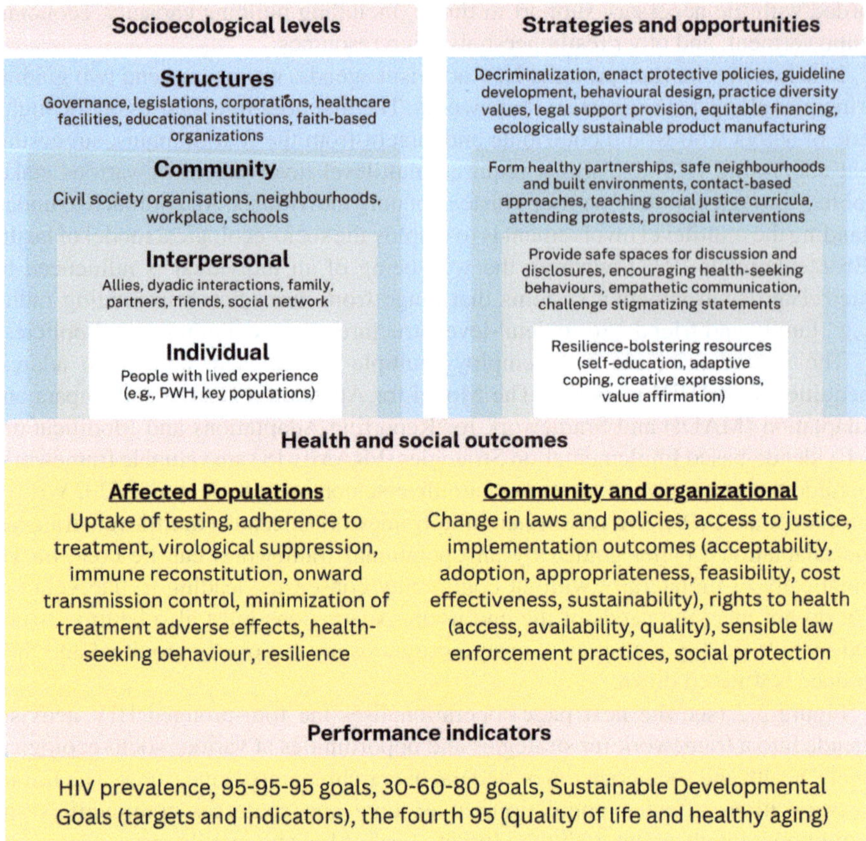

Fig. 3.2 The HIV activism framework for the Asia–Pacific region

3.4 Discussion of Implications

The four-fold HIV activism agenda and HIV activism framework for Asia serve as an empowerment-based agenda as opposed to conventional deficit-focused perspectives in negotiating for change and challenging HIV-related stigma. Applicable across Asian countries, the four-fold HIV activism agenda highlights four main ingredients that could move the dial on the HIV response in Asia, narrow healthcare disparities, and improve how countries in Asia manage a range of health conditions and diseases. Additionally, it is also pertinent to consider the applicability of the agenda and framework as well as the need to increase mobility for HIV activism in various countries in Asia, particularly in countries with a comparatively higher burden of HIV, significant disparities in the HIV treatment and prevention cascade, population size of key populations, within-country presence of HIV activism and resource mobilisation. Based on the Global AIDS Monitoring Report 2024, countries with the heaviest HIV burden and disparities in the HIV response are low to middle-income countries as well as countries with stringent punitive laws against people living with and affected by HIV. This includes but not limited to Indonesia, Philippines, Viet Nam, Malaysia, Sri Lanka, Iran, Myanmar, and Pakistan. An increased mobility of HIV activism is also due in countries where available and updated data are limited (e.g., India, Singapore, Nepal, Laos, and Timor-Leste).

3.5 Future Directions

Researchers, policymakers, and program implementers may leverage the HIV Activism Framework to guide discourse around research priorities and the development of innovative, adaptable, and scalable solutions. The framework and agenda can help community-based organisations and program implementers make informed decisions about the best combination and level of interventions to address healthcare inequities. Policymakers can benefit from the agenda and framework by enhancing funding efficiency and ensuring implementation efforts align with community needs while safeguarding individual liberties. The agenda underscores time sensitivity, particularly in making the case for urgency and specificity in driving change, considering the high costs of inaction and apathy.

The HIV Activism Framework for Asia also includes strategies and opportunities that can be employed across socio-ecological levels to strengthen and sustain HIV activism, particularly in resource-limited settings. On the individual level, resilience-bolstering resources (e.g., self-education, adaptive coping, creative expressions, and value affirmation) may improve the mental health of HIV activists, particularly in the wake of strong resistance or slow progress. This may enhance engagement and continuity of HIV activism, especially for individuals living with and affected by HIV. On the interpersonal level, safe spaces for discussions and disclosures may encourage health-seeking behaviours while encouraging empathetic communication between individuals with the lived experience and allies. The practice of allyship can also be enhanced by challenging stigmatising statements.

On the community level, we emphasise the importance of healthy partnerships between communities. The pathway to ending HIV as a public health threat necessitates a collaborative environment, engaging as many affected stakeholders as possible toward a unified vision. This can be enhanced through contact-based approaches, including social justice in teaching curricula and various prosocial interventions.

On the structural level, decriminalisation of individuals living with and affected by HIV and enacting protective policies should be at the forefront of the HIV activism agenda. Institutions and governing bodies can develop guidelines grounded in behavioural design and practice diversity values. Funding organisations on the local and international level would need to ensure equitable financing supporting community-based initiatives. Collaborations between countries, particularly between the Global North and South, ensure that initiatives are sustainable. Australia, where HIV disproportionately affects overseas-born key populations, is a good example for a focus on between-countries collaborations. The successes in addressing disparities in other Asia–Pacific countries through between-countries collaborations, may improve the HIV response in Australia for overseas-born individuals affected by HIV.

3.6 Conclusion

Noteworthily, this conceptual article makes clear that a multi-stakeholder approach is most effective in narrowing the gaps in the HIV response in Asia, particularly in addressing structural stigma, insufficient endorsement of community-centred HIV response, and the widening of the HIV funding gap. In guiding the progress in the HIV response, we introduce four time-sensitive action items for HIV activism and an operational framework.

Although this conceptual article provides valuable insights into the successes and gaps in the HIV response in Asia, we recognise several limitations that may open avenues for future research. The four-part agenda and framework for HIV activism proposed here are based on our experience and understanding of the evolving HIV epidemic. As such, while the propositions are intended to be broadly applicable, they may not be generalisable to all contexts. We encourage researchers, healthcare professionals, practitioners, community leaders, and HIV activists to implement and adapt these propositions to their specific local contexts and continue to expand upon them.

Recommendations hold little weight until they are translated into concrete actions that yield positive results. The time is ripe to unite and empower communities, granting historically marginalised individuals greater control over health-related choices and resources. HIV activism transcends empowerment; it is a force of political mobilisation, fostering collective identity and strength beyond individualism.

Highlights

- Health disparities, if not addressed adequately, may continue to disrupt HIV-related services in the Asia–Pacific region.
- HIV activism is instrumental in ensuring access to HIV care and prevention is sustainable and community-centred.
- Adopting the Four-Fold HIV Activism Agenda can empower communities and drive policy changes needed to address structural inequities and improve the long-term well-being of people living with HIV in Asia.

Acknowledgements The authors would like first to acknowledge all the vocal, visible, silenced, and invisible HIV activists of the past, present, and future, who have led the way from the beginning of the AIDS epidemic.

References

Anam FR, Nkosi S, Sebayang M, Jokonya M, Dunaway K, El Alaoui T. Let us lead: community leadership in the AIDS response is its fundamental pillar for success. J Int AIDS Soc. 2023;26(12): e26196. https://doi.org/10.1002/jia2.26196.

Anderson MJ, Eskandari MF. From peace talks to pandemics: the continuum of feminist peace activism. Glob Stud Q. 2023; 3(2). https://doi.org/10.1093/isagsq/ksad025.

Azzeri A, Dahlui M, Mohamed R, McDonald SA, Jaafar H, Shabaruddin FH. Budget impact analysis of two treatment approaches for hepatitis C in Malaysia through the use of voluntary and compulsory licensing options. Front Public Health. 2023;11:1114560. https://doi.org/10.3389/fpubh.2023.1114560.

Baer HA. On the political economy of health. Med Anthropol Newsl. 1982;14(1):1–17.

Barmania S. Successes and challenges in Thailand's HIV response. Lancet HIV. 2023;10(11):e699-700. https://doi.org/10.1016/S2352-3018(23)00270-9.

Bong SA. Becoming queer and religious in Malaysia and Singapore. 1st ed. London: Bloomsbury Academic; 2020. https://doi.org/10.5040/9781350132764.

Bekker LG, Das M, Abdool Karim Q, Ahmed K, Batting J, Brumskine W, Gill K, Harkoo I, Jaggernath M, Kigozi G, Kiwanuka N, Kotze P, Lebina L, Louw CE, Malahleha M, Manentsa M, Mansoor LE, Moodley D, Naicker V, Naidoo L, PURPOSE 1 Study Team. Twice-yearly Lenacapavir or daily F/TAF for HIV prevention in cisgender women. N Engl J Med. 2024; 391(13):1179–92. https://doi.org/10.1056/NEJMoa2407001

Bowleg L. Evolving intersectionality within public health: from analysis to action. Am J Public Health. 2021;111(1):88–90. https://doi.org/10.2105/ajph.2020.306031.

Brashers D, Haas S, Neidig J, Rintamaki L. Social activism, self-advocacy, and coping with HIV illness. J Soc Pers Relat. 2002;19:113–33. https://doi.org/10.1177/0265407502191006.

Bronfenbrenner U. Toward an experimental ecology of human development. Am Psychol. 1977;32(7):513–31. https://doi.org/10.1037/0003-066X.32.7.513.

Bunge K, Balkus JE, Fairlie L, Mayo AJ, Nakabiito C, Mgodi N, Gadama L, Matrimbira M, Chappell CA, Piper J, Chakhtoura N, Szydlo DW, Richardson B, Hillier SL. DELIVER: a safety study of a dapivirine vaginal ring and oral PrEP for the prevention of HIV during pregnancy. J Acquir Immune Deficiency Syndr. 2024;95(1):65–73. https://doi.org/10.1097/QAI.0000000000003312.

Carvalho M. Groups want HIV prevention treated as a health issue. The Star. 2022. https://www.thestar.com.my/news/nation/2022/12/08/groups-want-hiv-prevention-treated-as-a-health-issue. Accessed 30 Mar 2024

Corning AF, Myers DJ. Individual orientation toward engagement in social action. Polit Psychol. 2002;23(4):703–29.

DOC2US. Understanding PrEP, a preventative medicine for all Malaysians. 2023. https://www.freemalaysiatoday.com/category/leisure/health/2023/02/16/understanding-prep-a-preventative-medicine-for-all-malaysians/. Accessed 30 Mar 2024

Dyer O. Covid-19: Many poor countries will see almost no vaccine next year, aid groups warn. BMJ. 2020;371: m4809. https://doi.org/10.1136/bmj.m4809.

Earnshaw VA, Bogart LM, Dovidio JF, Williams DR. Stigma and racial/ethnic HIV disparities: moving toward resilience. Am Psychol. 2013;68(4):225–36. https://doi.org/10.1037/a0032705.

Earnshaw VA, Jin H, Wickersham J, Kamarulzaman A, John J, Altice FL. Exploring intentions to discriminate against patients living with HIV/AIDS among future healthcare providers in Malaysia. Trop Med Int Health. 2014;19(6):672–9. https://doi.org/10.1111/tmi.12306.

Earnshaw VA, Jin H, Wickersham JA, Kamarulzaman A, John J, Lim SH, Altice FL. Stigma toward men who have sex with men among future healthcare providers in Malaysia: would more interpersonal contact reduce prejudice? AIDS Behav. 2016a;20(1):98–106. https://doi.org/10.1007/s10461-015-1168-x.

Earnshaw VA, Rosenthal L, Lang SM. Stigma, activism, and well-being among people living with HIV. AIDS Care. 2016b;28(6):717–21. https://doi.org/10.1080/09540121.2015.1124978.

Freudenberg N, McDonough J, Tsui E. Can a food justice movement improve nutrition and health? A case study of the emerging food movement in New York City. J Urban Health. 2011;88(4):623–36. https://doi.org/10.1007/s11524-011-9598-x.

Gilson LL, Goldberg CB. Editors' comment: so, what is a conceptual paper? Group Organ Manag. 2015;40(2):127–30. https://doi.org/10.1177/1059601115576425.

Gilster ME. Comparing neighborhood-focused activism and volunteerism: psychological well-being and social connectedness. J Community Psychol. 2012;40(7):769–84. https://doi.org/10.1002/jcop.20528.

Hatzenbuehler ML. Structural stigma: research evidence and implications for psychological science. Am Psychol. 2016;71(8):742–51. https://doi.org/10.1037/amp0000068.

Head JKN. Thailand moves to legalise same-sex marriage. BBC. 2024. https://www.bbc.com/news/world-asia-68672318. Accessed 30 Mar 2024

Naning H, Kerr C, Kamarulzaman A, Osornprasop S, Dahlui M, Ng CW, Wilson DP. Return on investment and cost-effectiveness of harm reduction program in Malaysia. Washington, DC; 2014. https://www.aidsdatahub.org/sites/default/files/resource/roi-and-cost-effectiveness-harm-reduction-programme-malaysia.pdf. Accessed 2 August 2024

Integrated Biological and Behavioural Surveillance (IBBS) Survey 2022. HIV/STI/Hep C sector disease control division. Ministry of Health Malaysia: Malaysia. 2022. https://www.moh.gov.my/moh/resources/Penerbitan/Laporan/Umum/Final_IBBS_Report_2022.pdf. Accessed 24 Nov 2024

International AIDS Society (IAS), Me and My Healthcare Provider, East Asia Pacific. Me and MyHealthcare Provider. 2024. https://www.iasociety.org/me-and-my-healthcare-provider/2023-east-asia-pacific. Accessed 30 Mar 2024

Jaakkola E. Designing conceptual articles: four approaches. AMS Rev. 2020;10(1):18–26. https://doi.org/10.1007/s13162-020-00161-0.

Jin H, Earnshaw VA, Wickersham JA, Kamarulzaman A, Desai MM, John J, Altice FL. An assessment of health-care students' attitudes toward patients with or at high risk for HIV: implications for education and cultural competency. AIDS Care. 2014;26(10):1223–8. https://doi.org/10.1080/09540121.2014.894616.

John S. Societal enablers targets and the cost of inaction. In: International AIDS Conference 2022; 2022; Montréal, Canada.

Kavanagh MM, Agbla SC, Joy M, Aneja K, Pillinger M, Case A, Erondu NA, Erkkola T, Graeden E. Law, criminalisation and HIV in the world: have countries that criminalise achieved more or less successful pandemic response? BMJ Glob Health. 2021;6(8): e006315. https://doi.org/10.1136/bmjgh-2021-006315.

Kirk MA, Moore JE, Wiltsey Stirman S, Birken SA. Towards a comprehensive model for understanding adaptations' impact: the model for adaptation design and impact (MADI). Implement Sci. 2020;15(1):56. https://doi.org/10.1186/s13012-020-01021-y.

Klar M, Kasser T. Some benefits of being an activist: measuring activism and its role in psychological well-being. Polit Psychol. 2009;30(5):755–77. https://doi.org/10.1111/j.1467-9221.2009.00724.x.

Kraus S, Breier M, Lim WM, et al. Literature reviews as independent studies: guidelines for academic practice. Rev Manag Sci. 2022;16:2577–95. https://doi.org/10.1007/s11846-022-00588-8.

Krieger N. Epidemiology and the web of causation: has anyone seen the spider? Soc Sci Med. 1994;39(7):887–903. https://doi.org/10.1016/0277-9536(94)90202-x.

Krieger N. Theories for social epidemiology in the 21st century: an ecosocial perspective. Int J Epidemiol. 2001;30(4):668–77. https://doi.org/10.1093/ije/30.4.668.

Laverack G. Health activism foundations and strategies. 2013. https://doi.org/10.4135/9781446270004

Lertpiriyasuwat C, Jiamsiri S, Prommali P. Thailand's commitment to support community-led HIV services. Lancet HIV. 2024;11(3): e141. https://doi.org/10.1016/S2352-3018(24)00012-2.

Lim SH, Brown SE, Shaw SA, Kamarulzaman A, Altice FL, Beyrer C. "You have to keep yourself hidden": perspectives from Malaysian Malay-Muslim men who have sex with men on policy, network, community, and individual influences on HIV risk. J Homosex. 2020;67(1):104–26. https://doi.org/10.1080/00918369.2018.1525946.

Ling CY. Malaysia's experience in increasing access to antiretroviral drugs: exercising the 'government use' option; 2006.

Link BG, Phelan J. Social conditions as fundamental causes of disease. J Health Soc Behav. 1995; Spec No:80–94.

Link BG, Phelan JC. Understanding sociodemographic differences in health–the role of fundamental social causes. Am J Public Health. 1996;86(4):471–3. https://doi.org/10.2105/ajph.86.4.471.

Logie CH, Earnshaw V, Nyblade L, Turan J, Stangl A, Poteat T, Baral S. A scoping review of the integration of empowerment-based perspectives in quantitative intersectional stigma research. Glob Public Health. 2021;17(8):1451–66. https://doi.org/10.1080/17441692.2021.1934061.

Ministry of Health, Malaysia. The global AIDS monitoring report 2023. 2023. https://www.moh.gov.my/moh/resources/Penerbitan/Laporan/Umum/Laporan_Global_AIDS_Monitoring_2023.pdf. Accessed 30 Mar 2024.

Muhammed A, Amuda Y. LGBT: an evaluation of Shariah provisions and the laws of Malaysia and Nigeria. Glob J Al Thaqafah. 2018;8:15–29. https://doi.org/10.7187/GJAT072018-2.

Nguyen TT, Tran MTN, Le GM, Jauffret-Roustide M. Drug harm reduction in Vietnam: a review of stakeholders' perspectives and implications for future interventions. Am J Public Health. 2022;112(S2):S182-90. https://doi.org/10.2105/AJPH.2022.306764.

Ni Z, Shrestha R, Earnshaw VA, Tee YC, Altice FL, Azwa I, Kamarulzaman A, Zhou X, Wickersham JA. Exploring Malaysian physicians' intention to discriminate against gay, bisexual, and other men who have sex with men patients. LGBT Health. 2023;10(2):169–75. https://doi.org/10.1089/lgbt.2021.0452.

Nkengasong J, Ratevosian J. Legal and policy barriers for an effective HIV/AIDS response. Lancet. 2023;401(10386):1405–7. https://doi.org/10.1016/S0140-6736(23)00729-8.

Owens C, Voorheis E, Lester JN, Green HD, Herbenick D, Dodge B, Hubach RD. The lived experiences of rural HIV social workers. AIDS Care. 2021;35(1):48–52. https://doi.org/10.1080/09540121.2021.1981817.

Ozduzen O, Aslan Ozgul B, Ianosev B. 'Institutions of governance are all corrupted': anti-political collective identity of anti-lockdown protesters in digital and physical spaces. Soc Mov Stud. 2023:1–19. https://doi.org/10.1080/14742837.2023.2246920.

Parker R, Aggleton P. HIV and AIDS-related stigma and discrimination: a conceptual framework and implications for action. Soc Sci Med. 2003;57(1):13–24. https://doi.org/10.1016/s0277-9536(02)00304-0.

Pengnonyang S, Ramautarsing RA, Janyam S, Chaisalee T, Chanlearn P, Vannakit R, Phanuphak P, Phanuphak N. Certification of lay providers to deliver key population-led HIV services in Thailand's National Healthcare System: lessons learned. J Int AIDS Soc. 2022;25(7): e25965. https://doi.org/10.1002/jia2.25965.

Phanuphak N, Janyam S, Patpeerapong P, Sittikan S, Daramadhaj K, Pengnonyang S, Janamnuaysook R, Suya I, Mills S, Phanuphak P. Response to Thailand's commitment to support community-led HIV services. Lancet HIV. 2024;11(3):e141-2. https://doi.org/10.1016/S2352-3018(24)00013-4.

Rana DS. Promoting through consumer nostalgia: a conceptual framework and future research agenda. J Promo Manag. 2020; 27. https://doi.org/10.1080/10496491.2020.1829773.

Raphael D, Bryant T. Maintaining population health in a period of welfare state decline: political economy as the missing dimension in health promotion theory and practice. Promot Educ. 2006;13(4):236–42.

Reid G, Kamarulzaman A, Sran SK. Malaysia and harm reduction: the challenges and responses. Int J Drug Policy. 2007;18(2):136–40. https://doi.org/10.1016/j.drugpo.2006.12.015.

Revision of the Sentence of Death and Imprisonment for Natural Life (Temporary Jurisdiction of the Federal Court) Act. D.R. 7/2023; 2023.

Schwartz SR, Nowak RG, Orazulike I, Keshinro B, Ake J, Kennedy S, Njoku O, Blattner WA, Charurat ME, Baral SD, TRUST Study Group. The immediate effect of the Same-Sex Marriage Prohibition Act on stigma, discrimination, and engagement on HIV prevention and treatment services in men who have sex with men in Nigeria: analysis of prospective data from the TRUST cohort. Lancet HIV. 2015; 2(7):e299-306. https://doi.org/10.1016/S2352-3018(15)00078-8.

Siraprapasiri T, Ongwangdee S, Benjarattanaporn P, Peerapatanapokin W, Sharma M. The impact of Thailand's public health response to the HIV epidemic 1984–2015: understanding the ingredients of success. J Virus Erad. 2016;2(Suppl 4):7–14. https://doi.org/10.1016/S2055-6640(20)310 93-1.

Stangl AL, Earnshaw VA, Logie CH, van Brakel W, Simbayi LC, Barré I, Dovidio JF. The health stigma and discrimination framework: a global, crosscutting framework to inform research, intervention development, and policy on health-related stigmas. BMC Med. 2019;17(1):31. https://doi.org/10.1186/s12916-019-1271-3.

Tee YC, Earnshaw VA, Altice FL, Jin H, Kamarulzaman A, Wickersham JA. Evaluating physicians' intention to discriminate against patients living with HIV in Malaysia. AIDS Behav. 2019;23(4):1039–47. https://doi.org/10.1007/s10461-018-2362-4.

The Lancet HIV. Anti-LGBT laws continue to hinder the HIV response. Lancet HIV. 2022;9(10): e667. https://doi.org/10.1016/S2352-3018(22)00265-X.

The Lancet. American chaos: standing up for health and medicine. Lancet 2025;405:10477–439. https://doi.org/10.1016/S0140-6736(25)00237-5

Tricco AC, Lillie E, Zarin W, O'Brien KK, Colquhoun H, Levac D, et al. PRISMA extension for scoping reviews (PRISMA-ScR): checklist and explanation. Ann Intern Med. 2018;169(7):467–73. https://doi.org/10.7326/M18-0850.

UNAIDS. In danger: UNAIDS global AIDS update 2022. https://www.unaids.org/sites/default/files/media_asset/2022-global-aids-update-summary_en.pdf. Accessed 30 Mar 2024.

Versteegh L, Amatavete S, Chinbunchorn T, Thammasiha N, Mukherjee S, Popping S, et al. The epidemiological impact and cost-effectiveness of key population-led PrEP delivery to prevent HIV among men who have sex with men in Thailand: a modelling study. Lancet Reg Health Southeast Asia. 2022;7: 100097. https://doi.org/10.1016/j.lansea.2022.100097.

Vi VTT, Diep NTN, Han VHN, Van Dung D, Nguyen NLH, Binh LQ, Balhara YPS. Experience and insights from the emergency multiday take-home dose methadone dispensing in Ho Chi Minh City in context of COVID-19. J Addict Dis. 2023;41(4):317–21. https://doi.org/10.1080/10550887.2022.2116254.

Wiltsey Stirman S, Baumann AA, Miller CJ. The FRAME: an expanded framework for reporting adaptations and modifications to evidence-based interventions. Implement Sci. 2019;14(1):58. https://doi.org/10.1186/s13012-019-0898-y.

Wong J, Macikunas A, Manduric A, Dawkins J, Dhunna S. Reaching the hard to reach in Thailand: eliminating mother-to-child HIV transmission. Health Syst Reform. 2019;6(1). https://doi.org/10.1080/23288604.2019.1625498.

Norman Chong , MD MMedSc. Norman is interested in leveraging mathematical modelling to understand the associations between social determinants of health and health outcomes across the life span. Norman's work focuses on supporting decision-making processes that emphasiwe the highest return on investment for HIV prevention and care. Norman started a young people-led initiative for the HIV response (https://www.weaskthewhy.com.my) as the recipient of the International AIDS Society Youth Hub Seed Grant. Through the initiative, they have trained 30 young leaders, most of whom were young people with and affected by HIV, to design sustainable and people-powered HIV programs, including projects that leverage technology and social media.

Valerie A. Earnshaw , Ph. D. Valerie is a social psychologist and an associate professor at the Department of Human Development and Family Sciences at the University of Delaware. Her research focuses on stigma associated with substance use disorders, HIV, and mental illness. She aims to understand and address associations between stigma, health inequities, and health outcomes across the lifespan and what moderates these relationships in protective ways. She also aims to contribute to interventions to improve the well-being of stigmatised children, youth, and families.

Iskandar Azwa , MBChB, MRCP,. Iskandar is an associate professor and senior consultant in Infectious Diseases at the University of Malaya, Kuala Lumpur and he is the director of the Centre of Excellence for Research in Infectious Diseases and AIDS (CERIA). His research interests include HIV pre-exposure prophylaxis (PrEP) implementation research among key populations, differentiated HIV service delivery models, HIV resistance, and evaluation of newer HIV treatment strategies. He is the site Principal Investigator for several international collaborative HIV clinical trials, including the TAHOD, START, and D2EFT studies. He is also the president of the Malaysian AIDS Council.

Chapter 4
Navigating the Known Unknown: A Comparative Study of Pandemic Preparedness and Border Management in Southeast Asia (Thailand, Cambodia, Myanmar)

Clusters of Future Studies: Corporate Foresight; Post-Normality and Complexity; Technological Trends

Khathaleeya Liamdee

Abstract This chapter presents a comparative analysis of pandemic preparedness and response strategies in Thailand, Cambodia, and Myanmar. It is a combination of ethnographic fieldwork, in-depth interviews with multiple stakeholders including border crossers, frontline officers, representatives from health-related authorities between June 2022 and May 2023, and desk-review methodology to analyse official documents, media reports, and scholarly journals. The study examines the impacts of political, social, and economic factors on the effectiveness of the cross-border management strategies in the region. Border closures and travel restrictions were common but varied in implementation and effectiveness across the three countries. To illustrate, Thailand presented effective pandemic preparedness and response through its robust health surveillance system. Cambodia managed to contain the disease, yet cybercrime issues emerged out of the pandemic. While Myanmar's response was hindered by political instability leading to the collapse of health system. It underscores the importance of regional interconnectedness and cooperation in pandemic preparedness and responses, the need for dedicated border taskforces to enhance health security. Lastly, this chapter suggests that political stability and robust public health infrastructure are crucial for effective management and regional health security.

Keywords Pandemic preparedness · Border health · Health surveillance · COVID-19 · Health security

K. Liamdee (✉)
Department of Sociology and Anthropology, Faculty of Political Science, Chulalongkorn University, Bangkok, Thailand
e-mail: khathaleeya.l@chula.ac.th

R. Xiang Ng et al. (eds.), *Future of Healthcare in Asia*,
https://doi.org/10.1007/978-3-032-03183-9_4

4.1 Introduction

At the beginning of 2020, the world faced the emergence of a new contagious disease, COVID-19, which was initially unknown. This chapter, grounded in medical anthropological studies of border permeability and pandemic preparedness, aims to navigate this 'known unknown' by revisiting whether state authorities in Southeast Asia were prompt and prepared to take responsibility for the pandemic.

The discussion emphasises what COVID-19 means for government policies on pandemic preparedness and the future of cross-border mobility flows and health surveillance in mainland Southeast Asia, particularly focusing on Thailand and its neighbouring countries, Myanmar and Cambodia. The goals of this chapter are twofold: first, to highlight the challenges and successes of Thailand's proactive intervention and strong health surveillance system, especially the role of the Department of Disease Control, Ministry of Public Health; and second, to draw attention to the political and social factors that shifted practices of cross-border migration and border management in response to COVID-19.

As disease control practices require transnational cooperation, this study further explores Cambodia and Myanmar's healthcare systems and pandemic responses to indicate regional interconnectedness on various issues such as border health, migration flows, human trafficking, illegal border crossings, and political uncertainty in Myanmar. This chapter invites a re-evaluation of the future of biosecurity and public health in the subregion, taking political, economic, and social factors into account to enhance the efficacy of disease controls and biosecurity measures, especially for vulnerable populations affected by the pandemic and state responses.

4.2 Literature Review

According to Ong (2016), 'the known unknowns'[1] mean 'the combination of anticipated catastrophic events and the uncertain capacities to prepare for managing them'. It is a history of communicable disease that represent infection as the human bodies are exposed to the unpredictable shapeshifting through the accounts of interspecies intimacies (Ahuja 2016). The importance to point out the emerging infectious disease (EID) as the known unknown is to remind how we must stay in prompt and prepared to take responsibility for any outbreak that could eventually evolve to a disastrous event which is capable of haunting the future of science itself and the society as a whole.

---

[1] This term became widely discussed when Donald Rumsfeld, a former US Secretary of State for Defence, mentioned in 2002 during his speech regarding the War on Iraq stating that 'There are known knowns; there are things we know we know. We also know there are known unknowns; that is to say we know there are some things we do not know. But there are also unknown unknowns—the ones we don't know we don't know.' (Graham 2014).

In the time of COVID-19, 'pandemic preparedness' is becoming more important than ever not only in the studies of biological science but also in the social studies of science. According to World Health Organization (WHO), a pandemic is announced when a communicable disease's growth is exponential and covers a wide area, affecting several countries and populations (Mailman School of Public Health 2021). The situation falls into a category of the known unknown that has invoked the scientific prophecy about the anticipated spillover event. Now, what does the pandemic preparedness mean as we are in the middle of the COVID-19? Caduff (2015) argues that pandemic preparedness must be situated in relation to the sensibilities and anxieties to which they respond. Thus, it needs to understand the architecture behind these effects in order to promote the sense of urgency that is driven by mobilizing public persuasion for cooperation. In other words, to find the right balance between fear and panic, anxiety and alarm, suspicion and paranoia while continue with the daily routine and maintain steady vigilance at the same time.

Historically, pandemic preparedness has been a part of a political necessity for the logics of biosecurity and biowarfare dominated by the United States especially after the 9/11 incident. It is also an opportunity for innovation and pharmaceutical product development as it is crucial to remind that preparedness is 'not how to avoid the event but how to respond to it when it occurs' (Caduff 2015). Preparedness should be exercised to identify the vulnerabilities of the public health systems, infrastructures, and technologies (ibid 2015). It is not only the way to prepare the emergency drills as a speculative mode for generating knowledge of catastrophic risks to the nations and humankind (Ahuja 2016), but also the way to constantly document state's limitation on knowledge for the known unknowns. By this definition, even in the middle of the pandemic, preparedness is much needed to actively recalculate plans and scenarios by the experts; the allocation of resources and fundings for staff trainings and vaccine production in order to provide more efficient and applicable responses to the current trend of COVID-19.

Oftentimes, pandemic preparedness requires scaling up capacities and militarised strategies to prevent the emergence of the biothreats (Ahuja 2016; Kelly et al. 2019). Border closure is one of the first strategies being authorised by the nation-states around the world as the way to control national space and intervene, speed and frequency of microbes' mobility that travel with border crossers. Kenwick and Simmons (2020) discusses the historical and contemporary use of border closures and restrictions as means of responding to pandemics. The authors argue that the pervasive use of border controls during the COVID-19 reflects growing anxieties about border security. The measures that are influenced by public opinion and preexisting commitments to border security, rather than clear scientific consensus reflect the complex interplay between domestic politics, public opinion, and international health regulations in shaping state responses to pandemics. They underscore the tendency of states to resort to border controls as a visible and politically expedient measure, despite their limited effectiveness in controlling disease spread. This reliance on border security measures reflects broader anxieties about border security and the influence of domestic political considerations in public health policy (Worsnop 2017; Lindholt et al. 2021). To put in simply, it is the surface level of disease management on various kinds of flows namely humans, data, animals, and microbes to stop the disease 'there' before it came 'here' (Lowe 2010).

Moreover, border surveillance is generated through the logic of tertiary spatialization[2] and knowledge of science. It is enacted by the governmental techniques which already existed in the bordering practices for political borders to contain the spread of disease carriers (Foucault 1994 cited in Ong 2016). Such intervention which is not always in a sense of US-centric definition of biosecurity but surely in the same manner of politics of inclusion-exclusion which may focus on political controls rather than emergency preparedness. The differential patterns of mobility and new common immobility need to be addressed in order to avoid harmful decision on border surveillance for those who are unable to follow the health measurement guidelines and eventually excluded from the healthcare system and intensify stigmas around the idea of disease always coming from the others—migrants, refugees, and tourists—and then becoming a subject of control and a manifestation of national purification rather than focusing on the capacity of healthcare system (Ennis-McMillan and Hedges 2020; Laocharoenwong 2020; Wille and Kanesu 2020).

In addition, Chuengsatiansup and Limsawart (2019) offers a thought-provoking point on the fact that although the permeability of the border seems to hinder the efficacy of surveillance and control, it is also important to highlight that such attempts require political authorities to intervene and search for technologies to differentiate between the healthy and the infected. In The intermediacy of borders and complex entanglements of microbes and human will keep changing not only the conditions of pathogenic possibility but also the prevailing biocontainment model for the EID controls. It leads to the new set of questions on how to develop a more flexible conception of border surveillance that can embrace 'the fluidity of pharmaceuticals, microbes, and human relations.' This approach will surely affect the calculation of surveillance and preparedness measures that could offer the new perspective on the way we are living with the more-than-human creatures like the viruses that do not carry passports and know no borders (Ong 2016; Ennis-McMillan and Hedges 2020; Ticktin 2020). The surveillance mechanism by sovereign states became more autonomous in decision-making as WHO believes that controlling the national borders and domestic boundaries may do little to stop disease spread (Ennis-McMillan and Hedges 2020). Meaning that what is the right balance between being the leader of global health governance and being solely a reliable distributor for calculative information regarding the development of the diseases and surveillance guidelines (Hooker 2007). Beyond the border surveillance, it is important to recognise the concept of PHEIC—making Public Health Emergency of International Concern—in order to overcome the idea of biothreats to a single nation but rather a singular event that the world is facing altogether (Lakoff 2017).

---

[2] Tertiary spatialization refers to a space for curing 'a disease is circumscribed, medically invested, isolated, divided up into closed, privileged regions, or distributed throughout cure centers, arranged in the most favorable way' (Foucault 1994).

4.3 Methodology

This chapter is based on an ethnographic fieldwork to the international border passes of Thailand including Aranyaprathet-Poipet international border pass and Klong Yai-Koh Kong international border pass on the Thai-Cambodian border. These field visits were not only for directly observing the border health situations but also for the purpose of conducting the in-depth interviews regarding the national strategies on pandemic preparedness and responses during the COVID-19 with multiple stakeholders between June 2022 and May 2023. The targeted population samples were curated through both purposive and accidental samplings from different stakeholders namely border crossers, frontline officers, the health authorities at the Ministry of Public Health, Nonthaburi, Thailand, and a representative from WHO Thailand. Snowball sampling was also used for demonstrating the extended network of border health management through the suggestion from the purposive sampling population. In addition, this research utilises desk review methodology to analyse the official documents archived and released by the government agencies, media outlets, World Health Organization (WHO), and International Organisation for Migration (IOM) regarding health and border security policies for both air and ground border controls.

For ethical considerations, the research project for this chapter was approved by Chulalongkorn University's Ethical Review Board before conducting the ethnographic fieldwork along with either verbal or written consent from all informants. Particularly the government officers from the public health sectors, names and positions will not be disclosed but rather stated as one of the officers in specific department to protect the privacy of the informants. In doing so, this is arguably considered a limitation of doing ethnographic approach such as potential bias. However, the selected informants were curated with the careful consideration and close discussion together with the health authorities by a purposive sampling method. There were also the formal letters of invitation to participate in the research were sent to ask for the permission to share their experiences with the research team. Not to mention, one of the goals for this project is to illustrate the government's narratives on pandemic preparedness and border management which are open to scholarly debate in the future (Andrews et al. 2008).

4.4 Findings

4.4.1 Thailand's Border Health Management in a Time of COVID-19

The first reported COVID-19 patient in Southeast Asia was a passenger from China to Thailand who was detected by Dr. Rome Buathong, who was assigned to lead the detection team of the International Communicable Disease Control and Quarantine Division at Suvarnabhumi International Airport (World Health Organization

Thailand 2020). The point-of-entry screening techniques used by Dr. Rome and the detection team was figuring out how to sort out the first patient as quickly as they were informed the potential outbreak that could be traveling to Thailand through the commercial flights during the 2020 New Year holidays, when there were over ten flights between Wuhan and Bangkok. They established a thermal scanning system to screen the passengers with the cautionary note that the symptoms of the mysterious pneumonia at that time were not easy to detect with regular protocols if the passengers had low fever. The team set up new protocols by installing the thermal screening system combining thermal scanners and ear thermometers that could detect febrile passengers through the calculation of the arriving group's average body temperature. With this technique, they were able to detect the sick passengers that could have been cleared as healthy passengers and proceeded to immigration check and possibly spread the virus to others if the team only proceeded with the usual screening protocol. What was happening behind such a swift response informed by one of the officers from the Department of Disease Control provides insightful details on what happened before January 8, 2020, as described below:

> The routine in the surveillance room (of the Department of Disease Control), we have been constantly monitoring the information from several parties. On that day (January 8th) we first heard of unknown pneumonia from the Reuters news agency. So, we started to retrieve the options for surveillance mechanism that is not only monitored within Thailand (domestic level) but also at the international level. Since it is considered our responsibility to raise awareness for the unusual case. Our team was researching for the place of origin for the potential outbreak. As Wuhan, China was mentioned, we checked how much the city has a close relationship with Thailand. At that time, there were more than ten flights between Bangkok and Wuhan per day Wuhan considered one of the most popular tourist destinations in China for Thai people. it is apparently that Thailand has a close contact with Wuhan through an air gateway. With this information, the department decided to closely monitor the situation with a plan for disease x.[3]

Besides what appeared on the media outlet about the story of Dr. Rome, the informant demonstrates how Thailand's Public Health system was capable of dealing with the 'unknown pneumonia' or the known unknown even five days before the World Health Organization (WHO) publicly confirmed the cases of the genetic sequence of SARS-CoV-2 in China (World Health Organization 2020a). With the effort of the COVID-19 Task Force, Thailand succeeded in limiting local transmission in the first half of the year 2020 until the arrivals of the mutated variants that shaken and challenged the national capacity not only for public health system but also collective responses by its citizens.

Two months after the first detection and numerous countries became aware of the emerging outbreak, WHO finally announced COVID-19 to be characterised as a 'pandemic' on March 11, 2020. Though, countries could simply follow the guidelines based on the 2005 International Health Regulation (IHR). Many countries including Thailand, Cambodia and Myanmar decided to apply travel bans which was understood as border closure. It indicates that every official point of entry (PoE) shall

[3] Representative from Disease Control Department, Ministry of Public Health of Thailand, interviewed in April 2023. Interviewed on April 24, 2023.

restrict any traveller entering the country to stop the spread of the disease from elsewhere before hitting the local populations (Lowe 2010).

Travel restriction does not mean the country fully reject all means of international border crossing. It means there are entry requirements and health measurement for travellers to complete through the process of border health surveillance to make sure travellers are healthy and virus-free before entering the country. As prior to the pandemic, in Thailand, there are three major screening stations integrated as QIC system starting from the 'Q'—quarantine station, then proceed to the 'I'—immigration and 'C'—customs before entering the country legally. This practice has been implemented to all official points of entry of Thailand and most of the countries around the world. It signifies the importance of health measurement as part of border control and management. To be precise, health security and biosecurity has always been significant to national security. The COVID-19 pandemic has underlined how 'Q' screening unit has never been less critical than the other two units but rather functioned in the same direction.

Besides the promptly responses by the Department of Disease Control, the Ministry of Public Health of Thailand, the government also acknowledged the seriousness of COVID-19 as suggested by the public health sector. The government introduced mandatory quarantine for people entering the country through airports. Whereas ground checkpoints were forced to fully close after the massive return of migrant workers to their home countries, particularly Myanmar and Cambodia. However, as the situation developed and there were domestic cases that required further actions at the national level. The one of the public health officers provides the analogy of the water flow in the river for the reason why the COVID-19 is beyond the responsibility of public health stating that:

> In the Ministry of Public Health, we look at matters of what people may call it 'downstream'. Meaning that if there are patients, we treat them, if there are emerging diseases, we prevent them. But for the upstream where the Public Health sector does not have power to implement and control such as travel regulation and restriction, the economic policies and pandemic relief packages as it is under supervision of other ministries. Therefore, the pandemic indeed needs cooperation from other agencies beyond public health sector.

Accordingly, the Office of Prime Minister declared a state of emergency on March 26, 2020, and established the Centre of COVID-19 Situation Administration (CCSA) to be a key coordinator and streamline the special taskforce between different government ministries and several pieces of legislation to be under the Emergency Decree on Public Administration in Emergency Situations B.E. 2548.[4] The integration between sectors enacted several health surveillance tactics that were used to control the movements of people not only within the country but also cross-border activities which is the focus of this chapter.

During the first year of COVID-19, Thai health authorities together with the CCSA designed the applicable Quarantine system to facilitate international travellers who

[4] See full translation by the Ministry of Foreign Affairs, Thailand here: https://image.mfa.go.th/mfa/0/mkKfL2iULZ/migrate_directory/news3-20200326-211539-804409.pdf.

express their will to enter Thailand during the travel ban period. Since the International Health Regulation and World Health Organization suggested that countries should not implement measures that restrict travel and trade without solid evidence of effectiveness of such restriction (World Health Organization 2020b), State Quarantine system thus became a solution for Thailand and many countries as a screening technique that not entirely reject international travels, especially for the returnees from abroad who had an impending trip to return to their homelands.

On the one hand, the returnees by air travels were required to submit pre-departure documents and proceeded to the free-of-charge state-provided quarantine facilities for fourteen days. On the other hand, it was more difficult for the returnees by ground travels to proceed through the mandatory quarantine in the border provinces as not all points of entry were able to provide adequate quarantine sites. It often depended on the provincial governor's orders and the outbreak situation in the province.[5]

Thailand's border health screening techniques changed throughout the period of pandemic depending on the mutability of the viruses as well as the attempt to gradually resume international travels and tourism which has been the income resources of the country. The government-led CCSA announced reducing the period of quarantine, sandbox model, and test-and-go options for air travellers. In other words, changes in quarantine periods and new measurements for international travellers indeed reflect the uncertainties and the need for revising preparedness measures.

For instance, Thailand invented the quarantine scheme called 'Phuket Sandbox' started on July 1, 2021, as a campaign to slowly reopen the country and resume tourism business for the vaccinated tourists back on the island province before extending the program to other provinces across the country (*Tourism Authority of Thailand* (TAT) Newsroom 2021). According to the health authority, choosing Phuket for a sandbox model was considered effective strategy because it was aligned and well accepted from different sectors. First, the location of the province is an island isolating from the mainland of the country which deemed safe for health security principles. Second, Phuket has always been renowned destination for tourists. It was undeniably attractive campaign to resume international travels back for the country (see Table 4.1).

However, this was not the case for migrant workers from Thailand's neighbouring countries who wished to return to Thailand for work amidst travel restriction during the second half of the year 2020. The cases of human trafficking and illegal crossing found in the border provinces were increasingly reported to the authorities. Because at the beginning of the COVID-19, the groups of migrant workers were forced to go back home as they were unsure how they would be treated by employers and the Thai authorities if remaining in Thailand during the lockdowns. When the situation in Thailand was getting better and the business needed migrant workers back meanwhile the borders were not fully re-opened yet, illegal crossing became the fastest route to return to work.

[5] Combined sources from the interviews with the representatives from National Security Council, the International Communicable Disease Control and Quarantine Division and the Department of Disease Control, Ministry of Public Health, Thailand in 2023.

Table 4.1 Summary of COVID-19 responses on border management of Thailand, Myanmar, and Cambodia

Border-related health regulations	Thailand	Myanmar	Cambodia
National emergency response unit	Centre of COVID-19 Situation Administration (CCSA)	Central Committee led by Aung San Suu Kyi (Pre-Coup era)	Inter-Ministerial Committee to combat COVID-19
Dates of border closure (2020)	*March 26* (via a declaration of State Emergency)	Multiple dates for different borders *March 22* (Thailand) *April 1* (China) **April 13** (Bangladesh)	*March 17*—travel bans from Germany, Italy, Spain, France, and the U.S *March 30*—fully closed
Required documents upon arrival (2020)	• Fit-to-fly/travel certificate • Certificate of entry • Quarantine consent form	Laboratory evidence of absence of COVID-19 infection issued no more than 72 h prior to the date of travel (foreigners only)	• COVID-19 health certificate • Proof of medical insurance for $50,000 • Letter of confirming hotel reservation
Initial mandatory quarantine policy	14 days in state quarantine (SQ) facilities (free service)	• 21-day facility quarantine for Myanmar nationals • 7-day facility quarantine for foreign nationals[6]	14 days at designated hotels (paid service)
Innovative quarantine	Phuket sandbox	N/A	N/A
Dates of fully reopening the border	October 1, 2022	February 20, 2023 (Border with Thailand)	November 1, 2021

Eventually, the CCSA made the decision to resume MOU recruitment of migrant workers. The Ministry of Public Health along with the Ministry of Labour and Border Provincial Administrations thus teamed up to operate the quarantine system specialised for the migrant workers who wished to return to Thailand. The operation also received the assistance from the Border Patrol Police to facilitate the screening procedures at the field quarantine facilities in the border provinces before the business owners would directly recruit the screened migrant workers for their businesses (International Organisation for Migration 2021).

[6] See more details here: https://tourism.gov.mm/travelling-to-myanmar-covid-19-faqs-moht/.

4.5 Discussion of Implications

4.5.1 *Pandemic in Southeast Asia and the Question of Regional Interconnectedness*

Thailand, Cambodia, and Myanmar have gone through the period of political transitions in their own routes. Whether it is relevant to the disease or not, political changes did occur in the span of the pandemic crisis. Chronologically, as discussed in the previous section, Thailand may have demonstrated the effective responses to the pandemic. However, the country also experienced the mass protest during the first year of the pandemic due to the dissatisfaction of COVID-19 management by Prayuth's administration which had added on to the grudge of people toward the corrupted government.

Followed by the political situation in Myanmar, the military led by Min Aung Hlaing has seized the power by the coup which resulted in the political turmoil and armed fights across the country. Subsequently, public health system has been collapsed by the interference of the military and the breach of medical neutrality. It caused the unsafe conditions for health professionals which led many of them to flee the country.

While in Cambodia, there was no significant change in national politics and no major crisis in public health dealing with COVID-19. However, the country's regulation on travel restriction and along with the cooperation with Chinese government on fighting alleged online casinos. It led to the proliferation of cybercrime operation across the country due to increasing number of people from ASEAN countries attracted to the promise of quick-earned high income after being affected economic disruption by the lockdowns (Faulder 2022).

In terms of public health infrastructure, according to the Global Health Security (GHS) Index 2021, Myanmar and Cambodia ranked 85th and 126th respectively whereas Thailand ranked 5th place in with the highest score on detecting and reporting tasks.[7] The swift detects and responses by Thailand's health authority illustrate that the country's capacity have been internationally recognised for its robust healthcare system with a strong health surveillance system for both national and subnational levels. For example, Thailand received the GHS's full scores for real-time surveillance and reporting, epidemiology workforce, emergency preparedness and response planning, linking public health and security authorities, and infection control practices.

In short, the lack of cross-border health cooperation and the prolongation of political conflicts can have a profound impact on public health in the long run. Hence this study intends to highlight the regional connectedness beyond demonstrating the substantial differences in health capacity of Thailand and its neighbouring countries.

[7] See Global Health Security Index (Year 2019) here: Thailand; Myanmar; and Cambodia.

4.5.1.1 Myanmar's Coup, COVID-19, and the Collapse of Healthcare System

At the same time, not all the border health management could be solved as planned. The political situation in Myanmar has brought the new challenges for Thailand even after the reopening of the Thai-Myanmar border checkpoints. Myanmar's response to the COVID-19 back in 2020 was much loose and slow comparing to Thailand because Myanmar's structural condition and its vulnerable health infrastructure prevented an effective response to the pandemic itself and the large amount of the returnee population from Thailand and Malaysia. The response was thus mostly relied on the community-based mobilisation along with international aids.

At first, the civilian government operated the national COVID-19 Response through the Central Committee which led by the de facto head of the government, Aung San Suu Kyi based on two major laws namely the Prevention and Control of Communicable Diseases law and the Natural Disaster Management Law.[8] They also offered the COVID-19 Fund and Economic Relief Plan (CERP) to help create resilient recovery through tax relief, credit for businesses, and food and cash for households (Ostwald and Myint 2021; Perera et al. 2022).

The deep political fractions and mistrusts toward the central government further underscores the ethnic conflicts and also urban and rural divides beyond the fear of the pandemic itself. With travel restrictions and the mass return of migrant workers from Thailand and Malaysia, there was confusion and frustration because of substantial inconsistencies in health and quarantine management which led to the misconduct of illegal crossing to avoid the mandatory quarantine (Sangsuradej 2022).

Despite the ongoing pandemic, the military declared a coup d'état on February 1, 2021, and formed the State Administration Council (SAC). They also took the advantage of the COVID-19 situation as part of the reasons for the country reform after the general election in November 2020. Moreover, the SAC also used the pandemic as an excuse to implement the restriction on free speech and mass protest while ignoring the task for strengthening public health capacity to fight against COVID-19 (Wicaksana et al. 2023).

For example, the SAC forced to restrict oxygen delivery which caused long queues for oxygen cylinders during the Delta-variant wave (July–August 2021). Renowned doctors were house arrested, some were killed by unidentified assailants at the private clinics or died in prisons. The country became an epidemiological blind spot as there have been no reliable reports for health situation inside Myanmar. The attacks on health workers and health facilities have become one of the prominent features of the 2021 coup d'état (Physicians for Human Rights 2022).

The escalation of armed violence across the country not only affected the healthcare system that has been fragile disorganised and collapsed as the country failed to achieve solidary in a pandemic time when health and safety should have been a priority (Rocha et al. 2023). But it also led to the new wave of migration including

[8] The official titles of the law: The State Law and Order Restoration Council Law No.1/95 (1995) and The Pyidaungsu Hluttaw Law No.21, 2013 (Ostwald and Myint 2021).

health professionals fleeing the country as an act against the SAC. The influx of the displaced and the injured to the Thai border means the increasing needs of humanitarian aids and border health concerns which undoubtedly shaken the health security of Thailand as well.

4.5.1.2 Cambodia's COVID-19 Response and Cybercrimes

On the other side of the Thai border, Cambodia had a relatively better record on strong will and high level of cooperation from the public in containing the spread of COVID-19 throughout the period of travel restriction and also quickly covered the vaccination rollout program in 2021. Most of the infected cases were detected at the border and quarantine facilities. There have been no fatalities and mass local transmission. Cambodia was the fastest country in mainland Southeast Asia to resume international travels in November 2021. (Asian Development Bank (ADB) 2020; Chhim et al. 2020).

The key contribution to COVID-19 Response was the Cambodia's Communicable Disease Control Department (CCDC) that has recently collaborated with World Health Organization to establish 'Cambodia Early Warning System (CamEWARN)' and Cambodian Emergency Operations Centre and Incident Management System Standard Operating Procedures in 2019, a year prior to the pandemic, as a national surveillance system. Moreover, the Hun Sen government created the National COVID-19 Response Committee to mobilise resources and the technical working group and to provide technical advice on the COVID-19 emergency response for all related government sectors (Lan et al. 2022).

Still, the shortage of health workers remained significant challenges for the country. In addition, financial support for low-income population was limited by the registration system. The eligible households must register the IDPoor[9] to receive food and cash support during the lockdowns. There were groups of people that did not receive support as they were mostly people who have been working in Thailand or along the Thai border that may have not been part of the programme.

For cross-border mobility flows, border closure had impacted on regular border traders, workers, and travelers who were making a living by cross-border trade. Only trucks transporting commercial goods were allowed to travel across the border with the same set of required documents for air travels when returning to Thailand. According to World Health Organization (2020c), border screening and quarantine procedures at the points of entry between Thailand and Cambodia have been well-organised as well as for the coordination and cooperation among the agencies involved in border affairs for both sides. The border market areas were required to set up disease control measures in order to keep business running safely.

Regarding illegal crossing activities, there were significant numbers of news report concerning the illegal crossing from Thailand to Cambodia and Myanmar

[9] Identification of Poor Households Programme. See more details about the programme here https://idpoor.gov.kh/en/about/.

for working in organised crime compounds. This phenomenon could be labelled as 'scamdemic' as the pandemic has intensified the online platforms for cybercrimes across the Mekong region due to travel restriction and rapid emergence of technologies whilst the legal investigations remain limited. The organised crime activities have also targeted the groups of people who have been affected by COVID-19 such as the unemployed who are in desperation to get work. The network of digital criminal activities reveals the interconnectedness between countries in Southeast Asia in full circle as we have been witnessing since the beginning of the pandemic (Faulder 2022; Nikkei 2022; Strangio 2023; United Nations Office for Drugs and Crime (UNODC) 2023).

4.5.2 Implications to the Post-pandemic Era

Based on the empirical findings, the significant differences in the pandemic responses of Thailand, Cambodia, and Myanmar, driven by their unique political, social, and healthcare contexts. Thailand's robust health surveillance system and innovative quarantine measures, such as the 'Phuket Sandbox,' demonstrate the importance of a proactive and well-coordinated public health response. In contrast, Myanmar's response was severely hindered by political instability and a collapsed healthcare system, underscoring the critical role of political stability in effective pandemic management. Cambodia's rapid vaccination rollout and strong public cooperation were effective, but the country faced other social challenges, such as the rise in cybercrime. It tackles the convergence among political, economic, social, and health issues all at the same time which have an impact far beyond the pandemic era itself.

4.5.2.1 Policy Implications

- Enhancing Cross-Border Health Cooperation:
  - Establish dedicated border taskforces and data-sharing mechanisms to improve the detection and management of health threats.
  - Develop mutual acceptance of health documents to streamline travel and avoid unnecessary disruptions.
  - Improve quarantine management to ensure the right to freedom of movement and prevent illegal crossings.
- Investing in Public Health Infrastructure:
  - Prioritise investments in healthcare infrastructure, including hospitals, laboratories, and surveillance systems, to enhance the region's capacity to respond to future health emergencies.
  - Address challenges in attracting and retaining healthcare workers, particularly in rural and underserved areas.

- Ensuring Political Stability:
  - Recognise the impact of political stability on public health responses and work towards creating stable governance structures that can support effective health measures.

4.5.2.2 Practical Implications

- Healthcare Providers and Communities:
  - Train healthcare workers to adapt to new health surveillance and quarantine measures.
  - Improve communication strategies to ensure clear and consistent messaging about health guidelines and measures.
  - Ensure equitable access to healthcare services, especially for vulnerable populations affected by the pandemic and state responses.
- Addressing Health Inequities:
  - Implement policies to reduce health disparities and ensure equitable access to healthcare for marginalised and vulnerable populations.

4.6 Future Directions

4.6.1 Future of Cross-Border Health Cooperation

Based on the comparative analysis in previous section, the apparent pain point appeared in this study is about different epidemiological logics that each state deployed in decision making on COVID-19 Response, border closure/reopening dates, and health surveillance practices were not fully seamless and depending on its political stability. According to the proposed amendments of the 2005 International Health Regulation, the delegates recommended that States Parties sharing common borders to consider bilateral or multilateral agreements in the event of a public health emergency of international concern (PHEIC) concerning prevention or control of international transmission of disease at ground crossings (World Health Organization 2023).

There are three recommendations for the future of border health management as learned from the COVID-19 Responses between Thailand and neighbouring countries that face the challenges because of the huge difference of public health infrastructure namely:

4.6.1.1 Dedicating Border Taskforce

States parties beyond the public health and security sectors are encouraged to collaborate at the border level to enhance capacity to detect, assess, notify, and report events at the designated points of entry including airports, ports, and ground crossings through the channels of the Committee of Border Control (CBC). The committee may consider the potential role of screening technology, such as AI-powered surveillance and predictive analytics, in enhancing border security and public health for the management of quarantine mandates and facilities especially for the ground crossings where the information of migration patterns is crucial for designing practical quarantine mechanism.

4.6.1.2 Border Health Measurement Strategies

It is not only about advancing screening technologies but also data collection and required documents that travellers are obliged to provide to be permitted to travel amidst the pandemic. There should be mutual acceptance of health documents to avoid unnecessarily disruption or being used for political purposes beyond the purpose of disease containment.

4.6.1.3 The Management of Quarantine Mandates and Facilities

The task is for containing the spread of the disease and confirming everyone has the right to freedom of movement and prevent illegal crossings, especially for the ground crossings where the information of migration patterns is crucial for designing practical quarantine mechanism.

4.6.2 Future of Public Health System

By focusing on these following areas, the region can build a resilient and equitable public health system capable of responding to future health challenges and ensuring the well-being of all its citizens.

4.6.2.1 Attracting and Retaining Healthcare Workers

Address the challenges of attracting and retaining healthcare workers, particularly in rural and underserved areas. This includes offering competitive salaries, providing professional development opportunities, and ensuring safe working conditions.

4.6.2.2 Training and Education

Invest in continuous training and education programs to keep healthcare workers updated on the latest medical practices and technologies. Especially on the development and regular updates on emergency response plans to ensure preparedness for future health crises. This includes conducting regular drills and simulations as well.

4.6.2.3 Investing in Public Health Infrastructure Through Resource Allocation Plans

To ensure adequate allocation of resources, including funding, medical supplies, and personnel, to respond effectively to health emergencies as for prioritizing investments in healthcare infrastructure, including hospitals, laboratories, and clinics, to enhance the region's capacity to respond to future health emergencies.

4.6.2.4 Addressing Health Inequities and Justice

Equitable Access to Healthcare: Implement policies to reduce health disparities and ensure equitable access to healthcare for marginalised and vulnerable populations. This includes expanding healthcare coverage and providing financial support for low-income households.

Community-Based Health Programmes: Strengthen community-based health programs to ensure that healthcare services reach remote and underserved areas. This includes mobile clinics and telemedicine services.

4.6.3 Future of Asia's Regional Interconnectedness

This study suggests that the importance of regional interconnectedness in health responses and the lessons learned from the COVID-19 pandemic can inform future health policies, particularly in enhancing border health surveillance and health security cooperation. Comparing to the other parts of Asia, Thailand shares strong and effective management through pre-existing institutional infrastructure of public health system with countries like Taiwan, Hong Kong, South Korea, and Singapore that had pre-established Centres for Disease Control and Prevention (CDC), which facilitated swift and coordinated responses. Not to mention, the collectivist culture both in East Asia and Southeast Asia supported public cooperation with stringent measures, such as mask mandates and social distancing, which were more readily accepted compared to Western countries (An and Tang 2020; Ho 2020; Chen et al. 2021).

While in Central Asia, the region's response has been shaped by the severity of the health crisis, fiscal capacity, and the need for international support. The economic recovery requires addressing structural challenges, improving connectivity, and supporting the private sector. For example, Kazakhstan, Mongolia, and Uzbekistan responded swiftly with strict confinement measures, border closures, and large support packages. Kyrgyzstan had delayed responses due to limited public capacities and fiscal constraints. Tajikistan and Turkmenistan reported few or no cases initially and had limited policy measures (Organisation for Economic Co-operation and Development (OECD) 2020; Shadmanov et al. 2020; Alfano 2022).

Therefore, this is crucial for addressing both public health challenges and socioeconomic issues that arise from migration and cross-border interactions including:

4.6.3.1 Post-pandemic Recovery

Analyse the long-term economic and social impacts of the COVID-19 pandemic on the region, predominantly on vulnerable populations. Explore strategies for building resilience and ensuring equitable recovery.

4.6.3.2 Emerging Infectious Diseases (EIDs)

Explore the potential impact of future pandemics, mainly in light of climate change and increased global interconnectedness.

4.6.3.3 Ensuring Political Stability in the Region

Recognise the impact of political stability on public health responses and work towards creating stable governance structures that can support effective health measures.

4.7 Conclusion

Border health is considered as one of non-traditional security issues which taking various perspectives such as public health, geopolitics, medical anthropology, and global health into account. The major prevalent challenges for border health in mainland Southeast Asia, especially for Thailand have been concerning about the accessibility to primary health care and disease prevention among the undocumented and migrant populations. The findings of this study demonstrate the interconnectedness between borders, migration, and health. 'Viruses do not carry passport'[10] but the politics of inclusion and exclusion persist and remain the focal point for health surveillance and border management. This is an opportunity for the collaboration among the scholars and practitioners from both the Global South and North that could join this endeavour through this chapter's recommendations in supporting border health initiatives that run by the community-based health organisations, especially on the Thai-Myanmar border where the ongoing armed conflict in Myanmar has disrupted the entire country's public health system.

Border surveillance thus remains crucial for disease controls. It requires better understanding on political, economic, and social conditions before making the decisions on disease control practices. One the one side, the phenomenon of the 'scamdemic' in Cambodia reflects the economic impact of the pandemic that has flourished cyber organised crime activities exploiting people who are desperate for work and money. It has shifted cross-border migration flows that could make the region more vulnerable in all aspects of life.

On the other side, political instability in Myanmar that has destroyed the public health system of the country and influenced the shifts in border health situation on the Thai side by the rising in irregular reasons for seeking healthcare in Thailand such as the injured from the armed clashes. The loss of right to health and human rights in Myanmar resulted in the urgent needs not only for humanitarian assistance but also for reconfiguring the missions for cross-border health cooperation. It is a reminder that there would not be such effective or feasible pandemic preparedness for the country that remains in conflict. And it will eventually have an impact on the public health system beyond national borders.

Highlights

- Known Unknowns: Enhance regional risk assessment and scenario planning to improve preparedness for future pandemics with unknown characteristics.
- Border Management: Strengthen cross-border health cooperation and establish dedicated border taskforces to improve the detection and management of health threats.
- Regional Interconnectedness: Foster collaboration among Southeast Asian countries to enhance public health responses and ensure coordinated efforts in managing cross-border health issues.

Acknowledgements This research project is supported by Ratchadapiseksompotch Fund Chulalongkorn University. I also would like to express my deepest gratitude to all informants from the borders of Thailand to the heart of the Ministry of Public Health of Thailand, Nonthaburi Province. All opinions, omissions, and errors remain my own.

[10] The chapter title in Ong (2016). Fungible Fife.

References

Ahuja N. Bioinsecurities: disease interventions, empire, and the government of species. Durham: Duke University Press; (ANIMA (Duke University Press)); 2016.

Alfano V. COVID-19 in Central Asia: exploring the relationship between governance and non-pharmaceutical intervention. In: Health Policy and Planning. 2022.

An BY, Tang SY. Lessons from COVID-19 responses in East Asia: institutional infrastructure and enduring policy instruments. Am Rev Publ Admin. 2020;50(6–7):790–800.

Andrews M, Squire C, Tamboukou M. Doing narrative research. SAGE Publications Ltd.; 2008.

Asian Development Bank (ADB). Cambodia: COVID-19 active response and expenditure support program first quarterly monitoring report. https://www.adb.org/projects/documents/cam-54195-001-dpta. Accessed July-September 2020

Caduff C. The pandemic perhaps: dramatic events in a public culture of danger. University of California Press; 2015. P. 270.

Chen S, Guo L, Alghaith T, Dong D, Alluhidan M, Hamza MM, et al. Effective COVID-19 control: a comparative analysis of the stringency and timeliness of government responses in Asia. Int J Environ Res Public Health. 2021;18(16):8686.

Chhim S, Ku G, Mao S, Put WVD, Van Damme W, Ir P, et al. Descriptive assessment of COVID-19 responses and lessons learnt in Cambodia, January 2020 to June 2022. BMJ Glob Health. 2023; 8(5).

Chuengsatiansup K, Limsawart W. Tuberculosis in the borderlands: migrants, microbes and more-than-human borders. Palgrave Commun. 2019;5(1):1–10.

Ennis-McMillan MC, Hedges K. Pandemic perspectives: responding to COVID-19. Open Anthropol. 2020; 8(1). https://www.americananthro.org/StayInformed/OAArticleDetail.aspx?ItemNumber=25631

Faulder D. Asia's scamdemic: how COVID-19 supercharged online crime. Nikkei Asia. 2022. https://asia.nikkei.com/Spotlight/The-Big-Story/Asia-s-scamdemic-How-COVID-19-supercharged-online-crime

Foucault M. The birth of the clinic: an archaeology of medical perception. New York: Vintage; 1994. p. 240.

Graham DA. Rumsfeld's knowns and unknowns: the intellectual history of a quip. The Atlantic. 2014. https://www.theatlantic.com/politics/archive/2014/03/rumsfelds-knowns-and-unknowns-the-intellectual-history-of-a-quip/359719/

Ho HK. COVID-19 pandemic management strategies and outcomes in East Asia and the Western World: the scientific state, democratic ideology, and social behavior. Front Sociol. 2020.

Hooker C. Drawing the lines: danger and risk in the age of SARS. In: Bashford A, editor. Medicine at the border: disease, globalization and security, 1850 to the present. London: Palgrave Macmillan UK; 2007. p. 179–95. https://doi.org/10.1057/9780230288904_10

International Organisation for Migration. Impact of Covid-19 on migrants in thailand situation report [Internet]. 2021. https://thailand.iom.int/sites/g/files/tmzbdl1371/files/documents/unnm-covid19-sitrep_november.pdf.

Kelly AH, Keck F, Lynteris C, editors. The anthropology of epidemics. London: Routledge; 2019. 194 p.

Kenwick MR, Simmons BA. Pandemic response as border politics. Int Organ. 2020; 74(S1):E36–58.

Lakoff A. A fragile assemblage: mutant bird flu and the limits of risk assessment. Soc Stud Sci. 2017;47(3):376–97.

Lan CS, Sok S, Chheang K, Lan DM, Soung V, Divi N, et al. Cambodia national health hotline—participatory surveillance for early detection and response to disease outbreaks. Lan Region Health West Pac. 2022; 29. https://www.thelancet.com/journals/lanwpc/article/PIIS2666-6065(22)00199-7/fulltext

Laocharoenwong J. Center for social development studies. 2020. critical nature: COVID-19, health borders, and the purity of the Thai Nation. https://www.csds-chula.org/publications/2020/6/2/critical-nature-covid19-health-borders-and-the-purity-of-thai-nation

Lindholt MF, Jørgensen F, Bor A, Petersen MB. Support for border security during the COVID-19 pandemic: evidence on levels and predictors from eight Western democracies in 2020. J Elect Publ Opin Part. 2021;31(sup1):1–14.

Lowe C. Viral clouds: becoming H5N1 in Indonesia. Cult Anthropol. 2010;25(4):625–49.

Mailman School of Public Health. Epidemic, endemic, pandemic: what are the differences?. Columbia University Mailman School of Public Health. 2021. https://www.publichealth.columbia.edu/news/epidemic-endemic-pandemic-what-are-differences

Nikkei. Cambodia holds back Vietnam and Thailand in trafficking prevention—Nikkei Asia. https://asia.nikkei.com/Spotlight/Asia-Insight/Cambodia-holds-back-Vietnam-and-Thailand-in-trafficking-prevention. Accessed 16 Aug 2022.

Ong A. Fungible life experiment in the Asian city of life. Duke University Press; 2016.

Organisation for Economic Co-operation and Development. COVID-19 crisis response in Central Asia. 2020. https://www.oecd.org/en/publications/covid-19-crisis-response-in-central-asia_5305f172-en.html

Ostwald K, Myint T. Myanmar: Pandemic in a time of transition. In: Ramraj VV, editor. Covid-19 in Asia: law and policy contexts. Oxford University Press; 2021. https://doi.org/10.1093/oso/9780197553831.003.0023

Perera S, Maung C, Hla S, Moo Moo H, Than Lwin S, Bruck C, et al. Access to community-based reproductive health services and incidence of low birthweight delivery among refugee and displaced mothers: a retrospective study in the Thailand-Myanmar border region. BMJ Open. 2022;12(1): e052571.

Physicians for Human Rights. "Our health workers are working in fear": after Myanmar's military coup, one year of targeted violence against health care. PHR. 2022. https://phr.org/our-work/resources/one-year-anniversary-of-the-myanmar-coup-detat/

Rocha IC, Cedeño TD, Pelayo MG, Ramos K, Victoria HOH. Myanmar's coup d'état and its impact on COVID-19 response: a collapsing healthcare system in a state of turmoil. BMJ Mil Health. 2023;169(2):103–4.

Sangsuradej P. Community struggles and the challenges of solidarity in Myanmar. LSE Press. 2022. https://press.lse.ac.uk/site/chapters/e/10.31389/lsepress.cov.s/

Shadmanov A, Boklage E, Storozhenko O, Jansen A, Uzbekistan C. 19 mission team for. COVID-19 in Central Asia: Uzbekistan's new approach to international cooperation. Lanc Reg Health Europe. 2020; 1:100007.

Strangio S. UN report details massive scale of Southeast Asia's 'Scamdemic.' 2023. https://thediplomat.com/2023/09/un-report-details-massive-scale-of-southeast-asias-scamdemic/

Ticktin M. No borders in the time of cOVID-19. 2020. http://www.americananthropologist.org/2020/07/02/no-borders-in-the-time-of-covid-19/

Tourism Authority of Thailand (TAT) Newsroom. Thai PM Prayut Chan-o-cha confirms 'Phuket Sandbox' starting 1 July 2021. TAT Newsroom. 2021. https://www.tatnews.org/2021/06/thai-pm-prayut-chan-o-cha-confirms-phuket-sandbox-starting-1-july-2021/

United Nations Office for Drugs and Crime. Casinos, cyber fraud, and trafficking in persons for forced criminality in Southeast Asia: policy report. 2023. https://www.unodc.org/roseap/uploads/documents/Publications/2023/TiP_for_FC_Policy_Report.pdf

Wicaksana IGW, Nauvarian D, Pramudia PS. ASEAN, COVID-19 and Myanmar crisis: dealing with critical juncture. Int Area Stud Rev. 2023;26(1):40–54.

Wille C, Kanesu R. Bordering in pandemic times: insights into the COVID-19 lockdown. 2020; https://ubt.opus.hbz-nrw.de/frontdoor/index/index/docId/1428

World Health Organization Thailand. Point of entry screening and quarantine systems enabled Thailand to control COVID-19. 2020. https://www.who.int/thailand/news/feature-stories/detail/point-of-entry-screening-and-quarantine-systems-enabled-thailand-to-control-covid-19

World Health Organization. Archived: WHO timeline—COVID-19. 2020a. https://www.who.int/news/item/27-04-2020-who-timeline---covid-19

World Health Organization. Thailand controlling spread of COVID-19 along important trading border with Cambodia. 2020c. https://www.who.int/thailand/news/feature-stories/detail/thailand-controlling-spread-of-covid-19-along-important-trading-border-with-cambodia

World Health Organization. WHO director-general's statement on IHR emergency committee on novel coronavirus (2019-nCoV). 2020b. https://www.who.int/director-general/speeches/detail/who-director-general-s-statement-on-ihr-emergency-committee-on-novel-coronavirus-(2019-ncov)

World Health Organization. Report of the review committee regarding amendments to the International Health Regulations (2005). 2023. https://apps.who.int/gb/wgihr/pdf_files/wgihr2/A_WGIHR2_5-en.pdf

Worsnop CZ. Domestic politics and the WHO's International health regulations: explaining the use of trade and travel barriers during disease outbreaks. Rev Int Organ. 2017;12(3):365–95.

Khathaleeya Liamdee is a lecturer in the Department of Sociology and Anthropology, Faculty of Political Science, Chulalongkorn University. She holds a PhD in anthropology from the University of Washington with a dissertation titled 'On the Move across Phnom Dangrek: Mobilities and Silences in the Thai-Cambodian Borderland' in 2020. Her subsequent research focuses on health humanitarian assistance and health security along the Thai-Myanmar border.

Chapter 5
Enhancing Health Insurance for Farmers in Vietnam's Mekong Delta: Influential Factors and Policy Directions

Clusters of Future Studies: Past and Futures; Humanity at the Limen

Nguyet Thi Thu Tran, Anja Reid, Carol Warren, and Kathryn Trees

Abstract This chapter focuses on the factors influencing Mekong Delta farmers' satisfaction with health insurance services in the context of agricultural practice. The study identifies three key factors affecting farmers' satisfaction with health insurance services: participation and payment procedures, hospital service quality, and the support of hospital and social insurance staff. Statistical analysis indicates that these factors significantly impact farmers' satisfaction levels, with women rating health insurance services higher than men. Moreover, the study underscores the importance of improving access to quality care and enhancing client satisfaction to address the challenges faced by farmers in accessing healthcare services. The research employs a mixed-method approach, including surveys, interviews, and focus group discussions, to gather empirical data from 180 farming households in An Giang, Can Tho, and Soc Trang provinces. The chapter recommends policy reforms to address the socio-economic and environmental changes in the Mekong Delta, advocating for the

Disclaimer This chapter emerges from the first author's Ph.D. thesis, completed at Murdoch University. The thesis is currently under embargo in the university's repository, i.e. Murdoch University thesis repository.

N. T. T. Tran (✉)
Faculty of Law and Political Science, An Giang University- Vietnam National University, Ho Chi Minh City, Vietnam
e-mail: tttnguyet@agu.edu.vn

A. Reid · K. Trees
College of Law, Arts and Social Sciences, Murdoch University, Murdoch, WA, Australia
e-mail: A.Reid@murdoch.edu.au

K. Trees
e-mail: K.Trees@murdoch.edu.au

C. Warren
Indo-Pacific Research Centre, Murdoch University, Murdoch, WA, Australia
e-mail: C.Warren@murdoch.edu.au

R. Xiang Ng et al. (eds.), *Future of Healthcare in Asia*,
https://doi.org/10.1007/978-3-032-03183-9_5

integration of technological innovations such as telemedicine and artificial intelligence (AI) diagnostic tools to enhance healthcare accessibility. The study's implications extend to the Vietnam Ministry of Health and Social Security Agency, providing insights to identify areas for improvement in health insurance policies and social security systems. The research contributes to the ongoing discourse on enhancing healthcare services for rural populations, with implications for policymakers, healthcare providers, and social security agencies.

Keywords Health insurance · Farmer · Influential factors · Policy direction · Mekong delta

5.1 Introduction

Agriculture accounts for 31.37% of the Mekong Delta's gross domestic product (GDP). With 39,734 km2 or 12.2% of the country's total area carrying approximately 19% of the total population, this region contributes about 12.8% of the national GDP (Le Minh Hoan 2022; Vietnam Chamber of Commerce and Industry (VCCI) and Fulbright 2022). In the Mekong Delta—Vietnam's 'rice bowl'—37.5% of the 17.3 million people living there are engaged in agriculture, forestry and fisheries (General Statistics Office 2023). Agricultural production is one of the fields facing the most risks today, from weather and crop diseases to market instability, price, and health risks. The overuse of fertilisers and pesticides in agriculture negatively affects the health of many farmers (Damalas 2009; Yazd et al., 2019). The higher the volume of pesticide used, the greater the medical costs from exposure to these substances. Farmers' abuse of pesticides in rice production in the Mekong Delta has many impacts, including reducing their income and increasing health and environmental costs (Kongtip et al., 2017; Yazd et al., 2019). Further, the health of the agricultural workforce can be affected by direct exposure to extreme temperatures due to the impacts of changing weather (Mbow et al., 2019). Therefore, participating in health insurance is essential because it reduces the financial burden for farmers in the face of disease and, in the long term, supports national development and wellbeing.

Health insurance (HI), implemented by the State, is compulsory for all Vietnamese citizens of every age. HI is a form of insurance in which the participant is covered for part or the full cost of medical examinations and treatment. Vietnam's Law on Health Insurance No. 25/2008/QH12 promulgated on November 14, 2008 and the Law amending and supplementing a number of articles of the Law on Health Insurance No. 46/2014/QH13 promulgated on June 13, 2014 are the legal basis for the Vietnamese government's national health insurance program, which aims to institute universal health insurance. Under the law, the State is responsible for subsidising the full cost of health insurance for children under six years old, the elderly, the poor and ethnic minorities, as well as subsidising a portion of the cost of health insurance for near-poor people and students.[1]

In principle, Vietnam's health insurance scheme applies a unified package of benefits to a single payer from a common fund (Somanathan et al. 2014). In 2017, Resolution No. 20 was approved by the 6th Conference of the 12th Party Central Committee, stating that Social Health Insurance is the financial mechanism to implement Universal Health Insurance (The Central Committee of the Communist Party of Vietnam 2017).

Health insurance coverage in the Mekong Delta was reported as 89.3% in 2019; according to the Vietnam Ministry of Health report, evidently, health insurance coverage has not yet achieved the 'universal' expectation. In part, this is because of the difficulties people face regarding their living conditions, they are not financially able to participate in health insurance. The inability to pay health insurance premiums was also mentioned in some research of Boateng and Awunyor-Vitor (2013); Adebayo et al. (2015); Ogundeji et al. (2019) and Le et al. (2023). In addition, the policies and services provided by health insurance do not meet users' medical examination and treatment needs, and the process of making a claim is still complicated (Loan and Ban 2013; Le et al. 2023). Moreover, customers' intention to participate and satisfaction with medical services provided by health insurance are significantly influenced by the continuous changes in health insurance policies (Vietnam Ministry of Health 2020b; Nguyen et al. 2020; Nguyet 2022).

A mixed-method approach was deployed to gather data in three provinces in the Mekong Delta (An Giang, Can Tho, and Soc Trang). Building on the study findings, we argue that improving health insurance policy plays an essential role in stabilising farmers' lives and adapting to the fluctuating local socio-economic and environmental conditions. This analysis provides critical insights into how farmers respond to health insurance regulations and services and how the policy could be further improved. More importantly, this study advances knowledge of farmers' understanding of the health insurance policy and its benefits and provides a general context for policymakers and other government bodies to amend the health insurance policy, toward enhancing the effectiveness of Vietnam's social security programs.

This chapter aims to: identify the factors affecting the satisfaction of farmers in the Mekong Delta region with health insurance; propose recommendations to help improve farmers' satisfaction and contribute to finding solutions to reform health insurance policy for Vietnamese farmers in general. Moreover, the challenges faced by Vietnamese farmers, including geographic barriers, low educational attainment, and financial hardships, are common across rural regions in many other Asian countries. By examining Vietnam's healthcare system, we can gain a deeper understanding of the factors influencing health outcomes and patient satisfaction in similar contexts.

[1] Clause 3, Article 12 of The National Assembly's June 13, 2014 Law No. 46/2014/QH13 Amending and Supplementing a Number of Articles of the Health Insurance Law which takes effect on January 1, 2015.

5.2 Literature Review

5.2.1 Health Insurance for Farmers and Their Satisfaction with Healthcare Services

This study found that farming practices, including chemical use and personal protection, have been slow to change to the detriment of the health of farmers and the environment. Farmers in developing nations face poor health outcomes due to climate change, physical labour-intensive farming practices, use of pesticides and other hazardous substances, or subpar living conditions (O'Brien et al. 2004; Hu et al. 2015). Antle and Pingali (1994) found that pesticides had negative consequences for the health of rice farmers in the Philippines. They considered the humid tropical environment and farmers' lack of protective clothing as factors in their health. Nicolopoulou-Stamati et al. (2016), more than 20 years later, in their study of pesticides or agrochemicals and farmers' health, had similar findings. These studies found that farmers have higher rates of disease (skin, respiratory, neurological, and kidney issues) and poorer mental health (anxiety/depression, memory loss, dizziness) according to research conducted in both developed and developing nations (Antle et al. 1994; Nicolopoulou-Stamati et al. 2016). Low technological infrastructure (Benavides-Vaello et al., 2013; Doan et al., 2018), geographic barriers, low educational attainment, poor quality of healthcare professionals, and financial hardships (Shaikh and Hatcher 2004; Arcury et al. 2005; Ager and Pepper 2005) are all contributing factors to this issue. These adverse health effects may lead to reduced productivity at work and a rise in household debt (Berry et al. 2011; Burgard et al. 2013; Kunde et al. 2017).

The accessibility and affordability of medical care are pivotal aspects influencing a patient's contentment and play key roles in judging a healthcare system's efficiency. Getting care refers to how simply individuals can reach essential medical services, including nearby clinics, transportation options, and prompt scheduling of appointments. On the other hand, affordability is tied to the fiscal struggles related to obtaining care, covering out-of-pocket costs, insurance plans, and the overall impact on persons and families (Arrow 1963). When services are both accessible and affordable, patients are more apt to seek well-timed and suitable treatment, resulting in improved outcomes and overall fulfilment. In contrast, limited accessibility and unaffordable care can lead to postponed or deficient treatment, negatively impacting patient satisfaction and health consequences. The importance of medical insurance in addressing these barriers is essential as it can alleviate monetary hurdles and boost access to services, thereby contributing to heightened patient satisfaction levels and the overall effectiveness of healthcare systems (Arrow 1963; Hsiao and Shaw 2007; Baeza and Packard 2006). Health and access to the health system strongly affect farmers' quality of life and economic viability. Mafimisebi and Oguntade (2010) studied the accessibility and affordability of health services for agrarian people in Nigeria. They found that low accessibility to health care leads to farmers' dependence on traditional medicine based on socio-cultural and religious practices, which leads to using herbal remedies to manage farmers' health. The socio-economic and political environment surrounding farmers plays a crucial role in shaping their lives, particularly in evaluating insurance coverage options relative to other pressing priorities within their farm operations and households (Inwood 2017). This context affects their overall health and economic well-being. Furthermore, the ability to access a health care system—whether public or private—and secure affordable health care and insurance is a significant concern for individuals in the agricultural sector.

The multifaceted healthcare outcome for those, including farmers, who use them is influenced by various factors related to the patient, the doctor, and the system (Thiedke 2007). Patients are key partners and recipients of well-being protection plans. Their opinions are vital for affecting well-being protection approaches, giving feedback on the responsiveness and quality of protection programs, and bringing responsibility and straightforwardness to the decision-making for protection arrangements (Geng et al. 2021). Haile et al. (2022) identified several factors that affect beneficiary satisfaction, including family size, awareness of the benefit packages, the demeanour of healthcare providers, privacy and confidentiality, accessibility of prescribed medications, availability of laboratory services, perceived cleanliness of health facilities, waiting times, staff responsiveness to inquiries, and alignment with the benefit packages offered by the Community-Based Health Insurance Program. Getaneh et al. (2023) found similar results from their research in North-East Ethiopia.

Various factors, including individual and household socio-demographic characteristics, influence health insurance uptake and utilisation. A qualitative study conducted in the six southern provinces of Vietnam reported reasons for involvement in health insurance as its: importance for good health care, compulsory nature, role in protecting people from financial risk during illness, and its ability to share financial risk (Thanh 2015). The factors significantly associated with the level of satisfaction are age, gender, marital status, education, occupation, self-perceived health status, area of residency, and type of household plan, as found in studies by Ali Jadoo et al. (2012). These researchers also found that to raise overall patient satisfaction, more information must be provided about resource availability, technical excellence, as well as higher standards of personal attention and care (Ali Jadoo et al. 2012). Adequate access to and continuity of care were linked to higher patient satisfaction. A decade later, Fatimah et al., in their study of the health system and insured and non-insured users in Indonesia, have found that insured users report higher satisfaction (2022).

Consequently, the efficacy of the health system relies on enhanced access to high-quality care and customer satisfaction. Conducting satisfaction surveys is crucial for addressing issues related to access and performance. The satisfaction of patients is a pivotal consideration intertwined with strategic decisions regarding health services. Understanding beneficiary satisfaction improves access to quality care and enables informed decisions about health insurance policies. It also highlights the need for health insurance agencies and health facilities to work together to improve beneficiaries' care and satisfaction levels (Mohammed 2011).

This chapter focuses on satisfaction with healthcare services when using health insurance among Mekong Delta farmers and provides critical insights into the effectiveness and accessibility of health insurance schemes in rural and agricultural communities. Similar studies in countries like India and Indonesia have highlighted significant disparities in healthcare access and satisfaction between urban and rural populations (World Health Organization (WHO) 2020). In India, research has shown that rural residents often face greater challenges in accessing healthcare due to financial constraints and distance to facilities, despite having health insurance (Sakamoto et al. 2018). Similarly, in Indonesia, the uneven distribution of healthcare professionals has led to varying satisfaction levels and health outcomes across different regions (Harimurti et al. 2017). Examining farmers' healthcare in the Mekong Delta enables identification of unique obstacles and facilitators to treatment in Vietnam, providing a useful point of comparison for more comprehensive regional strategies. These understandings are indispensable for creating targeted policies guaranteeing equitable access to care and enhancing the overall performance of insurance programs throughout Asia, contributing to the region's health and economic strength.

5.2.2 Conceptual Framework

The chapter analysis of farmers' satisfaction with health insurance using the Donabedian model, through adapting the criteria for access to social security, includes coverage, eligibility, affordability, participant and information, and physical access (United Nations 2007). The analysis involves examining the intricate linkages between structure, process, and outcome. Donabedian categorised the evaluation of healthcare quality into three key components: structure, process, and outcome of care (Donabedian 2005). This three-part framework for quality assessment emphasises the interconnection among each element, suggesting that a solid structure enhances the likelihood of an effective process, which in turn boosts the chances of achieving a good outcome (Donabedian 1988). In this study, the structure dimension encompasses foundational elements such as infrastructure and medical equipment, which are essential for delivering high-quality healthcare services. Additionally, the accessibility (to policy and hospital), participation rates, information dissemination, and physical access—play a crucial role in determining how effectively farmers can utilise these services. The process dimension focuses on the interactions and communication between healthcare providers and farmers, the thoroughness and quality of examinations and treatments, and the efficiency of administrative procedures. Effective communication and streamlined administrative processes are vital for ensuring that farmers receive timely and appropriate care. Finally, the outcome dimension assesses farmers' satisfaction with the health services they receive under their health insurance plans. Satisfaction is a critical indicator of the success of the health insurance system, reflecting whether the services meet the farmers' needs and expectations. By linking these three dimensions, this study provides a comprehensive framework

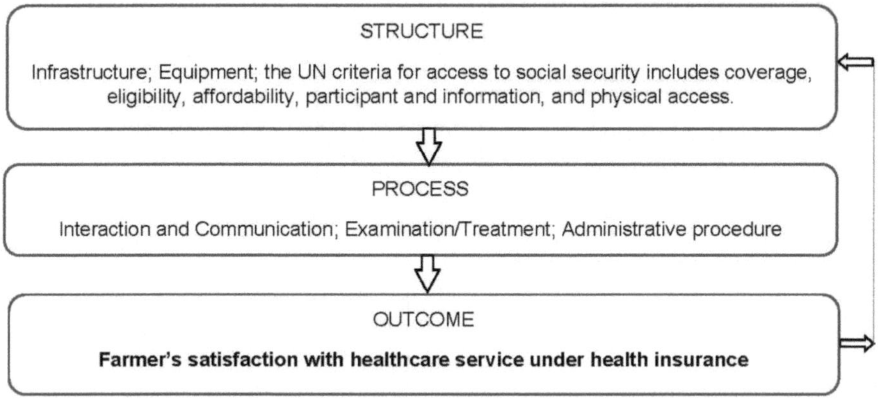

Fig. 5.1 Conceptual framework of farmers' satisfaction with health insurance by using the Donabedian model

for understanding and improving the factors that contribute to farmers' satisfaction with health insurance, ultimately guiding policymakers in enhancing the overall effectiveness and sustainability of health insurance programs for rural populations (Fig. 5.1).

5.3 Methodology

5.3.1 Ethical Considerations

Fieldwork for this project was carried out after receiving approval from the Human Research Ethics Committee. The survey team attended a one-day group briefing on Murdoch University's regulations on Human Research Ethics, the research background, and the general requirements for conducting the survey. All participants received a participant information letter and consent form. Before starting the survey, for those participants who had limited capacities in reading and understanding, interviewers were required to read these to them and only start interviewing when oral consent had been given by saying "I agree". The fieldwork strictly followed the human research ethics guidelines, with no adverse reactions from the participants.

5.3.2 Research Hypotheses

The hypotheses aim to provide a framework for understanding the multifaceted factors that contribute to farmers' satisfaction with health insurance, encompassing both the procedural and service-related aspects and the influence of socio-demographic variables (Fig. 5.2).

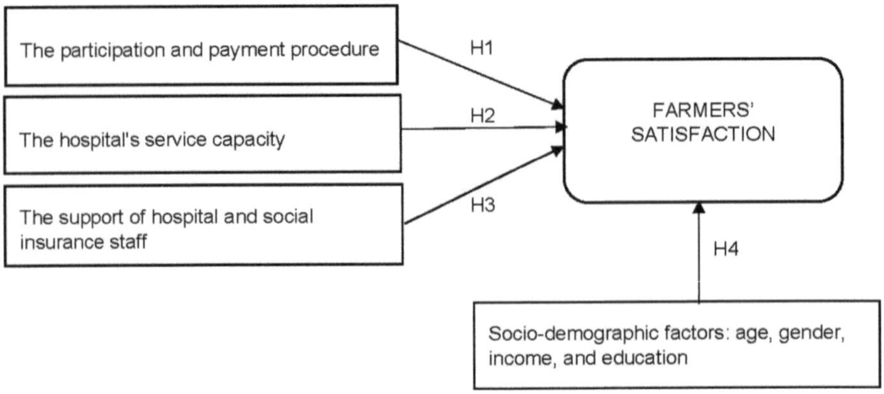

Fig. 5.2 Model of hypotheses on factors affecting farmers' satisfaction with health insurance

H1: The ease of participation and the transparency of payment procedures positively impact farmers' satisfaction with health insurance.
H2: The availability of quality services, medical facilities, and efficient healthcare delivery at hospitals positively impacts farmers' satisfaction with health insurance.
H3: The support services, such as the responsiveness, helpfulness, and professionalism of hospital and social insurance staff, positively impact farmers' satisfaction with health insurance.
H4: Variations in age, gender, income, and education levels impact farmers' satisfaction with health insurance, with different demographic groups exhibiting varying satisfaction levels.

5.3.3 Recruitment and Data Collection

A purposive selection of 180 farming households to conduct a survey and 54 respondents (including six commune officials) for interviews and focus group discussions was carried out in six communes spread over the three provinces of An Giang, Can Tho, and Soc Trang (Fig. 5.3). Since households are directly impacted by health insurance policy, they were used as the unit of analysis in this study. Heads of households were the target respondents for the survey. Because they have the authority to speak for the entire household, household heads are the appropriate respondents, according to Bookwalter et al. (2006). Thus, information supplied by the heads of the home can serve as a reasonable stand-in for the entire household. The household head (or another knowledgeable adult if the household head was not at home) was interviewed using the household questionnaire. A pre-test survey was undertaken before using the questionnaire to collect data to limit any problems with the interpretation of the wording of survey questions that might lead to biased answers.

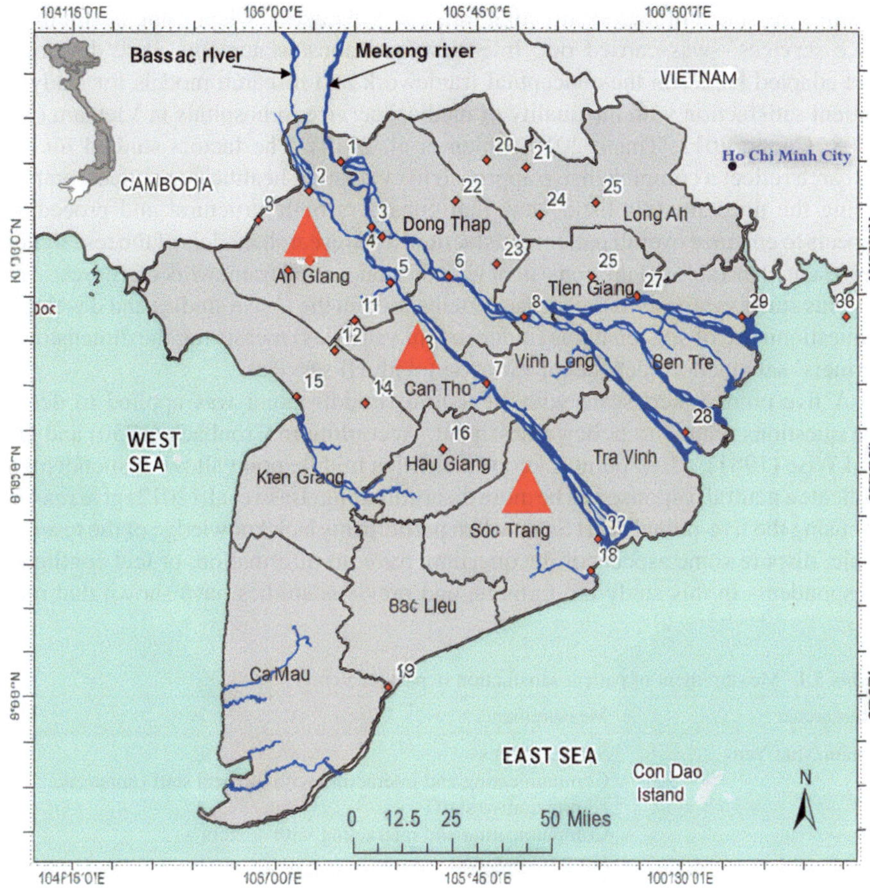

Fig. 5.3 Location of Soc Trang, An Giang, and Can Tho provinces in the Mekong Delta. *Source* A base map adapted from Vu et al. (2014)

5.3.4 Measures

Many potential items were generated from the literature, relevant elements of previous empirical research were examined, and relevant statements were extracted for consideration. This research adapted the United Nations (2007) accessibility criteria to social security, as outlined in the literature review, to focus on affordability, participation, information, and physical access. The questions in the questionnaire were also developed based on these criteria. For example, the question: 'How did you hear about health insurance?' provides data to analyse respondents' access to information on HI. Household income was collected and compared with insurance costs to assess whether a farmer could afford the insurance premium.

An assessment of factors affecting farmers' satisfaction when using health insurance services—was carried out. Interview questions/ statements were developed and adapted based on the conceptual framework and research models for studying patient satisfaction with the quality of medical service in hospitals in Vietnam (Yen 2008; Cuong 2012; Thanh 2013; Chien et al. 2014). The factors studied for this research reflect a comprehensive approach to evaluating healthcare services, emphasizing the need for a holistic view that considers both structural and procedural aspects to enhance overall patient satisfaction. The research models of the researchers represented in Table 5.1 are consistent with the conceptual framework of this research.

This study synthesised the measurements used in the above studies and developed a questionnaire of sixteen items (independent variables) measuring the dimension of farmers' satisfaction (dependent variables) with HI services.

A five-point Likert scale with a 'neutral' middle point was applied to design the questions related to policy assessment. According to Cronbach (1950) and Ory and Wise (1981), a five-point Likert scale with a middle point allows respondents to indicate a neutral response and be more discriminating. Baka et al. (2012) give reasons for using the five-point Likert Scale when participants lack knowledge of the research topic, dispute some aspects of the question, have no information, or feel conflicted. Respondents in this study are farmers, and previous studies have shown that most

Table 5.1 Measurement of patient satisfaction in past studies

Researcher	Measurements
Pham Nhat Yen	Service Access Communicating and interacting with medical staff (nurse and administrative staff) Communicating and interacting with doctors Hospital facilities Health care results Quality of medical care
Chu Hong Cuong	Waiting time for service Communicating and interacting with medical staff (nurse and administrative staff) Communicating and interacting with doctors Hospital facilities and environment Medical examination and treatment results
Nguyen Quang Thanh	Waiting time for service access Attitude and quality of care by nursing staff Attitude and quality of treatment by doctors Infrastructure Service price Patient's needs
Le Thanh Chien	Infrastructure Administrative procedures Doctor's quality of care Nurse's quality of care Hygiene and security issues of the hospital

farmers in the Mekong Delta have a low level of education and lack knowledge of policy (Anh 2010; Hoa et al. 2014; Tien et al. 2018); this potentially led some respondents to express uncertain opinions about the effect of policy implementation. In this study, neutral was interpreted as 'neither bad nor good', and to ensure that all respondents understood the options, the research applied a face-to-face survey, in which an interviewer was physically present to ask the survey questions and to assist the respondent in answering them.

5.3.5 Data Analysis

Quantitative data collected from the survey was analysed using SPSS software. Descriptive statistics are used to analyse data on demographic background and farmers' perceptions of social security policies in terms of frequency, percentage, mean or standard deviation. In order to identify factors affecting participants' satisfaction with health insurance services, Exploratory Factor Analysis and Regression Analysis were carried out. Factor analysis is a method of shortening data by identifying continuous variables and separating them from less relevant ones. It is important to separate independent and dependent variables for better prediction (Hair et al. 2010). The appropriate criteria for conducting exploratory factor analysis are that the minimum KMO (Kaiser–Meyer–Olkin) as an indicator to consider the sample size is reasonable, the correlation between variables is 0.5, and the Bartlett test has a p-value less than 0.05 (Kaiser 1970; Dziuban and Shirkey 1974; Cerny and Kaiser 1977). The minimum Cronbach's alpha coefficient of 0.6 (Hair et al. 2010), and the minimum total correlation coefficient of the variables of 0.3 (Nunnally and Bernstein 1994; Maltby et al. 2007; Brzoska and Razum 2010) are the criteria for assessing the reliability of the scale in this study. For this study, a statistical significance of 0.05 was used. Content and ethnographic analyses were used to analyse qualitative data. Content analysis allows data to be systematically coded by organising information into categories to discover patterns that cannot be detected simply by listening to tapes or reading transcripts (Yin 1989; Robson 1993).

This study employed a convergent parallel mixed-method design, which involves collecting both quantitative and qualitative data simultaneously and then merging the results for a comprehensive interpretation (Morse 1991; Creswell et al. 2003; Johnson and Christensen 2008).[2] The study prioritizes quantitative results to address the research questions, following a quantitatively driven concurrent design (Morse 1991; Johnson and Christensen 2008). Data integration occurs during the discussion phase, synthesizing complementary quantitative and qualitative results to better understand the research topic (Andrew and Halcomb 2009).

[2] This approach, also known as simultaneous triangulation, allows for the confirmation, cross-validation, enrichment, and verification of findings by integrating different types of data (Creswell 2012).

5.4 Findings

5.4.1 *Characteristics of Sampled Participants*

Survey data was gathered for 180 farmers from three research regions (equally dispersed with 60 respondents from each province). The percentages of men and women were roughly equal in the sample, making up 48.3% and 51.7%, respectively. About 70% of respondents were heads of households, and the results indicate that males have historically been designated to head homes in the rural Mekong Delta, accounting for 44.4% of household heads, about twice as many as female household heads (23.3%).

The age group 40–49 comprised the majority of informants (36.1%), followed by 50–59 (27.2%). The proportion of respondents under 30 years old (3.9%) was the lowest. Most farmers had completed primary and lower secondary school (46.7% and 34.4%, respectively), followed by senior high school graduates (10.6%); 8.3% of farmers were illiterate. Out of the three provinces, An Giang had the largest percentage of illiterate farmers (5%) and the highest percentage of high school graduates (4.4%). The three provinces' educational status and household size differ statistically ($p = 0.044 < 0.05$ and $p = 0.026 < 0.05$, respectively), according to the findings of ANOVA and Chi-square tests.

When comparing provinces, Can Tho had the greatest income gap (USD 52 and USD 1789). Farming households in Soc Trang had higher expenditure rates than An Giang and Can Tho. The average household spending of each province/city was less than their income. Can Tho was the province with the highest average household income (USD 429), followed by An Giang (USD 324) and Soc Trang (USD 287). The difference in average household expenditure was similar. However, only the difference in income ($p = 0.002 < 0.05$) between provinces was statistically significant. Based on average income and expenditure, most farming households could afford HI. According to the regulations, the health insurance premium for a household of four members is about 8 USD/month (USD 96/year), which can be paid in terms of three months, six months or one year, depending on the household's choice. The HI program also covered some poor farmers with no productive land, who are eligible for full State sponsorship.

5.4.2 *Factors Affecting Farmer's Satisfaction*

Exploratory factor analysis (EFA) was used to identify the latent factors that represent the concept (farmers' satisfaction). In Table 5.2 (see the next page), the percentage of explained variance means component 1 explains 33.55% of the variance, whereas component 2 explains 12.24%, and component 3 explains 10.56% of the variance. The total percentage of explained variance of the three factors in this study is 56.35%. Some criteria say that the total variance explained by all components should be between 70 and 80%, which in this case would mean four to five components. Peterson (2000) and Hair (2010) also comment that this may be untenable for social science research where extracted factors usually explain only 50–60% (Peterson 2000; Hairc et al. 2010) of variance. Thus, the total variance explained in this study accounts for 56.35% with three components considered appropriate.

The results from Table 5.3 show that KMO = 0.651 > 0.5, Bartlett's test has p-value = 0.000 < 0.05, variance extracted = 61% > 50%, the variables grouped into 1 component with factor loading of dependent variables are greater than 0.5, thus using EFA was appropriate.

Then, a Pearson correlation analysis was conducted to evaluate the correlation between independent and dependent variables. The research results show that there is a correlation between customer satisfaction with components (factors), which has a high correlation with factor SUP 0.672 (Sig. = 0.000) and a lower correlation with PRO factor 0.466 (Sig. = 0.000). There is also a correlation between factors as well (see Table 5.4).

This study conducted regression analysis with the dependent variable—'member satisfaction' (SAT) and three independent variables or factors as the process and procedure (PRO); the hospital's service capacity (HOS); the support of the hospital and social insurance staff (SUP); which was extracted after factor analysis to assess the impact of factors on farmers' satisfaction when using health insurance services.

The Adjusted R Square of the model is 0.551, showing that three independent variables explain 55.1% of the variance of the dependent variable—member's satisfaction. The regression analysis results show that all three factors impact the satisfaction of Mekong Delta farmers due to Sig. index less than 0.05 and VIF less than 2. Table 5.5 shows the relationship between the control variables and the factors PRO, HOS and SUP to farmers' satisfaction when using HI.

A one-way ANOVA test was used to see if there were differences in satisfaction among HI members across age groups, income groups and education levels. The quantitative variable used was the SAT, and the qualitative variables (control variables) were age, education, and income.

Analysis results of Levene's test with Sig. = 0.795 > 0.05, and ANOVA results show that Sig. = 0.182 > 0.05, it can be concluded that there is no difference between ages regarding satisfaction with health insurance services. Similarly, there are no differences between education[3] and income groups[4] regarding the satisfaction of HI members.

With the gender control variable, having only two groups of values, male and female, this study used the Independent Sample T-test. The mean value of the two groups of male/female is 3.50 (SD = 0.614) and 3.68 (SD = 0.569), representing a moderate difference. The result from the Sig. F-test equals 0.183 > 0.05, and there is no difference in variance between the two groups, men and women. However, the Sig. of the T-test is equal to 0.039 < 0.05, meaning there is a difference in the member's satisfaction (SAT) between sexes defined as male and female.

The research results show that there were three factors influencing farmers' satisfaction when using HI services: (1) PRO- the participation and payment procedure

[3] Levene's test with Sig. = 0.923 > 0.05, and ANOVA with Sig. = 0.426 > 0.05.

[4] Levene's test with Sig. = 0.561 > 0.05, and ANOVA with Sig. = 0.621 > 0.05.

Table 5.2 Factor loadings from principal components analysis for independent variables (i.e. the process and procedure (PRO), hospital's service capacity (HOS); support of hospital and social insurance staff (SUP))

Variables	Factor loading		
	Component 1	Component 2	Component 3
You can participate in health insurance easily and simply	0.811		
You are provided with complete information about benefits when participating in health insurance	0.733		
The cost of buying health insurance compared to your income is reasonable	0.711		
The procedures for payment of health insurance are made quickly and simply	0.566		
The equipment for medical examination and treatment of the hospital is full and modern		0.756	
List of medicines granted to health insurance users meet your requirements		0.686	
You believe in the quality of hospitals that accept health insurance		0.686	
The doctors' professional qualifications meet your medical needs		0.668	
Your benefits when examining with health insurance are guaranteed		0.532	
Doctors and nurses are always interested and willing to help, solve your problems			0.831
You are given full answers about voluntary health insurance			0.785
Hospital admission, hospital transfer, and discharge procedures are done quickly and simply			0.710
You find hospital staff have a positive attitude in service			0.516
Kaiser–Meyer–Olkin	0.809		
Bartlett's test of sphericity	Sig. = 0.000		
Number of items retained	4	5	4
Eigenvalue	4.36	1.59	1.37
Percentage of explained variance	33.55	12.24	10.56
Cronbach's alpha (α)	0.731	0.666	0.748

Eigenvalues greater than 1 selected.
Factors retained with loading values greater than 0.5.
Four items were dropped from the scale.

Table 5.3 Component Matrix of dependent variable, member satisfaction

Variables	Factor loading
	Component 4
You will continue to participate in health insurance for next years	0.818
You are satisfied with the participation and payment procedures of the health insurance	0.793
You are satisfied with medical services using health insurance	0.731
Kaiser–Meyer–Olkin	0.651
Bartlett's Test of Sphericity	**Sig.** = 0.000
Percentage of explained variance	61.078
Eigenvalue	1.832

Eigenvalues greater than 1 selected.
Factors retained with loading values greater than 0.5.

Table 5.4 Pearson Correlations among the factors of farmers' satisfaction (i.e. the process and procedure (PRO), hospital's service capacity (HOS), support of hospital and social insurance staff (SUP), and farmers' satisfaction with health insurance)

		PRO	HOS	SUP	SAT
PRO	Pearson correlation	1	0.449**	0.393**	0.466**
	Sig. (2-tailed)		0.000	0.000	0.000
	N	180	180	180	180
HOS	Pearson correlation	0.449**	1	0.403**	0.549**
	Sig. (2-tailed)	0.000		0.000	0.000
	N	180	180	180	180
SUP	Pearson correlation	0.393**	0.403**	1	0.672**
	Sig. (2-tailed)	0.000	0.000		0.000
	N	180	180	180	180
SAT	Pearson correlation	0.466**	0.549**	0.672**	1
	Sig. (2-tailed)	0.000	0.000	0.000	
	N	180	180	180	180

** Correlation is significant at the 0.01 level (2-tailed)
Note: PRO: the process and procedure; HOS: the hospital's service capacity; SUP: the support of hospital and social insurance staff; SAT: the member's satisfaction

of HI; (2) HOS- the hospital's service capacity; (3) SUP- the support of hospital and social insurance staff. The farmers' evaluations of each factor varied, with SUP receiving the highest average score of 3.6236, and PRO receiving the lowest score of 3.2444. Additionally, the standard deviations of the responses among the variables differed, with HOS demonstrating the smallest deviation ($\partial = 0.60848$), and PRO showing the largest deviation ($\partial = 0.68501$).

Table 5.5 Factors Affecting Farmers' Satisfaction When Using Health Insurance

Model		Unstandardised coefficients		Standardised coefficients	*t*	Sig	Collinearity statistics	
		B	Std. error	beta			Tolerance	VIF
1	(Constant)	0.443	0.215		2.061	0.041		
	PRO	0.123	0.051	0.142	2.442	0.016	0.745	1.343
	HOS	0.278	0.057	0.283	4.861	0.000	0.738	1.355
	SUP	0.490	0.055	0.502	8.862	0.000	0.781	1.280

Dependent Variable: the member satisfaction (SAT).

Thus, the model fits the research data, and the results show that the first three proposed hypotheses are accepted, while hypothesis 4 is accepted only for gender differences in satisfaction. Recommendations obtained from open-ended questions through the questionnaire survey were summarised into a word cloud, and recommendations from respondents all focused on health insurance. The comments focused on 'improving hospital service quality', 'improving service attitudes of medical staff', 'continuing to provide free health insurance cards' and 'reducing incurred medical expenses' (Fig. 5.4; see the next page).

Fig. 5.4 Farmers' recommendation on health insurance policy. *Source* Nguyet (2022)

5.5 Discussion of Implications

Among contribution-based policies, health insurance is the most prevalent welfare policy, and Mekong Delta farmers are more knowledgeable about it than other policies. This finding is similar to the study of the International Labour Organisation (ILO) (2016), which indicated that informal and self-employed workers are more likely to have health insurance and feel protected by this policy than by other social protection schemes. The farming households in this study have good access to the State health insurance (HI) policy, including access to information, participation in and receipt of the benefits from HI. Survey results show that 30.6% of the beneficiaries have free health insurance cards, and 69.4% purchased them.

Similarly, the qualitative results show that farmers highly appreciate the role of health insurance in reducing the financial burden on the family's medical expenses. At the same time, many farmers commented that they had difficulties with administrative procedures because of discrimination by medical staff. Some respondents sought medical services at higher-level medical facilities such as provincial or city hospitals because they believed they could access better doctors and facilities there. Forty-one-point-one per cent (41.1%) of farmers had yet to decide whether to continue participating in health insurance in the coming years because the quality of the service did not meet their needs.

Farmer uncertainty is exacerbated by a limited understanding of risk sharing in health insurance. Risk sharing is one of the five principles of Vietnam's Health Insurance Law; health insurance works best when both healthy and sick people participate. If only sick people sign up, then the risk probability is high for everyone—participants, insurers, and medical professionals—and everyone must pay more. If a good distribution of healthy and sick people contributes to health insurance, the risk probability is lower, and costs stay down (Vietnam Ministry of Health 2020b). This study found that Mekong Delta farmers require an enhanced understanding of risk sharing to improve their willingness to continue participating in the health insurance program.

Mekong Delta farmers were generally satisfied with the health care services provided by health insurance, although the level of satisfaction was different among the three surveyed provinces. Survey results indicate a difference between provinces in understanding the regulations on choosing medical facilities for health examination and treatment. Can Tho farmers understand this regulation best, followed by An Giang and Soc Trang. This result corresponds to the level of education: in Can Tho, the number of graduates from primary and secondary schools was higher than in the other two provinces, where literacy was low.

The three factors influencing HI acceptability to farmers were (1) procedures for participating in health insurance and payment for health care services; (2) the hospital's range of health service capacity; and (3) the support of hospital and social insurance staff. As the analysis indicates, the third factor—the support of hospital and social insurance staff—had the highest impact on farmers' satisfaction.

Administrative procedures and requirements (for example, registration identification) may prevent people (especially the poor) from participating in social protection programmes (Basaza et al. 2008; Odeyemi 2014). The qualitative data also showed that complicated administrative procedures (transfer, payment) caused negative reactions by farmers. Thus, this study suggests that the Social Security Agency should focus on customer support in its service improvement strategy.

Men and women in this study had statistically significant differences in satisfaction when using HI services; on average, women rated the services provided by HI at 3.68 points, while men rated them at 3.50 points, 0.18 units lower (on a five-point scale). The provision of maternity care services can be a feature that women appreciate in medical services covered by HI, as observed by the United Nations Population Fund (UNFPA) (UNFPA and Ministry of Health 2017). Participants' satisfaction level with medical services in An Giang was higher than Can Tho and Soc Trang. Statistically, income, age group and education level did not affect farmers' satisfaction with HI services. However, many studies have shown that low education and economic status hinder social security (Perry 2007; World Bank 2014; ILO 2016; Allieu and Ocampo 2019). By analysing what factors affect farmers' satisfaction when using HI services, the Vietnam Ministry of Health and Social Security Agency can identify factors that need improvement to increase their members' satisfaction.

5.6 Future Directions

5.6.1 Suggestions from the Study's Results

This study used data on farmers' assessment of the State's policy implementation and their expectations, which were collected through interviews and surveys. Comments on policy implementation by local officials were also used for this discussion. Farmers proposed that the Vietnam Social Security agency should increase the availability of health insurance information, improve administrative procedures, and provide financial support to health insurance participants. Healthcare providers were also expected to improve service quality and invest more in community infrastructure and equipment. The requirement to provide timely feedback to beneficiaries of the emergency assistance policy was also noted from focus group discussions and in-depth interviews. Summarised opinions from the survey results focused on three expectations: improve administrative procedures, improve hospital service quality and the service attitude of medical staff, and invest more in community infrastructure and equipment. This finding shows that if these issues are improved, it will contribute to greater satisfaction with health services, thereby improving users' access to social service support.

In this study, opinions like 'continue to provide free health insurance cards', '[provide] financial support to health insurance participants', and 'reduce incurred medical expenses' are grouped into a category called 'financial issues of HI'. The surveys found that 149 (82.8%) households had annual health expenditures below 5% (below USD 208/year), and 31 households (17.2%) spent 5–10% (between USD 208—416/year) of household income. The regulations on health insurance in Vietnam have provisions to support and reduce prices for middle-income households working in agriculture, forestry, fisheries and salt production. However, based on data on income groups in the quantitative study, most households in An Giang and Soc Trang fell into quintiles 1 and 2 (lowest and under-middle class). Low earnings can hinder farmers' participation in insurance policies (Anh 2010; Nhuong 2011; Liang et al. 2014; Li et al. 2015; Allieu and Ocampo 2019). In addition, regarding socio-economic and environmental changes in the Mekong Delta, this study suggests that authorities consider farmers' living and working conditions in policy reform.

In the rural areas in the Mekong Delta, where the facilities and staff of medical centres are limited in both quantity and quality, and farmers already face great risks from climate change, the addition of technological innovations such as telemedicine and AI diagnostic tools could be crucial to bringing affordable care to more people. These telemedicine platforms allow farmers to receive consultation and diagnosis as well as receive prescriptions from reputable medical providers or provincial or city hospitals, helping farmers save money on accommodation and travel costs. The Vietnamese government approved a remote medical examination and treatment scheme for 2020–2025 due to COVID-19 issues. Launched in September 2020, the program connects 1000 hospitals, including 20 central ones, aiming to reduce the burden, improve quality, save costs, and enhance patient satisfaction. Financial support is provided for both inpatients and outpatients; the Scheme is funded by the state budget, official development assistance (ODA) and other legal funding sources (Ministry of Health 2020a).

Furthermore, AI-powered applications can be used to monitor health and detect health problems early in farmers. They provide personalised recommendations and alert farmers and their healthcare providers about health hazards using Health data analytics. AI can help farmers take preventative measures and protect their health from diseases, which are anticipated to be increasingly affected by climate change.

Implementing telemedicine and AI-enabled tools in healthcare services provided to rural farmers may require time to strengthen infrastructure and balance budgets. However, this strategy has the potential to help redefine future health insurance policy directions. These technologies promote personalised and preventative healthcare, enabling a move towards lower long-term healthcare costs. With telemedicine tools, virtual consultations, and remote health monitoring devices, Vietnam's social insurance, the Ministry of Health, and relevant organisations, including private insurance, can promote increased health outcomes and cost-effectiveness in providing health care to farmers.

5.6.2 Policy Recommendation

Expanding health insurance coverage is the goal for all national health insurance systems. The World Health Assembly and the United Nations General Assembly have called on governments to develop their health systems to ensure that healthcare services are available to everyone, and that people do not suffer financial hardship in paying for them (Thanh 2015). The important issue for Mekong Delta farmers is not only a financial one but also a matter of adjusting and supplementing policies suitable for each household. The financial resources for implementing the social security system, including HI, are not highly sustainable.

In Vietnam, the master plan for universal health coverage was produced in 2012. However, there is a lack of research evaluating the level of coverage, benefit packages, and reimbursement of commercial health insurance schemes (Thanh 2015).

The Vietnamese government and relevant ministries have a plan to implement and develop private Health Insurance provided by commercial health insurance companies. Private health insurance is designed as medical benefit packages (especially packages suitable to farming characteristics) beyond current health insurance coverage, increasing patient benefits, reducing out-of-pocket expenses, and choosing a service provider.

According to the Vietnam Ministry of Health, if additional forms of health insurance (private health insurance package) are applied, the co-payment amount when people go for outpatient medical examination and treatment is expected to decrease by 25%. The reduction in inpatient medical examination and treatment will be 41%. Thus, patients will reduce the total cost (of co-payment) by about 20–31% (Vietnam Ministry of Health 2020a). However, this proposed plan is for all subjects and does not consider the socio-economic characteristics of farmers. For example, the premiums may outweigh these cost reductions for poor and near-poor farming households. If private Health Insurance is poorly regulated, the profit motive may outweigh any benefit. Lessons learned about health insurance for farmers in the Mekong Delta may be informative for other parts of Asia. Along with rural areas similar to the rest of Asia, farmers in the Mekong Delta typically encounter barriers such as low income, geographic isolation, and low awareness about insurance advantages (Organisation for Economic Co-operation and Development (OECD) and WHO 2022). Still, while facing challenges in the Delta, these specific challenges also provide opportunities—such as ready-made communities that might take up group insurance schemes and other local initiatives.

Meanwhile, more specifically, some Asian regions can require more adjusted infrastructural and economic contexts. For example, in Thailand and Indonesia, there have been examples of Universal Health Coverage, which covers farmers and rural populations (Harimurti et al. 2017). These arrangements frequently combine government funding with an element of employer payroll contributions and variable levels of individual premiums and could be tested in the Mekong Delta context.

The reliance exclusively on private health insurance is problematic regarding the adequacy and sustainability of health insurance; this is important as farmers in the Mekong Delta often have lower incomes. More sustainable models could, in fact, look like community-based health insurance (CBHI), which pools risks and resources within a community. Some examples of successful CBHI schemes have been reported in Asia, as they are designed context-specific and more feasible in terms of costs for the underprivileged (Liu et al. 2018). Furthermore, such a system could consider public–private partnerships to pay for the premiums of the poorest or to offer premium reductions and increased benefits (such as coverage for more preventative care) based on a competitive basis. CBHI—perhaps with complementary government support and public–private partnerships—might represent a more sustainable and equitable path toward improving health insurance in the region.

5.7 Conclusion

Mekong Delta farmers' awareness of health insurance is relatively good, although some farmers do not know all the regulations and benefits. A lack of knowledge may disadvantage them when using health care services. Health insurance is a policy that receives positive feedback from farmers for its benefits; the health insurance card has partly contributed to subsidising their health costs. Most people interviewed were satisfied with the services provided by the system. The factors affecting farmers' satisfaction when using health insurance include (i) the participation and payment procedure of HI, (ii) the hospital's service capacity, and (iii) the support of hospital and social insurance staff. Farmers' satisfaction with health insurance services differs among males and females. To generalise the findings and implement more effective policies, the current research aims to motivate comparable future studies focused on a larger sample of rural people.

From the policy perspective, this study suggests upgrading the infrastructure of commune and district health centres, improving the quality of medical services in hospitals that conduct medical examinations and treatment under health insurance, and improving some medical staff service attitudes. In addition, research programmes, policy training, and amendments to relevant legislation should be set up to encourage farmers to participate and maintain health insurance, which could contribute to raising awareness of their social security and the risks to their lives. In addition, studying, analysing and learning from experiences in health insurance policies from other countries in the Asian region plays an important role in improving and enhancing the effectiveness of this policy for Vietnamese farmers.

Highlights

- Participation, payment procedures, hospital service quality, and medical staff support affect farmers' satisfaction with health insurance

- Telemedicine and AI Tools are crucial for affordable, personalised healthcare in Vietnam and Asia's rural areas.
- Community-Based Health Insurance (CBHI) may be feasible and sustainable for low-income farmers
- Learning from other Asian countries' experiences is vital for enhancing Vietnam's health insurance policies.
- Essential healthcare infrastructure upgrade for rural healthcare access and quality

References

Adebayo EF, Uthman OA, Wiysonge CS, Stern EA, Lamont KT, Ataguba JE. A systematic review of factors that affect the uptake of community-based health insurance in low-income and middle-income countries. BMC Health Serv Res. 2015;15(1):543.

Ager A, Pepper K. Patterns of health service utilisation and perceptions of needs and services in rural Orissa. J Health Policy Plan. 2005;20:176–84.

Ali Jadoo SA, Wan Puteh SE, Ahmed Z, Jawdat A. Level of patients' satisfaction toward National Health Insurance in Istanbul City-Turkey. BMC Public Health. 2012;12(Suppl 2):A5. https://doi.org/10.1186/1471-2458-12-S2-A5.

Allieu AM, Ocampo A. On the path to universal coverage for rural populations: removing barriers of access to social protection. Rome: FAO;2019. http://www.fao.org/3/ca7246en/ca7246en.pdf.

Andrew S, Halcomb E. Mixed methods research for nursing and the health sciences. Chichester, U.K.: Wiley-Blackwell Pub;2009.

Anh MN. Social security for farmers in market economy conditions in Vietnam. Ha Noi: National Political Publishing House;2010.

Antle JM, Pingali PL. Pesticides, productivity, and farmer health: a Philippine case study. J Am J Agric Econ. 1994;76:418–30.

Arcury TA, Gesler WM, Preisser JS, Sherman J, Spencer J, Perin J. The effects of geography and spatial behaviour on health care utilisation among the residents of a rural region. J Health Serv Res. 2005;40:135–56.

Arrow KJ. Uncertainty and the welfare economics of medical care. Am Econ Rev. 1963;53(5):941–73.

Baeza C, Packard T. Beyond survival: protecting households from health shocks in Latin America. The World Bank;2006.

Baka A, Triga V, Figgou L. Neither agree, nor disagree. Int J Electron Gov. 2012;5(3/4):244–263. https://doi.org/10.1504/IJEG.2012.051306.

Basaza R, Criel B, Van der Stuyft V. Community health insurance in Uganda: why does enrolment remain low? a view from beneath. Health Policy. 2008;87 (2):172–184. https://doi.org/10.1016/j.healthpol.2007.12.008.

Benavides-Vaello S, Strode A, Sheeran BC. Using technology in the delivery of mental health and substance abuse treatment in rural communities: a review. J Behav Health Serv Res. 2013;40:111–20.

Berry HL, Hogan A, Ng SP, Parkinson A. Farmer health and adaptive capacity in the face of climate change and variability. Part 1: Health as a contributor to adaptive capacity and as an outcome from pressures coping with climate related adversities. J Int J Environ Res Public Health. 2011;8:4039–4054.

Boateng D, Awunyor-Vitor D. Health insurance in Ghana: evaluation of policy holders' perceptions and factors influencing policy renewal in the Volta region. Int J Equity Health. 2013;12(1):50.
Bookwalter J, Fuller B, Dalenberg D. Do household heads speak for the household? a research note. Soc Indic Res. 2006;79(3):405–19. https://doi.org/10.1007/s11205-005-4925-9.
Brzoska P, Razum O. Validity issues in quantitative migrant health research. The example of illness perceptions. Frankfurt am Main: Peter Lang; 2010.
Burgard SA, Lin KY. Bad jobs, bad health? how work and working conditions contribute to health disparities. J Am Behav Sci. 2013;57:1105–27.
Cerny BA, Kaiser HF. A study of a measure of sampling adequacy for factor-analytic correlation matrices. Multivar Behav Res. 1977;12(1):43–7. https://doi.org/10.1207/s15327906mbr1201_3.
Chien LT et al. Assessment of patients' satisfaction with medical examination and treatment at some Grade I general hospitals under the Department of Health of Ho Chi Minh City. J MilY Med Pharmacol. Supplementary issue 2014:35–43.
Chu Hung Cuong. Evaluation of patient satisfaction about medical examination and treatment services at the medical examination department, Ninh Giang district general hospital, Hai Duong Province in 2012 [Master dissertation]. Ha Noi (Vietnam): University of Public Health; 2012.
Creswell JW. Educational research: planning, conducting, and evaluating quantitative and qualitative research. 4th ed. Boston, MA: Pearson; 2012.
Creswell JW, Plano Clark VL, Gutmann ML, Hanson WE. Advanced mixed methods research designs. In: Tashakkori A, Teddlie C, editors. Handbook of mixed methods in social and behavioral research. Thousand Oaks, CA: Sage; 2003. p. 209–40.
Cronbach LJ. Further evidence on response sets and test design. Educ Psychol Measur. 1950;10:3–31. https://doi.org/10.1177/001316445001000101.
Damalas C. Understanding benefits and risks of pesticide use. Sci Res Essays. 2009;4:945–9.
Doan DTT, Mirzoev T, Nguyen CC, Bui HTT. Utilization of services provided by village-based ethnic minority midwives in Vietnam: lessons from implementation research. J Public Health Manag Pract. 2018;24:S9–18.
Donabedian A. The quality of care. How can it be assessed? JAMA. 1988;260(12):1743–8. https://doi.org/10.1001/jama.260.12.1743. PMID: 3045356.
Donabedian A. Evaluating the quality of medical care. Milbank Q. 2005;83(4):691–729. https://doi.org/10.1111/j.1468-0009.2005.00397.x. PMID: 16279964; PMCID: PMC2690293.
Dziuban CD, Shirkey EC. When is a correlation matrix appropriate for factor analysis? Some Decis Rules Psychol Bulletin. 1974;81(6):358–61. https://doi.org/10.1037/h0036316.
Fatimah FS, Mars S, Sarwadhamana RJ, Mulyani R, Handayani PD. A Comparison of patient satisfaction when using the insured and non-insured in public health center (Puskesmas Kasihan 1) Bantul, Indonesia. Open Access Maced J Med Sci. 2022;10(T8):1–4. https://doi.org/10.3889/oamjms.2022.9454.
Fuller J, Edwards J, Procter N, Moss J. How definition of mental health problems can influence help seeking in rural and remote communities. J Aust J Rural Health. 2000;8:148–53.
General Statistics Office (GSO). Report of the Vietnam labour and employment survey 2022. Ha Noi: Statistical Publishing House;2023.
Geng J, Chen X, Shi J, Bao H, Chen Q, Yu H. Assessment of the satisfaction with public health insurance programs by patients with chronic diseases in China: a structural equation modeling approach. BMC Public Health. 2021;21:1886. https://doi.org/10.1186/s12889-021-11947-7.
Getaneh MM, Bayked EM, Workneh BD, Kahissay MH. Satisfaction of beneficiaries with community-based health insurance and associated factors in Legambo District, North-East Ethiopia: a cross-sectional study. Front Public Health. 2023;11:1127755. https://doi.org/10.3389/fpubh.2023.1127755.
Haile M, Hunduma F, Haile K. Clients' knowledge and satisfaction with utilizing healthcare services in community-based health insurance program and its associated factors among public health centers. Addis Ababa, Ethiopia: Research Square;2022.

Hair JF, Black B, Babin B, Anderson RE. Multivariate data analysis: a global perspective. 7th ed. New Jersey: Pearson Education; 2010.

Harimurti P, Prawira J, Hort K. The Republic of Indonesia health system review. Health Systems in Transition. WHO Regional Office for South-East Asia;2017. https://iris.who.int/bitstream/handle/10665/254716/9789290225164-eng.pdf?sequence=1.

Hoa, Dang Le., Li, E., Nuberg, I., and Bruwer, J. Farmers' assessments of private adaptive measures to climate change and influential factors: a study in the Mekong Delta, Vietnam. Nat Hazards. 2014;71(1):385–401.

Hoan LM. Report on agricultural restructuring and investment in irrigation systems in the Mekong Delta for the period 2022–2030. In: Conference to announce the Mekong Delta Regional Planning for the period 2021–2030;2022.

Holst J, Normann O, Herrmann M. Strengthening training in rural practice in Germany: new approach for undergraduate medical curriculum towards sustaining rural health care. J Rural Remote Health. 2015;15:3563.

Hsiao WC, Shaw RP. Social health insurance for developing nations. The World Bank. World Bank Institute, IBRD, Harvard University (Cambridge, Mass.);2007.

Hu R, Huang X, Huang J, Li Y, Zhang C, Yin Y, Chen Z, Jin Y, Cai J, Cui F. Long-and short-term health effects of pesticide exposure: a cohort study from China. PLoS ONE. 2015;10: e0128766.

ILO. Extending the social security coverage for farmers in the republic of Moldova: findings and recommendations based on the social security assessment survey. Budapest: ILO;2016. https://www.ilo.org/wcmsp5/groups/public/---europe/---ro-geneva/---sro-budapest/documents/publication/wcms_504710.pdf.

Inwood S. Agriculture, health insurance, human capital and economic development at the rural-urban-interface. J Rural Studies. 2017;54:1–14. https://doi.org/10.1016/j.jrurstud.2017.05.009.

Johnson B, Christensen LB. Educational research: quantitative, qualitative, and mixed approaches. Thousand Oaks, CA: Sage Publications, Inc.; 2008.

Kaiser HF. A second-generation little jiffy. Psychometrika. 1970;35(4):401–15.

Kongtip P, Techasaensiri B, Nankongnab N, Adams J, Phamonphon A, Surach A, Sangprasert S, Thongsuksai A, Srikumpol P, Woskie S. The impact of prenatal organophosphate pesticide exposures on Thai infant neurodevelopment. Int J Environ Res Public Health. 2017;14(6):570. https://doi.org/10.3390/ijerph14060570.

Kunde L, Kõlves K, Kelly B, Reddy P, De Leo D. Pathways to suicide in Australian farmers: a life chart analysis. J Int J Environ Res Public Health. 2017;14:352.

Le TXH, Le THT, Nguyen T. Expanding health insurance coverage in Vietnam: a view from a study of factors affecting the non-participation in health insurance of informal sector workers. J Econ, Financ Manag Stud. 2023;06 (05):2095–210. https://doi.org/10.47191/jefms/v6-i5-27.

Li H, Hung X, Kwan M, Bao H, Jefferson S. Changes in farmers' welfare from land requisition in the process of rapid urbanisation. Land Use Policy. 2015;42:635–41. https://doi.org/10.1016/j.landusepol.2014.09.014.

Liang Y, Lu W, Wu W. Chinese landless farmers really effective on health in the process of Chinese rapid urbanisation? a study on the effect of social security policies for Chinese landless farmers on their health-related quality of life. Int J Equity Health. 2014;13:5. https://doi.org/10.1186/1475-9276-13-5. PMID: 24433258; PMCID: PMC3896694.

Liu X, Zhu A, Tang S. Attraction and retention of rural primary health-care workers in the Asia pacific region. 2018. https://iris.who.int/bitstream/handle/10665/272663/9789290226499-eng.pdf?sequence=1&isAllowed=y. Accessed 15 Jun 2024.

Loan CTK, Ban NH. The reality of participating in voluntary health insurance in Ha Tinh city. J Sci Dev. 2013;115–124.

Mafimisebi TE, Oguntade AE. Preparation and use of plant medicines for farmers' health in Southwest Nigeria: socio-cultural, magico-religious and economic aspects. J Ethnobiol Ethnomed. 2010;6:1. https://doi.org/10.1186/1746-4269-6-1. PMID: 20089149; PMCID: PMC2821299.

Maltby J, Day L, Macaskill A. Personality, individual differences and intelligence. 2nd ed. London: Prentice Hall; 2007.

Mbow C, Rosenzweig C, Barioni LG, Benton TG, Herrero M, Krishnapillai M, Liwenga E, Pradhan P, Rivera-Ferre MG, Sapkota T, Tubiello FN, Xu Y. Food Security. In: Shukla PR, Skea J, Calvo Buendia E, Masson-Delmotte V, Pörtner H-O, Roberts DC, Zhai P, Slade R, Connors S, van Diemen R, Ferrat M, Haughey E, Luz S, Neogi S, Pathak M, Petzold J, Portugal Pereira J, Vyas P, Huntley E, Kissick K, Belkacemi M, Malley J, editors. Climate change and land: an IPCC special report on climate change, desertification, land degradation, sustainable land management, food security, and greenhouse gas fluxes in terrestrial ecosystems. Intergovernmental Panel on Climate Change;2019.

Mohammed S, Sambo MN, Dong H. Understanding client satisfaction with a health insurance scheme in Nigeria: factors and enrollees experiences. Health Res Policy Syst. 2011;9:20. https://doi.org/10.1186/1478-4505-9-20

Morse JM. Approaches to qualitative & quantitative methodological triangulation. Nurs Res. 1991;40:120–3. https://doi.org/10.1097/00006199-199103000-00014.

Nguyen TTN, Xiem CH, Anh BTM. Universal coverage challenges: Determinants of enrolment in family-based social health insurance. Int J Healthc Manag. 2020;14:1–7. https://doi.org/10.1080/20479700.2020.1752986.

Nguyen Quang Thanh. Description of inpatient satisfaction about medical examination and treatment services at Binh Phuoc Provincial General Hospital in 2013 [Master dissertation]. Ha Noi (Vietnam): University of Public Health; 2013.

Nguyet TTT. Beneficiary status and level of access to social security policies by farmers in the Mekong delta, Viet Nam [PhD thesis]. Perth (Australia): Murdoch University;2022.

Nhuong NV. Social security policy for farmers after land acquisition to develop industrial zones (study in Bac Ninh) [Ph.D. thesis]. Ha Noi (Vietnam): National Economics University;2011.

Nicolopoulou-Stamati P, Maipas S, Kotampasi C, Stamatis P, Hens L. Chemical pesticides and human health: the urgent need for a new concept in agriculture. J Front Public Health. 2016;4:148.

Nunnally JC, Bernstein IH. The assessment of reliability. Psychom Theory. 1994;3:248–92.

O'Brien K, Leichenko R, Kelkar U, Venema H, Aandahl G, Tompkins H, Javed A, Bhadwal S, Barg S, Nygaard L. Mapping vulnerability to multiple stressors: climate change and globalization in India. J Glob Environ Change. 2004;14:303–313.

Odeyemi, IA. Community-based health insurance programmes and the national health insurance scheme of Nigeria: challenges to uptake and integration. Int J Equity Health. 2014;13(20).https://doi.org/10.1186/1475-9276-13-20. PMID: 24559409; PMCID: PMC3941795.

OECD health statistics 2021. https://doi.org/10.1787/health-data-en.

OECD/WHO. Health at a Glance: Asia/Pacific 2022: measuring progress towards universal health coverage, OECD Publishing, Paris;2022. https://doi.org/10.1787/c7467f62-en.

Organisation Regional Office for the Western Pacific; 2020. Accessed 24 Feb 2024.

Ogundeji YK, Akomolafe B, Ohiri K, Butawa NN. Factors influencing willingness and ability to pay for social health insurance in Nigeria. PLoS ONE. 2019;14(8): e0220558.

Ory JC, Wise SL. Attitude change measured by scales with 4 and 5 response options. Paper presented at the meeting of the National Council on Measurement in Education, Chicago, IL;1981.

Perry GE, Maloney WF, Arias OS, Fajnzylber P, Mason AD, Saavedra-Chanduvi J. Informality: exit and exclusion. Washington D.C., World Bank;2007.

Peterson RA. A meta-analysis of variance accounted for and factor loadings in exploratory factor analysis. Mark Lett. 2000;11(3):261–75.

Pham Nhat Yen. Patient satisfaction about the quality of medical examination and treatment services at the medical examination and treatment department of Bach Mai hospital in 2008 [Master dissertation]. Ha Noi (Vietnam): University of Public Health; 2008.

Robson C. Real World Research. Oxford: A resource for social scientists and practitioner researchers. Blackwell Publishers Inc.; 1993.

Sakamoto H, Rahman M, Nomura S, Okamoto E, Koike S et al. Japan health system review. Health Syst Transit. 2018;8(1). World Health Organization. Regional Office for South-East Asia. https://iris.who.int/handle/10665/259941.

Shaikh BT, Hatcher J. Health-seeking behaviour and health service utilisation in Pakistan: challenging the policy makers. J Public Health. 2004;27:49–54.

Somanathan A, Tandon A, Dao HL, Hurt KL, Fuenzalida-Puelma HL. Moving toward universal coverage of social health insurance in Vietnam: assessment and options. Washington DC: World Bank; 2014.

Thanh ND. The impact of a health insurance program on the near-poor in Vietnam [PhD thesis]. Queensland (Australia): Queensland University of Technology;2015.

The Central Committee of the Communist Party of Vietnam. Resolution No. 20-NQ/TW of the sixth conference of the 12th party central committee on strengthening the protection, care and improvement of people's health in the new situation, issued on October 25 2017. Ha Noi;2017.

Thiedke CC. What do we really know about patient satisfaction? Fam Pract Manag. 2007;14(1):33–6 PMID: 17294978.

Tien DK, Young MD, Loch A, Thennakoon J. Mekong River Delta farm-household willingness to pay for salinity intrusion risk reduction. Agric Water Manag. 2018;200:80–9. https://doi.org/10.1016/j.agwat.2017.12.010.

UNFPA and Ministry of Health. Exploring barriers to accessing maternal health and family planning services in ethnic minority communities in Vietnam. Ha Noi;2017. https://vietnam.un.org/sites/default/files/2019-08/Web_Barriers%20to%20accessing%20maternal%20health%20and%20FP%20services%20in%20ethnic%20minority%20communities%20in%20VN.pdf. Accessed 26 Aug 2024.

United Nations 2007. General comment no. 191: the right to social security (art. 9). Economic and Social Council Distr. General E/C.12/GC/19. Geneva: UN; 2008. https://socialprotection-humanrights.org/wp-content/uploads/2015/06/CESCR-General-Comment-19.pdf.

http://vanban.chinhphu.vn/portal/page/portal/chinhphu/hethongvanban?class_id=1&_page=1&mode=detail&document_id=175349. Accessed 17 Jan 2024.

http://vanban.chinhphu.vn/portal/page/portal/chinhphu/hethongvanban?class_id=1&_page=1&mode=detail&document_id=81142. Accessed 17 Jan 2024.

VCCI and Fulbright. Annual economic report of the Mekong delta in 2022: transforming the development model and integrated planning. Can Tho University Publishing House;2022.

Vietnam Government. National assembly. Law No. 25/2008/QH12 on Health Insurance. Ha Noi;2008.

Vietnam Government. Central committee of communist party, sixth conference of central committee XII on 25th October 2017, resolution 20/NQ-TW on strengthening health protection, health care and health promotion for people in the new situation. Ha Noi;2017.

Vietnam Government. National Assembly. Law No. 46/2014/QH13 amending and supplementing a number of articles of the health insurance law. Ha Noi;2014.

Vietnam Ministry of Health. Summary report, evaluation of implementation of the law on health insurance for 2015–2020. Ha Noi;2020b. https://moj.gov.vn/qt/tintuc/Lists/ChiDaoDieuHanh/Attachments/3043/BC%20tong%20ket.pdf. Accessed 6 Feb 2025.

Vietnam Ministry of Health. Decision 2628/QD-BYT 2020 approving scheme for remote medical examination and treatment. Ha Noi;2020a. https://vanbanphapluat.co/decision-2628-qd-byt-2020-approving-scheme-for-remote-medical-examination-and-treatment. Accessed 18 Oct 2024.

Vu D, Trinh V, Nestmann F, Oberle P, Nam N. Land use based flood hazards analysis for the Mekong delta. Conference: IAHR;2014. https://doi.org/10.13140/2.1.5153.9842.

WHO: Health Workforce Country Profiles 2020. https://www.who.int/westernpacific/health-topics/health-workforce/country-profiles. World Health.

World Bank. Social safety nets and gender: learning from impact evaluations and World Bank projects. Washington, DC: World Bank;2014. https://ieg.worldbankgroup.org/Data/reports/ssn-gender-iefull-report.pdf.

Yazd SD, Wheeler SA, Zuo A. Key risk factors affecting farmers' mental health: a systematic review. Int J Environ Res Public Health. 2019;16(23):4849.

Yin R. Case study research: design and methods. 2nd ed. London, UK: Sage Publications Inc.; 1989.

Nguyet Thi Thu Tran is a lecturer at the Faculty of Law and Political Science, An Giang University- Vietnam National University HCMC, Vietnam. She received her Ph.D. degree in Asian Studies at Murdoch University (Perth, Australia). Her work focuses on teaching general sociology, rural sociology and social policy to undergraduates. Her research interests are mainly vulnerability groups in society, social security policy, and social issues in rural society.

Anja Reid is an Adjunct at the College of Law, Arts and Social Sciences and a Fellow of the Higher Education Academy, Murdoch University (Perth, Australia). She completed a Doctoral program in Art and Anthropology at Curtin University in Perth, Australia. Her areas of interest in Asian Studies are gender in Asian countries, religion in Asian society, cultural and social ecology in the Asian region.

Carol Warren is the Academic Chair of the Asian Studies program and a Fellow of the Indo-Pacific Research Centre (Murdoch University). She has a Ph.D. degree in Anthropology from the University of Western Australia. Her teaching and research focus on social, economic and environmental change in Asia, particularly Indonesia and Malaysia.

Kathryn Trees is a Senior Lecturer at the College of Law, Arts and Social Sciences, Murdoch University. She holds a Doctoral Degree that explores indigenous and non-Indigenous coexistence and storytelling. She teaches English and Creative Writing and supervises postgraduate candidates across English, Film, Media and Communication and, Law.

Part II
Health and Care: Advanced and Ethical Solutions for Physical and Mental Health

Chapter 6
Assessment of Mental Health Service Utilisation Among Indonesian Adults: Basis for Enhancing Future Services in Indonesia

Cluster of Future Studies: Corporate Foresight

Muhammad Wildan Rabbani Kurniawan, Benny Prawira, and Andrian Liem

Abstract A high number of people with mental disorders are in low- and middle-income countries (LMICs, including Indonesia) and they do not receive adequate treatment due to low service utilisation, evidence on factors impacting utilisation is inconsistent. This empirical study aimed to understand this phenomenon better by exploring mental health service use among Indonesian adults and identified contributing factors. A non-random-self-selection-sampling online survey was carried out between May and June 2021 including questions for service utilisation and potential factors affecting it. The participant inclusion criteria were Indonesian, aged 18 or older, and living in Indonesia. The measure used standardized questionnaires (i.e. Perceived Helpfulness of Interventions) and hierarchical logistic regression was used for statistical analysis. Out of 5,211 participants, only about 26.5% used mental health services in the past three years. Significant predicting factors ($p < 0.001$) including perceived helpfulness on the services (*OR* 2.82), disability status (*OR* 2.72), and adequate insurance literacy (*OR* 1.91). Based on these findings, suggestions for improving health financing and social health equity for future access to in Indonesia were recommended, such as the importance of expanding insurance coverage and the necessity to improve awareness of mental health services.

M. W. R. Kurniawan (✉)
Jakarta, Indonesia
e-mail: kurniawanmwr@gmail.com

B. Prawira
Into the Light Indonesia, Jakarta, Indonesia
e-mail: prawirabenny89@gmail.com

A. Liem
Faculty of Psychology, Universitas Sebelas Maret, Surakarta, Indonesia
e-mail: andrianliem@staff.uns.ac.id

R. Xiang Ng et al. (eds.), *Future of Healthcare in Asia*,
https://doi.org/10.1007/978-3-032-03183-9_6

Keywords Mental health · Health access · Health equity · Health financing · Indonesia · Health literacy

6.1 Introduction

In the last ten years, the prevalence of mental disorders has significantly increased globally (World Health Organization 2017). It was estimated the global cost of the two most common mental disorders, which are depression and anxiety, is around $1 trillion a year and is estimated to be more than $6 trillion by 2030 (The Lancet Global Health 2020). In Asia, the prevalence of mental disorders is indicated at approximately 11.8%, equating to 11,810.8 cases per 100,000 population (Chen et al. 2024). In Indonesia, data from 2019 shows that the prevalence of mental disorders is around 11%. These figures are comparable to those in other Asian regions such as Thailand, where the overall prevalence of mental disorders is 12.8%, Bangladesh with 11.8% prevalence, and India with 13.8% (World Health Organization, Regional Office for South-East Asia 2023).

Unfortunately, the rising burden of mental disorders does not receive the same level of attention as physical illnesses and remains significantly overlooked, especially in low- and middle-income countries (LMICs) (Sarikhani et al. 2021; Gilbert et al. 2015). Despite this high number of mental disorders, access to adequate mental health (MH) services—especially in LMICs are still limited (Bolton et al. 2023; Alonso et al. 2018). This drawback is closely related to inadequate provision (Sarikhani et al. 2021; Gilbert et al. 2015) and underutilisation of mental health (MH) services (Daliri et al. 2024). Focusing on the utilisation, the data shows that utilisation of MH services is still relatively low worldwide, particularly in the Global South including Indonesia, and data related to the influencing factors is not adequately addressed (Roberts et al. 2018). People with mental disorders often experience greater difficulty accessing MH services due to different attitudinal and structural barriers (Sarikhani et al. 2021; Mekonen et al. 2022; Roberts et al. 2018).

Therefore, the current study aims to explore the level and contextual factors influencing the MH service utilisation of Indonesian adults with greater emphasis on demographic variables, minority status (including disability and sexual orientation), perceived helpfulness of MH services, and ownership and understanding of the national insurance system. The findings are expected to contribute to improvements of MH care access both in Indonesia specifically and Asian country with similar cultural and demographic context in general.

6.2 Literature Review

There are several studies on MH services utilisation in Indonesia (e.g., Gronowski 2020; Munira et al. 2023; Andary et al. 2023). Most of these studies only focus on identifying barriers and facilitators in accessing MH services. They found that the primary barriers to accessing mental health services in Indonesia include widespread cultural stigmas and traditional beliefs, which significantly deter individuals from seeking mental health care. Additionally, there is a severe shortage of mental health professionals, compounded by high costs and limited mental health literacy, which further hinder the accessibility of mental health services. However, factors that are associated with the utilisation of mental health (MH) services in Indonesian adults are still relatively unknown. In addition, previous systematic review shows inconsistent evidence for factors predicting utilisation and there was a lack of research from LMIC contextual level factors, including Indonesia (Roberts et al. 2018). Identifying these factors would enable the stakeholders to plan tailored interventions and optimised the resource allocation, as different groups often have varying needs and barriers to accessing MH services. Individuals and groups who are likely to underutilise MH services can be specifically targeted for engagement interventions that potentially enhancing the utilisation and improving health outcomes.

Additionally, despite the introduction of the National Health Insurance Scheme (Jaminan Kesehatan Nasional/JKN) that aims to improve access to mental health services, uptake remains inadequate. For example, a study analysing data from 2015–2020 found that factors such as age, marital status and type of health facility have a significant impact on the use of mental health services, with age being the most influential factor, especially among the less productive age group (Rahmah and Indrawati 2024). However, it acknowledges the need for considering more contextual factors that could influence this utilisation. To the best of authors' knowledge, there is no data available on these additional factors in Indonesia. Furthermore, the study is limited to participants of the national health insurance program, overlooking the fact that the method of payment through health insurance or out-of-pocket may also significantly affect the use of MH services (Munira, Liamputtong, and Viwattanakulvanid 2023).

6.3 Methodology

6.3.1 Participants and Procedure

This research was a cross-sectional survey using Typeform online survey platform, in which the data were collected between 25 May and 16 June 2021 (Into The Light Indonesia/ITL 2021). The data was collected using non-randomized sampling methods from the inclusion criteria of Indonesian, aged 18 or older, and living in Indonesia. The online survey link was shared via the social media platforms and mailing lists of ITL and Change.org Indonesia.[1] To mitigate coverage bias, the survey was promoted to the social media accounts of marginalised groups, including Lesbian, Gay, Bisexual, Transgender, Intersex, and Queer (LGBTIQ) individuals, people with disabilities, and those living with HIV. To maintain the integrity of the data collected, we ensured secure data transmission, anonymised responses, implemented data validation and consistency checks, restricted one submission per email address, and enforced strict access to the collected data to authorised personnel only. The study received ethical approval from the Research Ethics Committee at the Research and Community Service Centre (LPPM) of Universitas Katolik Indonesia Atma Jaya (Approval No. 0519A/III/LPPM-PM.10.05/05/2021). At the end, 5,211 participants completed the survey, while incomplete responses were automatically removed by the system.

6.3.2 Measures

The questionnaire was developed using instruments from previous research, with several questions adapted to the socio-cultural context of Indonesian society, some adopted from other studies conducted in Indonesia, and a few developed by the team. The questionnaire was piloted among the research team and community members to check its clarity and reception. However, not all questions were statistically validated due to time constraints during the COVID-19 lockdown period. All survey items were presented in Bahasa Indonesia. It covered demographic information, including age groups, current provincial residence, ethnicity, educational attainment, employment status and type, income level, religious beliefs, and sex-gender identification. Additionally, three questions addressed participants' minority and marginalised status, including sexual orientation, HIV status, and both mental and physical disabilities. For analytical purposes, demographic variables were grouped based on the socio-cultural landscape of Indonesia.

The utilisation of MH services was defined as the history of accessing any MH services, both online and in-person, for any purpose at least once in the last three years. This question for individual MH service utilisation consists of two items, including the history of accessing MH services and the purpose of access. However, this study, only reported the first item where participants can select more than one of the following options: psychology unit at community health centre, psychology unit at hospital, psychology unit at university, private psychology service, psychiatric hospital, MH services on digital/online platforms, and never at all. Participants who have not accessed any MH services in the past three years will automatically be directed to the next section as they were grouped as not utilising MH services.

[1] Founded as Indonesia's first youth-led community for suicide prevention, Into The Light Indonesia addresses suicide prevention and mental health concerns through advocacy, research, and education initiatives. Change.org Indonesia is the Indonesian office of Change.org, an American petition website that offers the public the ability to promote the petitions they care about to potential signers.

The perceived helpfulness of MH service was measured using a modified version of Mental Health Literacy Survey to determine helpfulness of treatment interventions (Jorm et al. 1997; Slewa-Younan et al. 2014). The modified version used in this study only utilises one of the total three sub-sections of the instrument: activity and management, which consist of 12 choices of answers. Mental health services are perceived as helpful if participants chose '*Pergi ke layanan kesehatan maupun layanan daring untuk konseling/konsultasi*' (going to health services or online services for counselling/consultation) as one of their choices in the activity and management which they believe will be helpful for them when having psychological problems ('*jenis intervensi yang diyakini akan sangat membantu apabila partisipan memiliki masalah kesehatan mental*').

The insurance ownership question was created by the research team to understand whether or not the participant owns health insurance and then classified the insurance further into government, private or other insurance, and both government and private insurance. In addition to the ownership, participants were also be assessed regarding their literacy about government insurance (*BPJS Kesehatan*) utilisation in MH services. The instrument consists of one question '*Apakah pemeriksaan dan perawatan gangguan mental ditanggung BPJS*?' (Is the examination and treatment of MH issues covered by BPJS?). Participants were classified as good literacy only if they choose 'yes' as the answer to the question and the other answer will be considered as insufficient literacy.

6.3.3 Statistical Analysis

Descriptive analysis was performed first then followed by a Chi-Square test to examine the association between potential influencing variables and the utilisation of MH services. Hierarchical logistic regression analysis was then used to identify predictors of MH service utilisation. The analysis was structured into three blocks: demographic variables in Block 1, minority conditions (sexual orientation, HIV status, and disability status) in Block 2, and perceived helpfulness of MH services, insurance ownership, government insurance literacy, and proximity of MH services in Block 3. To find the optimal model, all variables from the three blocks were included using the "Backward-Stepwise" selection method, with the first category of each categorical variable set as the reference indicator. Adjusted odds ratios (ORs) and their 95% confidence intervals (CIs) were reported, and the significance threshold was set at a *p*-value of 0.05 (two-sided). All analyses were conducted using the Statistical Package for the Social Sciences (SPSS) version 26.

6.4 Findings

The descriptive analysis indicated that only 26.5% of respondents ($n = 1{,}382$) utilised MH services in the past three years and varied across respondent categories. Higher utilisation rate was observed among females (30.54%) and intersex/transgender/non-binary peoples (37.21%), 18–24 and 25–34 age groups (30.42% and 31.28%, respectively), non-heterosexual (42.28%), respondents with disability (43.15%), respondents who perceived MH services as helpful (44.1%), respondents with adequate insurance literacy (43.13%), and those aware of nearby services (34.98%) (see Table 6.1 for complete results). Significant association to utilisation of MH services was found with age group, current provincial residence, ethnicity, educational attainment, employment status and type, income level, gender, sexual orientation, disability status, perceived helpfulness of MH services, insurance status, insurance literacy, and awareness of nearby MH services using Chi-square ($p < 0.05$). The only two variables which did not show significant association were religion/belief ($p = 0.81$) and HIV status ($p = 0.11$).

The final predictive model for MH services utilisation (Model 3 in Table 6.2) demonstrated a significant goodness-of-fit with the observed data, as supported by the Omnibus test result ($\chi^2(25) = 1055.31, p < 0.001$). The model's C-statistic also suggested effective discrimination between participants who utilised MH services and those who did not, with an accuracy rate of 77%.

Factors significantly associated with increased odds and are significant predictors of utilizing MH services included being 25–34 years old ($OR = 1.72$, $p < 0.001$), residing in Java Island ($OR = 1.28, p < 0.05$), obtaining a diploma or higher education ($OR = 1.5, p < 0.001$), having an average monthly income over IDR 12,000,000 ($\approx$ USD 750) ($OR = 1.85$, $p < 0.05$), identifying as female ($OR = 1.67$, $p < 0.001$), having a disability ($OR = 2.27$, $p < 0.001$), perceiving MH services as helpful ($OR = 2.82$, $p < 0.001$), possessing adequate knowledge of insurance coverage for MH services ($OR = 1.91, p < 0.001$), and being aware of accessible services nearby ($OR = 1.55$, $p < 0.001$), with perceiving MH services as helpful, disability status, and adequate insurance knowledge being the most reliable predictor with $OR > 1.9$. (see Table 6.2). The final predictive model can be seen in Model 3 with adjusted odd ratio for each variable.

Table 6.1 Relationship between dependent variables and participant's mental health service utilisation in the past three years

Variables	*n* (%)	Mental health service utilisation in the past three years		*p* value*
		No	Yes	
		n (%)	*n* (%)	
All participants	5,211 (100%)	3,829 (73.48%)	1,382 (26.52%)	
Age Group				<0.001
18–24	2,393 (45.92%)	1,665 (69.58%)	728 (30.42%)	
25–34	1,771 (33.99%)	1,217 (68.72%)	554 (31.28%)	
>34	1,047 (20.09%)	947 (90.45%)	100 (9.55%)	
Province				<0.001
Outside Java Island	1,259 (24.16%)	988 (78.47%)	271 (21.53%)	
Within Java Island	3,952 (75.84%)	2,841 (71.89%)	1,111 (28.11%)	
Ethnicity				0.001
Non-Javanese	2,898 (55.61%)	2,180 (75.22%)	718 (24.78%)	
Javanese	2,313 (44.39%)	1,649 (71.29%)	664 (28.71%)	
Educational background				0.04
High school and lower	2,375 (45.58%)	1,777 (74.82%)	598 (25.18%)	
Diploma and higher	2,836 (54.42%)	2,052 (72.36%)	784 (27.64%)	
Job status/type				<0.001
Unemployed/student/retired	2,142 (41.11%)	1,481 (69.14)	661 (30.86%)	
Freelancer	728 (13.97%)	533 (73.21%)	195 (26.79%)	
Private employee/blue collar/ driver/domestic worker	1,436 (27.56%)	1,095 (76.25%)	341 (23.75%)	
Civil servants/police/military	290 (5.57%)	232 (80%)	58 (20%)	
Entrepreneur	403 (7.73%)	331 (82.13%)	72 (17.87%)	
Others	212 (4.07%)	157 (74.06%)	55 (25.94%)	
Average monthly income				<0.01
Other/have no income	509 (9.77%)	384 (75.44%)	125 (24.56%)	
<IDR 3,000,000	2,530 (48.55%)	1,802 (71.23%)	728 (28.77%)	
IDR 3,000,000–IDR 5,999,999	1,107 (21.24%)	824 (74.44%)	283 (25.56%)	
IDR 6,000,000–IDR 8,999,999	461 (8.85%)	342 (74.19%)	119 (25.81%)	
IDR 9,000,000–IDR 11,999,999	223 (4.28%)	180 (80.72%)	43 (19.28%)	
>IDR 12,000,000	381 (7.31%)	297 (77.95%)	84 (22.05%)	
Religion/Belief				0.81
Non-Islam	1,518 (29.13%)	1,112 (73.25%)	406 (26.75%)	
Islam	3,693 (70.87%)	2,717 (73.57%)	976 (26.43%)	
Sex-gender				<0.001

(continued)

Table 6.1 (continued)

Variables	*n* (%)	Mental health service utilisation in the past three years		*p* value*
		No	Yes	
		n (%)	*n* (%)	
Male	1,661 (31.87%)	1,366 (82.24%)	295 (17.76%)	
Female	3,507 (67.3%)	2,436 (69.46%)	1,071 (30.54%)	
Intersex/transgender/non-binary	43 (0.83%)	27 (62.79%)	16 (37.21%)	
Sexual orientation				<0.001
Non-heterosexual	544 (10.44%)	314 (57.72%)	230 (42.28%)	
Heterosexual	4,667 (89.56%)	3,515 (75.32%)	1,152 (24.68%)	
HIV status				0.11
Negative/do not know	5,168 (99.17%)	3,802 (73.57%)	1,366 (26.43%)	
Positive	43 (0.83%)	27 (62.79%)	16 (37.21%)	
Disability status				<0.001
Without disability	4,050 (77.72%)	3,169 (78.25%)	881 (21.75%)	
With disability	1,161 (22.28%)	660 (56.85%)	501 (43.15%)	
Perceived Helpfulness of Mental Health Services				<0.001
Not helpful	3,379 (68.84%)	2,805 (83.01%)	574 (16.99%)	
Helpful	1,832 (35.16%)	1,024 (55.9%)	808 (44.1%)	
Insurance status				<0.01
Did not own insurance	1,368 (26.25%)	1,042 (76.17%)	326 (23.83%)	
Own insurance	3,843 (73.75%)	2,787 (72.52%)	1,056 (27.48%)	
Insurance literacy regarding mental health services				< 0.001
Illiterate	3,669 (70.41%)	2,952 (80.46%)	717 (19.54%)	
Literate	1,542 (29.59%)	877 (56.87%)	665 (43.13%)	
Mental health services nearby				<0.001
Not available/did not know	3,041 (58.36%)	2,418 (79.51%)	623 (20.49%)	
Available	2,170 (41.64%)	1,411 (65.02%)	759 (34.98%)	

* chi-square test

Table 6.2 The predictive model for mental health service utilisation using hierarchical logistic regression

Variables	Model 1		*p*	Model 2		*p*	Model 3		*p*
	Adj OR	95% CI		Adj OR	95% CI		Adj OR	95% CI	
Age Group									
18–24	1 (ref)			1 (ref)			1 (ref)		
25–34	1.95	1.66–2.99	***	1.82	1.54–2.15	***	1.72	1.45–2.05	***
> 34	1.00			1.00			1.00		
Province									
Outside Java Island	1 (ref)			1 (ref)			1 (ref)		
Within Java Island	1.43	1.22–1.69	***	1.36	1.14–1.61	***	1.28	1.07–1.53	**
Ethnicity									
Non-Javanese	1 (ref)			1 (ref)			1 (ref)		
Javanese	1.14	0.99–1.31	0.07	1.18	1.03–1.37	0.02	1.09	0.94–1.27	0.23
Educational background									
High school and lower	1 (ref)			1 (ref)			1 (ref)		
Diploma and higher	1.59	1.36–1.88	***	1.72	1.45–2.03	***	1.50	1.26–1.80	***
Job status/type									
Unemployed/ student/retired	1 (ref)			1 (ref)			1 (ref)		
Freelancer	0.82	0.66–1.02	0.08	0.79	0.63–0.98	0.03	0.81	0.64–1.03	0.08
Private employee/ blue collar/driver/ domestic worker	0.59	0.48–0.73	***	0.61	0.49–0.76	***	0.65	0.52–0.82	***
Civil servants/ police/military	0.56	0.39–0.81	**	0.61	0.42–0.88	**	0.49	0.33–0.73	***
Entrepreneur	0.63	0.46–0.85	**	0.64	0.47–0.88	**	0.73	0.53–1.02	0.07
Others	0.77	0.53–1.1	0.15	0.79	0.55–1.15	0.22	0.84	0.57–1.23	0.37

(continued)

Table 6.2 (continued)

Variables	Model 1		*p*	Model 2		*p*	Model 3		*p*
	Adj OR	95% CI		Adj OR	95% CI		Adj OR	95% CI	
Average monthly income									
Other/have no income	1 (ref)			1 (ref)			1 (ref)		
<IDR 3,000,000	1.42	1.12–1.78	**	1.40	1.11–1.77	**	1.24	0.97–1.58	0.09
IDR 3,000,000–IDR 5,999,999	1.58	1.19–2.12	**	1.56	1.16–2.10	**	1.35	0.99–1.85	0.06
IDR 6,000,000–IDR 8,999,999	1.91	1.34–2.71	***	1.78	1.24–2.55	**	1.62	1.11–2.37	0.01
IDR 9,000,000–IDR 11,999,999	1.31	0.84–2.03	0.23	1.25	0.79–1.97	0.32	1.18	0.73–1.89	0.50
> IDR 12,000,000	1.93	1.32–2.81	**	1.93	1.31–2.84	**	1.85	1.24–2.78	**
Religion/Belief									
Non-Islam	1 (ref)			1 (ref)			1 (ref)		
Islam	0.86	0.74–1.01	0.06	0.93	0.79–1.09	0.38	1.04	0.88–1.23	0.66
Sex-gender									
Male	1 (ref)			1 (ref)			1 (ref)		
Female	1.68	1.44–1.96	***	1.86	1.58–2.19	***	1.67	1.41–1.99	***
Intersex/transgender/non-binary	2.44	1.26–4.73	**	1.50	0.75–2.99	0.24	1.66	0.81–3.40	0.17
Sexual orientation									
Non-heterosexual				1 (ref)			1 (ref)		
Heterosexual				0.46	0.37–0.57	***	0.54	0.43–0.68	***
HIV status									
Negative/do not know				1 (ref)			1 (ref)		
Positive				1.42	0.71–2.84	0.31	1.39	0.67–2.91	0.38

(continued)

Table 6.2 (continued)

Variables	Model 1		*p*	Model 2		*p*	Model 3		*p*
	Adj OR	95% CI		Adj OR	95% CI		Adj OR	95% CI	
Disability status									
Without disability				1 (ref)			1 (ref)		
With disability				2.38	2.04–2.75	***	2.27	1.95–2.66	***
Perceived Helpfulness of Mental Health Services									
Not helpful							1 (ref)		
Helpful							2.82	2.45–3.25	***
Insurance status									
Did not own insurance							1 (ref)		
Own insurance							1.09	0.93–1.29	0.29
Insurance literacy regarding mental health services									
Illiterate							1 (ref)		
Literate							1.91	1.65–2.22	***
Mental health services nearby									
Not available/did not know							1 (ref)		
Available							1.55	1.34–1.79	***
Constant	0.16		***	0.23		***	0.1		***
c-statistic	0.66			0.71			0.77		
Δ c-statistic	0.66			0.05			0.06		
Omnibus test/ Chi-square (*df*)	381.14 (18)***			593.19 (21)***			1055.31 (25)***		

Adj *OR*: Adjusted Odds Ratio; *CI*: Confidence Interval; *df*: degree of freedom; IDR: Indonesia Rupiah; ref: references

** p < 0.05

*** p < 0.001

6.5 Discussion of Implications

This study aimed to explore the MH service use among Indonesian adults and identified contextual contributing factors. It was found that only 26.5% of participants utilised MH services in the past three years. This result is similar with the findings from the Indonesia National Adolescent Mental Health Survey (I-NAMHS), which only a small percentage of adolescents seek treatment from healthcare providers. Specifically, 24.3% visit doctors or nurses, 3.7% consult community health centre (*PUSKESMAS*) staff, and just 2.9% seek help from specialists, including psychiatrists (Queensland Centre for Mental Health Research 2022). This number is higher than the estimations on the utilisation of both primary and secondary level services in Indonesia a decade ago, which indicates that the scope of MH services is still < 10% (Mamnuah et al. 2016). This change may happen because of the improvement of mental health promotion and mental health awareness within the society.

This finding is also in line with the low rate of MH service utilisation in Asia, with a treatment gap as high as 90% in some countries (Maddock et al. 2021). For example, in Singapore, only 9.3% of the general population and 32% of those with mental disorders had ever sought help, with the highest utilisation among those with anxiety and mood disorders seeking help from MH professionals (Shafie et al. 2021). Similar survey conducted in Northwestern China found a significantly lower utilisation of MH service in which only 3.21% and 6.53% of participants utilised MH service in the past 12 months and lifetime use subsequently (Liu et al. 2018).

The final predictive factors indicated that several demographic factors were significantly impacting the utilisation of MH services (see Table 6.3), with gender played a significant role. Females were 1.67 times more likely to use these services than males, while being intersex, transgender, or non-binary did not significantly predict service utilisation. Prior research indicates that cisgender females are more likely to utilise healthcare services for MH issues compared to males because of different cultural value and expectation related to the specific gender. (Fleury et al. 2014; Vasiliadis et al. 2013). Women are more likely than men to seek help from MH professionals than males often due to their more positive attitudes towards expressing psychological concerns (Mackenzie et al. 2006). The finding in which intersex, transgender, or non-binary being non-significant predictor is an interesting novel finding, as previous research usually found that non-cisgender being significant predictor for MH utilisation (Ben Khallouq et al. 2022). This non-significant finding in our study could be the very different sample sizes that led to low statistical power to detect an effect.

In terms of minority status, individuals with disabilities and non-heterosexuals are 2.27 times and 1.85 times more likely to use MH services, respectively. These findings shared similarities with people with disability status in the United States who had a higher proportion of utilising MH services compared to persons without disability (Xie et al. 2022). Also, people with disabilities tend to experience higher rates of mental illness than the general population because the "burden of disability" can lead to heightened discriminatory attitudes towards individuals with disabilities from co-workers, community members, and healthcare professionals (Lund and Ayers

Table 6.3 Demographic factors impacting mental health service utilisation: comparison with previous studies and potential explanation on how demographic factor affecting mental health service utilisation

Demographic Factors	Study Result	Comparison to previous studies	Potential explanation
Age	Individuals aged 25–34 were 1.72 times more likely to utilise MH services compared to both younger (18–24 years old) and older groups (over 34 years old)	Complements the previous study where productive age (15–69 years old) in Indonesia being a significant predictor of MH utilisation (Rahmah and Indrawati 2024)	This phenomenon might be because 25–34 age group may: • Have better MH literacy and fewer stigmatizing attitudes compared to older adults (Putri et al. 2021) • More likely to be employed and have financial means to access services • Increased responsibilities (e.g., starting families, careers) and motivate individuals to prioritize their mental well-being and seek support when needed (Juliansen et al. 2024) In contrast, among younger adults, the low utilisation rate of MH services may be caused by inadequate information about these services, the belief that they can manage the problem without professional help or scepticism about the efficacy of the services (Cadigan et al. 2019)
Residential Area	Residents within Java Island were also 1.28 times more likely to utilise these services	–	Attributed to infrastructure availability and access disparities within Java Island compared to other regions (Ministry of Health Indonesia 2019) • Only 60% out of 10,000 primary-care facilities nationwide feature operational MH programs, predominantly located within Java • Six out of Indonesia's 34 provinces lack a psychiatric hospital, which are all located outside Java

(continued)

Table 6.3 (continued)

Demographic Factors	Study Result	Comparison to previous studies	Potential explanation
Education	Those with at least a diploma were 1.5 times more likely to utilise MH services	Consistent with previous studies that higher education levels are generally associated with better MH literacy and reduced stigma towards mental illness (Putri et al. 2021)	Individuals with higher education may have greater awareness and understanding of MH issues, making them more likely to recognize the need for professional help and seek services
Job status and type	Private employees or those in blue-collar jobs, and civil servants, police, or military personnel were 35% and 51% less likely, respectively, to utilise MH services compared to those unemployed, students, or retired	Similar findings were found in Steele, et al. where employed participants were 36% less likely to use MH services compared to unemployed (Steele et al. 2007)	This finding may be caused by persistence of explicit forms of discrimination and stigma in the workplace (Elraz 2018)
Income	Participants earning an average monthly income of more than IDR 12 million/month ($\approx$ USD 750/month) were 1.85 times more likely to utilise MH services	Supported previous studies that have found a positive correlation between high income and MH service Utilisation (Fleury et al. 2014)	This behaviour might be fuelled by the fact that out-of-pocket costs for psychotherapy and psychotropic medications are common in Indonesia, even with national insurance coverage so wealthier individuals can more easily afford these expenses than people with low economic resources (Cipta and Saputra 2022)

2022). In the long run, these experiences could further exacerbate MH challenges of people with disabilities. Likewise, socially vulnerable non-heterosexual groups also face distinct stressors due to their marginalised status within societies dominated by heteronormative power structures (Filice and Meyer 2018). These additional stressors may be attributed to a higher prevalence of MH disorders among them, which could explain their greater use of MH services (Li et al. 2016).

Another finding, people who utilise MH services were 2.82 times more likely to view the services as useful, 1.91 times more likely to be well-informed regarding government MH insurance, and 1.55 times more likely to be aware of local services. It could be because by accessing MH services, individuals gained direct exposure to information about MH, available treatment options, insurance coverage, and local facilities (Anindyajati et al. 2022; Rahmah and Indrawati 2024). Hence, this increased awareness and knowledge compared to those who did not utilise services. Moreover, seeking treatment and having positive experiences helped reduce stigma and increased comfort with MH issues, which made people more open to discussing MH and sharing information with others (Tristiana et al. 2018; Susanti et al. 2024). Assessing insurance literacy for utilising mental health (MH) service is crucial, as emerging evidence suggests that health insurance literacy plays a significant role in facilitating effective healthcare use, including access to primary care and preventive services. A systematic review by Yagi et al. examining the link between health insurance literacy and healthcare utilisation supports this perspective. Their findings indicate that in ten out of thirteen studies on preventive care, individuals with higher health insurance literacy were more likely to utilise primary and preventive healthcare services. Similarly, eight out of nine studies on care avoidance demonstrated that those with lower health insurance literacy were more prone to delaying or forgoing necessary care. These findings highlight the importance of improving insurance literacy to enhance healthcare access and utilisation, albeit general and not specified to MH services (Yagi et al. 2022). As for MH services, a prior study highlighted that insufficient information regarding the cost and availability of MH services, along with a lack of recognition of professional help as a legitimate support option, has led to poor awareness among the Southeast Asian population, perpetuating barriers to accessing these services (Andary et al. 2023).

Based on this study, the most impactful facet of mental health services for Indonesians and other Asian countries with similar contexts might involve tailored support for specific populations, particularly males, younger individuals, and those without disabilities, coupled with comprehensive mental health education and awareness. Young population and males, often underrepresented in mental health initiatives, face unique challenges and stigma that hinder their access to care. By providing targeted support and resources, their specific needs can be addressed and mental well-being can be improved (Sagar-Ouriaghli et al. 2019). Furthermore, enhancing mental health education and awareness is crucial. This involves educating the public about nearby mental health services, the utility of insurance for mental health care, and the perceived helpfulness of these services. By increasing awareness and understanding, the stakeholders can reduce stigma, encourage the use of mental health services, and ensure that individuals know how to access and utilise available resources effectively (Tsai et al. 2021). One way to implement these results is by encouraging individuals to bring friends to mental health services or training them as peer educators, similar to the HIV Voluntary Counselling and Testing (VCT) model. Thus, it could significantly increase the utilisation of mental health services in Indonesia by reducing stigma and enhancing support networks. These combined efforts will have a significant impact on improving mental health outcomes and require the most availability to meet the diverse needs of the population.

In the Indonesian context, previous studies on mental health utilisation have often overlooked the specific needs and challenges of underserved populations, particularly those with minority statuses. Our research team, comprising individuals with lived experiences of minority status in Indonesia, was uniquely positioned to approach this study with a deep curiosity and a nuanced understanding of these populations. This positionality allowed us to design a study that addressed these gaps, reflected in our relatively large and socially diverse sample. Additionally, by engaging community members to assist in disseminating the survey link, we sought to mitigate power dynamics between researchers and participants, ensuring a more equitable and inclusive research process.

6.6 Future Directions

This study highlights the urgent need to improve the availability and quality of MH services in Indonesia. It emphasizes the necessity of enhancing awareness about these services, not only regarding the availability of services but also the potential use of health insurance and their coverage. The findings indicate that information on payment and availability of MH services are significant predictors of utilisation. Despite the establishment of BPJS Kesehatan, which covers MH service payments, low utilisation rates among marginalised populations, such as those with lower incomes and those living outside Java Island, highlight the need for interventions addressing financial literacy and social access barriers. This study also emphasizes the importance of broadening the information of insurance coverage to facilitate access to MH services. By expanding and improving how BPJS Kesehatan disseminate information about the MH services covered, Indonesia can progress towards more equitable healthcare.

This study also highlights a concerning trend among less marginalised groups, older adults, heterosexuals, males, and professionals such as private employees, civil servants, military, and police, who show a lower likelihood of utilising MH services that might be caused by stigmas around seeking MH care and mental illness. This resistance underscores the urgent need for targeted educational programs that specifically address and aim to reduce MH stigma within these populations. By focusing on targeted MH education that promotes both preventive measures and the benefits of service utilisation, these initiatives can foster a greater understanding and acceptance of MH challenges, ultimately encouraging more individuals from these groups to seek necessary support without fear of judgment or discrimination in the future.

Recognizing that individuals are more likely to use MH services if they believe these services are beneficial, it becomes imperative to enhance public perceptions through targeted educational campaigns and improved service quality. By actively promoting the effectiveness and benefits of MH care, as well as providing clear, accessible information on how these services can assist individuals, Indonesia can encourage a higher uptake of these vital resources. On the other hand, it is essential for MH professionals to enhance public trust in their services as a legitimate aid for MH issues, thus can encourage more people to seek help if needed. These approaches not only address the immediate need for increased service utilisation but also contribute to the broader goal of reducing MH stigma, thus developing a more supportive environment for MH care across the country.

To ensure the sustainability and replicability of these proposed measures for enhancing mental health care services, it is essential to establish robust frameworks that can be adapted to diverse contexts within Indonesia and other Asian nations. This could include developing scalable models for mental health education tailored to local cultural contexts, such as community-based peer educator programs similar to VCT initiatives or culturally specific awareness campaigns. Additionally, strengthening regional and multilateral cooperation, such as fostering collaborations among mental health professionals across Southeast Asia through joint clinical and research initiatives, can help minimise inequities in mental health services and promote cultural humility among participating countries. Regular monitoring and evaluation should be conducted to assess the effectiveness of these initiatives, allowing for continuous improvement and adaptation. By building strong partnerships and fostering community involvement, these measures can achieve long-term sustainability and serve as a model for other countries facing similar mental health challenges.

Moreover, these findings may direct potential collaboration with the Global North, especially in the fields of disability studies, sexual minority health, and public mental health services. Future research collaboration could involve learning exchanges about barriers and facilitators in both contexts, providing insights on how to improve utilisation among these groups. Such collaborations could also enable Global North scholars to better understand the utilisation of mental health services by Indonesian people, potentially leading to joint research or enhanced mental health services.

6.7 Conclusion

This study aimed to assess the level of mental health (MH) service utilisation among Indonesian adults and identify contextual predictive factors influencing this utilisation, which only about quarter of participants utilised MH services in the past three years. Key findings highlighted the significant impact with perceiving MH services as helpful, disability status, and adequate insurance knowledge being the most reliable predictor on MH service utilisation. Therefore, tailored support for specific populations, coupled with comprehensive mental health education and awareness to improve future MH utilisation, are needed as strategies to improve the use of MH services.

A notable limitation of this study is the use of a single-item measure to assess insurance literacy, specifically regarding the coverage of mental health services by BPJS Kesehatan. Due to time constraints and challenges posed by the COVID-19 pandemic, a more comprehensive and statistically validated instrument could not be developed. While this question provided an initial indication of participants' awareness of mental health coverage within the national insurance scheme, it may not fully capture the multidimensional nature of insurance literacy, which includes aspects such as knowledge of claims procedures, reimbursement policies, and service limitations. Consequently, this limitation should be considered when interpreting the findings, as the measure may underestimate or oversimplify the complexity of insurance literacy and its impact on mental health service utilisation. Future research should aim to develop and validate a more robust instrument that better encapsulates the breadth of insurance literacy in the Indonesian context.

Another main limitation of the current study is the reliance on self-reported data, which may be subject to biases. Future research should consider using mixed-method approaches, including qualitative interviews or clinical assessments, to validate self-reported data and gain deeper insights into MH service utilisation behaviours. Another limitation is the cross-sectional design of the study, which restricts the ability to establish causal relationships between predictive factors and MH service utilisation. Longitudinal studies are needed to explore how these factors influence MH service utilisation over time and to identify potential causal pathways. By addressing these limitations and exploring these future research directions, we can advance our understanding of MH service utilisation and develop more effective strategies to promote equitable and accessible MH care for all individuals.

Highlights

- Developing targeted interventions for specific populations, including people with disabilities and marginalised communities, will help address unique barriers to mental health care.
- Expanding mental health insurance coverage and literacy is essential to improving access and reducing financial barriers to care across Asia.
- Targeted stigma reduction initiatives can encourage help-seeking and normalise mental health service utilisation.

References

Alegria M, Bijl RV, Lin E, Walters EE, Kessler RC. Income differences in persons seeking outpatient treatment for mental disorders. Arch Gen Psychiatry. 2000;57:383. https://doi.org/10.1001/archpsyc.57.4.383.

Alonso J, Liu Z, Evans-Lacko S, Sadikova E, Sampson N, Chatterji S, Abdulmalik J, et al. Treatment gap for anxiety disorders is global: results of the world mental health surveys in 21 countries. Depress Anxiety. 2018;35(3):195–208. https://doi.org/10.1002/da.22711.

Andary S, Bassani J, Burrell G, Cole E, Evans R, Redman E, Kumar S. Barriers and enablers to access and utilization of mental health care services across Southeast Asia: a preliminary scoping review. Asia-Pac Psychiatry. 2023;15(4): e12549. https://doi.org/10.1111/appy.12549.

Anindyajati G, Mardiasmo DR, Sekarasih L, Susilaradeya D, Takwin B, Pelupessy DC, et al. The right to health: COVID-19 pandemic and the opportunity to transform mental health inequalities in Indonesia. Front Public Health. 2022;10: 844656. https://doi.org/10.3389/fpubh.2022.844656.

Ben Khallouq BA, Carter JS, Carter SK, Taliaferro LA. Race and gender matter: Insights into variation in health and patterns of health utilization for transgender and non-binary adolescents. In: Baumle AK, Nordmarken S, editors. Demography of Transgender, Nonbinary and Gender Minority Populations. Cham: Springer International Publishing; 2022. p. 91–116.

Bolton P, West J, Whitney C, Jordans MJ, Bass J, Thornicroft G, Murray L, et al. Expanding mental health services in low- and middle-income countries: a task-shifting framework for delivery of comprehensive, collaborative, and community-based care. Cambridge Prisms Glob Ment Health. 2023;10: e16. https://doi.org/10.1017/gmh.2023.5.

Cadigan JM, Lee CM, Larimer ME. Young adult mental health: a prospective examination of service utilization, perceived unmet service needs, attitudes, and barriers to service use. Prev Sci. 2019;20(3):366–76. https://doi.org/10.1007/s11121-018-0875-8.

Chen Q, Huang S, Xu H, Peng J, Wang P, Li S, Zhao J, et al. The burden of mental disorders in Asian countries, 1990–2019: an analysis for the global burden of disease study 2019. Transl Psychiatry. 2024;14(1):1–11. https://doi.org/10.1038/s41398-024-02864-5.

Cipta DA, Saputra A. Changing landscape of mental health from early career psychiatrists' perspective in Indonesia. JoGHNP. 2022;5: e2022011. https://doi.org/10.52872/001c.37413.

Daliri DB, Aninanya GA, Laari TT, Abagye N, Dei-Asamoa R, Afaya A. Exploring the barriers to mental health service utilization in the Bolgatanga Municipality: the perspectives of family caregivers, service providers, and mental health administrators. BMC Health Serv Res. 2024;24(1):278. https://doi.org/10.1186/s12913-024-10567-2.

Elraz H. Identity, mental health and work: how employees with mental health conditions recount stigma and the pejorative discourse of mental illness. Hum Relat. 2018;71(5):722–41. https://doi.org/10.1177/0018726717716752.

Filice E, Meyer SB. Patterns, predictors, and outcomes of mental health service utilization among lesbians, gay men, and bisexuals: a scoping review. J Gay Lesbian Ment Health. 2018;22(2):162–95. https://doi.org/10.1080/19359705.2017.1418468.

Fleury MJ, Ngamini Ngui A, Bamvita JM, Grenier G, Caron J. Predictors of healthcare service utilization for mental health reasons. Int J Environ Res Public Health. 2014;11(10):10559–86. https://doi.org/10.3390/ijerph111010559

Gilbert BJ, Patel V, Farmer PE, Lu C. Assessing development assistance for mental health in developing countries: 2007–2013. PLoS Med. 2015;12(6). https://doi.org/10.1371/journal.pmed.1001834.

Gronowski A. Indonesia's mental health services: Availability and current challenges [Master's Projects and Capstones]. San Francisco: University of San Francisco;2020. https://repository.usfca.edu/cgi/viewcontent.cgi?article=2224&context=capstone. Accessed 25 Sept 2024.

Harris MG, Kazdin AE, Chiu WT, Sampson NA, Aguilar-Gaxiola S, Al-Hamzawi A, Alonso J, et al. Findings from world mental health surveys of the perceived helpfulness of treatment for patients with major depressive disorder. JAMA Psychiat. 2020;77(8):830–41. https://doi.org/10.1001/jamapsychiatry.2020.1107.

Health TLG. Mental health matters. Lancet Glob Health. 2020;8(11): e1352. https://doi.org/10.1016/S2214-109X(20)30432-0.

Into The Light Indonesia. Seri Laporan ke-1: Laporan perilaku penggunaan layanan kesehatan mental di Indonesia 2021 - Hasil awal [Serial Report. No. 1: Use of mental health services in Indonesia 2021 - Preliminary report]. Jakarta, Indonesia: Into The Light Indonesia;2021. https://www.intothelightid.org/tentang-bunuh-diri/laporan-perilaku-penggunaan-layanan-kesehatan-mental-di-indonesia-2021-hasil-awal/. Accessed 25 Sept 2024.

Jorm AF, Korten AE, Jacomb PA, Christensen H, Rodgers B, Pollitt P. 'Mental health literacy': a survey of the public's ability to recognise mental disorders and their beliefs about the effectiveness of treatment. Med J Aust. 1997;166(4):182–6. https://doi.org/10.5694/j.1326-5377.1997.tb140071.x.

Juliansen A, Heriyanto RS, Muljono MP, Budiputri CL, Sagala YDS, Octavius GS. Mental health issues and quality of life amongst school-based adolescents in Indonesia. J Med Surg Public Health. 2024;2: 100062. https://doi.org/10.1016/j.glmedi.2024.100062.

Li W, Dorstyn DS, Denson LA. Predictors of mental health service use by young adults: a systematic review. Psychiatr Serv. 2016;67. https://doi.org/10.1176/appi.ps.201500280.

Liu L, Chen XL, Ni CP, Yang P, Huang YQ, Liu ZR, Wang B, Yan YP. Survey on the use of mental health services and help-seeking behaviors in a community population in Northwestern China. Psychiatry Res. 2018;262:135–40. https://doi.org/10.1016/j.psychres.2018.02.010.

Lund EM, Ayers KB. Ever-changing but always constant: 'Waves' of disability discrimination during the COVID-19 pandemic in the United States. Disabil Health J. 2022;15(4): 101374. https://doi.org/10.1016/j.dhjo.2022.101374.

Mackenzie CS, Gekoski WL, Knox VJ. Age, gender, and the underUtilisation of mental health services: the influence of help-seeking attitudes. Aging Ment Health. 2006;10(6):574–82. https://doi.org/10.1080/13607860600641200.

Maddock A, Blair C, Ean N, Best P. Psychological and social interventions for mental health issues and disorders in Southeast Asia: a systematic review. Int J Ment Health Syst. 2021;15:56. https://doi.org/10.1186/s13033-021-00482-y.

Mamnuah M, Nurjannah I, Prabandari YS, Marchira CR. Literature review of mental health recovery in Indonesia. GSTF J Nurs Health Care. 2016;3(2):20–5. https://doi.org/10.5176/2345-718X_3.2.115.

Mekonen T, Chan GCK, Belete T, Menberu M, Davidson L, Hides L, et al. Mental health service utilization in a low-resource setting: a qualitative study on perspectives of health professionals in Northwest Ethiopia. PLoS ONE. 2022;17(11): e0278106. https://doi.org/10.1371/journal.pone.0278106.

Ministry of Health Indonesia. Situasi kesehatan jiwa di Indonesia. Jakarta: Pusat Data Informasi Kementerian Kesehatan RI;2019. https://pusdatin.kemkes.go.id/article/view/20031100001/situasi-kesehatan-jiwa-di-indonesia.html. Accessed 1 Oct 2024.

Munira L, Liamputtong P, Viwattanakulvanid P. Barriers and facilitators to access mental health services among people with mental disorders in Indonesia: a qualitative study. Belitung Nurs J. 2023;9(2):110–7. https://doi.org/10.33546/bnj.2521.

World Health Organization, Regional Office for South-East Asia. Mental health conditions in the WHO South-East Asia Region. New Delhi: World Health Organization, Regional Office for South-East Asia;2023. Licence: CC BY-NC-SA 3.0 IGO. https://www.who.int/publications/i/item/9789290210788. Accessed 25 Sept 2024.

Pham TN, Pham TT, Nguyen DH, Duong HA, Bui DAT, Giang KB, Dang Huong GD, et al. Utilization of mental health services among university students in Vietnam. Int J Ment Health. 2021;50(2):113–35. https://doi.org/10.1080/00207411.2020.1816114.

Putri AK, Gustriawanto N, Rahapsari S, Sholikhah AN, Prabaswara S, Kusumawardhani AC, et al. Exploring the perceived challenges and support needs of Indonesian mental health stakeholders:

a qualitative study. Int J Ment Health Syst. 2021;15:81. https://doi.org/10.1186/s13033-021-00504-9.
Queensland Centre for Mental Health Research. Indonesia national adolescent mental health survey (I-NAMHS) [Internet]. Brisbane: QCMHR;2022 [cited 2024 Nov 26]. https://qcmhr.org/~teoqngfi/docman/reports/12-i-namhs-report-bahasa-indonesia/file. Accessed 30 Sept 2024.
Rahmah RH, Indrawati F. Analyzing utilization of mental health services for national health insurance participants in Indonesia. Unnes J Public Health. 2024;13(1):1–9. https://doi.org/10.15294/ujph.v13i1.68905.
Roberts T, Esponda GM, Krupchanka D, Shidhaye R, Patel V, Rathod S. Factors associated with health service utilization for common mental disorders: a systematic review. BMC Psychiatry. 2018;18(1):262. https://doi.org/10.1186/s12888-018-1837-1.
Sagar-Ouriaghli I, Godfrey E, Bridge L, Meade L, Brown JSL. Improving mental health service utilization among men: a systematic review and synthesis of behavior change techniques within interventions targeting help-seeking. Am J Mens Health. 2019;13(3):1557988319857009. https://doi.org/10.1177/1557988319857009.
Sarikhani Y, Bastani P, Rafiee M, Kavosi Z, Ravangard R. Key barriers to the provision and utilization of mental health services in low-and middle-income countries: a scope study. Community Ment Health J. 2021;57(5):836–52. https://doi.org/10.1007/s10597-020-00619-2.
Shafie S, Subramaniam M, Abdin E, Vaingankar JA, Sambasivam R, Zhang Y, et al. Help-seeking patterns among the general population in Singapore: results from the Singapore mental health study 2016. Adm Policy Ment Health. 2021;48:586–96. https://doi.org/10.1007/s10488-020-01092-5.
Slewa-Younan S, Mond J, Bussion E, Mohammad Y, Uribe Guajardo MG, Smith M, et al. Mental health literacy of resettled Iraqi refugees in Australia: knowledge about posttraumatic stress disorder and beliefs about helpfulness of interventions. BMC Psychiatry. 2014;14:320. https://doi.org/10.1186/s12888-014-0320-x.
Steele LS, Dewa CS, Lin E, Lee KL. Education level, income level and mental health services use in Canada: associations and policy implications. Healthc Policy. 2007;3(1):96–106.
Susanti H, Brooks H, Yulia I, Windarwati HD, Yuliastuti E, Hasniah H, et al. An exploration of the Indonesian lay mental health workers' (cadres) experiences in performing their roles in community mental health services: a qualitative study. Int J Ment Health Syst. 2024;18:3. https://doi.org/10.1186/s13033-024-00622-0.
Tristiana RD, Yusuf A, Fitryasari R, Wahyuni SD, Nihayati HE. Perceived barriers on mental health services by the family of patients with mental illness. Int J Nurs Sci. 2018;5(1):63–7. https://doi.org/10.1016/j.ijnss.2017.12.003.
Tsai CL, Lin YW, Hsu HC, Lou ML, Lane HY, Tu CH, Ma WF. Effects of the health-awareness-strengthening lifestyle program in a randomized trial of young adults with an at-risk mental state. Int J Environ Res Public Health. 2021;18(4):1959. https://doi.org/10.3390/ijerph18041959.
Vasiliadis HM, Gagné S, Jozwiak N, Préville M. Gender differences in health service use for mental health reasons in community-dwelling older adults with suicidal ideation. Int Psychogeriatr. 2013;25(3):374–81. https://doi.org/10.1017/S1041610212001895.
World Health Organization. Depression and other common mental disorders: Global health estimates. Geneva: WHO;2017. https://www.who.int/publications/i/item/depression-global-health-estimates. Accessed 25 Sept 2024.
Xie Z, Tanner R, Striley CL, Marlow NM. Association of functional disability with mental health services use and perceived unmet needs for mental health care among adults with serious mental illness. J Affect Disord. 2022;299:449–55. https://doi.org/10.1016/j.jad.2021.12.040.
Yagi BF, Luster JE, Scherer AM, Farron MR, Smith JE, Tipirneni R. Association of health insurance literacy with health care utilization: a systematic review. J Gen Intern Med. 2022;37(2):375–89. https://doi.org/10.1007/s11606-021-06819-0.

Muhammad Wildan Rabbani Kurniawan is an independent researcher, medical doctor, and global health professional with a history of eight years experience working in clinical practice, research, and the private sectors. He holds a medical doctor degree from Universitas Indonesia and a Master of Public Health in Global Health from Harvard University. He is currently working in the health policy sectors focusing on strategically aligning healthcare initiatives outside of government areas with government policies and public interests.

Benny Prawira is an Indonesian mental health and suicide prevention researcher, wellness facilitator, and lived experience engagement consultant whose work bridges academic scholarship with community-based interventions. He holds master's degrees in Social Health Psychology from Atma Jaya Catholic University of Indonesia and Medical Anthropology from University of Oxford. As founder of Into The Light Indonesia, the country's first youth-based suicide prevention community, Prawira integrates evidence-based practice, lived experience perspectives, and human rights approaches in developing mental health interventions. Throughout his career, he has been research and community engagement consultants for UN agencies, funding organizations, NGOs, and universities across Southeast Asia, Australia, Canada, the United States, and the UK to advance mental health promotion, suicide prevention, and meaningful lived experience engagement.

Andrian Liem is an Indonesian clinical psychology scholar earned his doctoral degree at the School of Psychology, the University of Queensland, Australia. After graduating, he worked as a postdoctoral fellow at the University of Macau, Macao (SAR), assisting a digital mental health intervention project for overseas Filipino workers; then continued his work at the Monash University Malaysia to explore the experiences of Indonesian migrant domestic workers in caring for older adults in Asian countries. He is now serving as an Assistant Professor at the Faculty of Psychology, Universitas Sebelas Maret, Indonesia.

Chapter 7
Future of Gut Microbiota as a Promising Treatment Target for Mental Illness in Asian Setting

Clusters of Future Studies: Past & Futures; Post-Normality & Complexity; Technological Trends

Arutchelvan Rajamanikam, Chandramathi Samudi, Freddy Franklin, and Suresh Kumar Govind

Abstract Most Asian countries report increasing trend of mental health problems among the younger populations. Advancements in mental health treatment is scarce and stigmatization often contributes to delayed clinical attention. Recent evidence suggests intricate linkage between gut health and mental illness. The gut microbiota plays an integral role from digestion and metabolism to modulation of the immune system. Diverse array of genes in gut microbiome gives a versatile gene function that impacts health. Recent evidence demonstrates unequivocal influence of gut microbiota on mental health. Restoration of compositional and functional derangement of gut microbiota can be a potential treatment target. However, microbiota targeted approach is still far-fetched in Asian setting. In this review, we identify recent publication from major scientific databases and search engines, highlight gut microbiota and mental health linkage and focus on gut microbiota-targeted precision treatment using probiotics, prebiotics, postbiotics and other potential modes. We also discuss the ethical considerations, guidelines and policy implications involved in the practice of treatment. This review intends to inform the clinicians on alternative and proactive intervention which can ultimately contribute to the reduction of national mental healthcare burden.

A. Rajamanikam (✉) · F. Franklin · S. K. Govind
Department of Parasitology, Faculty of Medicine, Universiti Malaya, Kuala Lumpur, Malaysia
e-mail: arun04@um.edu.my

S. K. Govind
e-mail: suresh@um.edu.my

C. Samudi
Department of Medical Microbiology, Faculty of Medicine, Universiti Malaya, Kuala Lumpur, Malaysia
e-mail: chandramathi@um.edu.my

R. Xiang Ng et al. (eds.), *Future of Healthcare in Asia*,
https://doi.org/10.1007/978-3-032-03183-9_7

Keywords Asian · Mental health · Gut microbiota · Gut-brain-axis · Targeted treatment · Future challenges

7.1 Introduction

The World Health Organization (WHO) has identified mental health condition as one of the health threats that are affecting the global population irrespective of age (Saxena and Davidson 2019). It is regarded as the largest contributor for the global non-fatal burden of disease. In 2019, it was estimated that 1 in every 8 people (approximately 970 million people) were living with a mental health condition. Especially after the recent coronavirus disease (COVID-19) pandemic, the estimates increased significantly by 25–28% in just one year. The worsening mental health condition is a concern particularly in countries with poor healthcare system (Bodeker 2020).

In Asia, the burden of diseases attributable to mental health has risen in major cities typically in the context of demographic, environmental, socio-political transitions (Bodeker 2020; World Health Organization 2017). In recent years, the number of disability-adjusted life of years (DALY) of mental disorder such as depressive disorder, anxiety disorder and schizophrenia increased from 43 to 63 million—at the rate of 37.18%, 21.52% and 13.75% respectively (Chen et al. 2024) with higher prevalence among middle-aged individuals for both male and female (see Fig. 7.1). The study, which has analysed data from 1990–2019 urged Asian governments to take proactive and urgent action before the situation turns catastrophic.

Past report demonstrated low mental health priority in some regions of Asia (Maramis et al. 2011). However, a recent infodemiological study on mental health in several countries like Malaysia, the Philippines and Singapore suggested otherwise

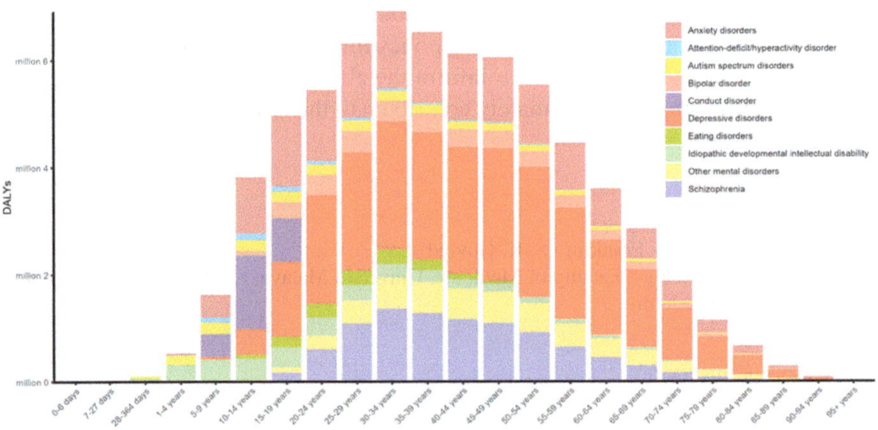

Fig. 7.1 Age-dependent variation of the number of DALY for mental health issues in Asia. The figure was adopted from Chen et al. (2024)

where increase in public interest on mental health was seen (Alibudbud 2022). In recent years, there was increase of public interest in mental health, partly due to the implementation of mental health laws in these regions implying the importance and attention given to mental health issues. Despite this, mental illness sufferers in Asia face a challenge of pervasive stigmatization contributed largely by supernatural, religious and magical approaches that are common in this region (Lauber and Rössler 2007). The stigmatization potentially leads to discrimination against this population leading to prevention of early intervention.

Factors contributing to mental illness are over-generalized based on the risk factors from specific western population. This has led to bias and often do not address the local population precisely. However, there are some risk factors that can be modified and play significant role in policy-making unique to Asia. In general, common risk factors in Asian and western are abuse in childhood, chronic diseases, drug use, hypertension, internet addiction and etc. (Rami et al. 2024). Those unique for Asian are female gender, harmful alcohol intake, less participation in social activities, light occupational physical activity, no religious activities, and work shift or at night (Rami et al. 2024). These factors are by large modifiable but requires specific and targeted approach. Moreover, the risk factors listed above are based on past reports and may fall short of other recently emerged risk factors or not sought-after risks such as diet and gut microbiota (consortium of bacteria that inhabits the gut) in mental illness.

Despite extensive understanding of mental illnesses and established market economics, burden of disease not only has persisted but increased significantly. This could potentially be due to either incorrect burden estimates (Vigo et al. 2016), lack of effective treatment (Exeter 2001) or non-compliance to treatment (Semahegn et al. 2020). The key question we should be asking, however, is whether the treatments are effective. There are a variety of treatment based on severity of the disorder. This treatment ranges from psychotherapy in mild to moderate cases to pharmacotherapy in severe cases. Despite standard of care treatment and continuous refinements in the past decades, the illness has not been addressed enough (Leichsenring et al. 2022).

It is becoming apparent that the treatment for mental illness requires integrated approach with more than a type of treatment option. It often requires a personalized approach as which treatment works for whom is currently unknown. Moreover, multitude of ways such as exercise, community engagement, engagement to nature that can help recover from mental illness is largely understudied (Cuijpers et al. 2020). The role of gut microbiota in mental health is increasingly becoming important but its potential as treatment, adjunct or alternative, has not gained traction yet (Berding and Cryan 2021). Advancement of technology has heightened the access to gut microbial composition and function which is deemed personalized. Whether modulation of bacterial composition and its underlying mechanism will shed light on personalized treatment potential of mentally ill is under-explored. In this narrative review, we highlight the role of gut microbiota in mental illness, the targeted therapeutic strategies, and its future direction, specific to Asian sentiments.

7.2 Review Methodology

Information used to write this review was collected from databases such as PubMed, Web of Science, Scopus and search engine like "Google Scholar" for any publication from year 2010 to 2024. The search was done using the keyword "mental illness", "Asian" "gut microbiota", "gut microbiome", "targeted treatment", "gut brain axis", "mental illness therapeutics".

7.3 Findings

There has been a complex association between gut microbiota and mental health. However, the potentials of targeting gut microbiota to treat mental illness is largely understudied. The following texts will explore the underlying mechanisms and how it can be exploited in the future for human mental wellbeing.

7.3.1 Gut Microbiota and Its Influence on the Brain

The diverse living microorganisms associated with human digestive tract are known as gut microbiota or gut flora. Trillions of microorganisms including bacteria, viruses, protists, and fungi coexist in symbiosis in the human intestine (Lloyd-Price et al. 2016). They collectively play a role in physiological functions as shown in Table 7.1 (see the next page). Illness has been attributed to a disruption in the balance of gut microbiota which can impair physical and mental health (Appleton 2018, Halverson and Alagiakrishnan 2020).

There are now increasing evidence on how gut and the brain interact between each other. This bidirectional interaction is coined as the gut-brain axis (Bienenstock et al. 2015, Foster and Neufeld 2013) (see Fig. 7.2 on the next page). There is increasing evidence on the role of gut microbiota in gut-brain axis. Thus far the axis is reported to link the nervous system and the gastrointestinal tract via various routes including neural (neurotransmitters), endocrine (hormones), immune (Hooper et al. 2012) (cytokines, antibodies) and metabolic signalling (Silva et al. 2020) (metabolites and short chain fatty acids). The bacteria in the gut produce crucial neuroactive molecules such as acetylcholine, gamma-aminobutyric acid (GABA) and precursors to other neurotransmitters such as serotonin that affects the brain chemistry.

Dysbiosis (imbalance in gut microbial composition) of the gut microbiota is linked several neuropsychiatric disorder such as stress, depression and anxiety, schizophrenia and bipolar disorders (Aizawa et al. 2016; Coello et al. 2019; Jiang et al. 2015; Naseribafrouei et al. 2014). When in dysbiosis, it induces alteration to systemic immune system through the release or inhibition of cytokine (O'Mahony et al. 2005) and activation of T cells (Sprouse et al. 2019). There have been multiple

Table 7.1 Physiological roles of gut microbiota

No	Physiological functions	Brief description	References
1	Nutrition/digestion	Gut microbiota plays crucial role in metabolism of dietary carbohydrate, proteins, plant polyphenols, bile acids, and vitamins into metabolites such as short chain fatty acids that are essential for health	Banerjee et al. (2022)
2	Inflammation	Taxa-based cytokines responses are seen in microbiota-host interactions and inflammatory modulations	Schirmer et al. (2016)
3	Mucosal immunity	Gut microbiota maintains integrity of intestinal epithelial barrier and stimulation of antimicrobial peptides as well as immune system maturation	Amoroso et al. (2020)
4	Protection against foreign pathogen	Commensal bacterial species confers protection against pathogens by colonization resistance to various pathogenic organisms	Ubeda et al. (2017)
5	Neurotransmitter production	Certain bacterial taxa contribute to the production and consumption of neurotransmitters that plays key role in gut-brain axis	Strandwitz (2018)

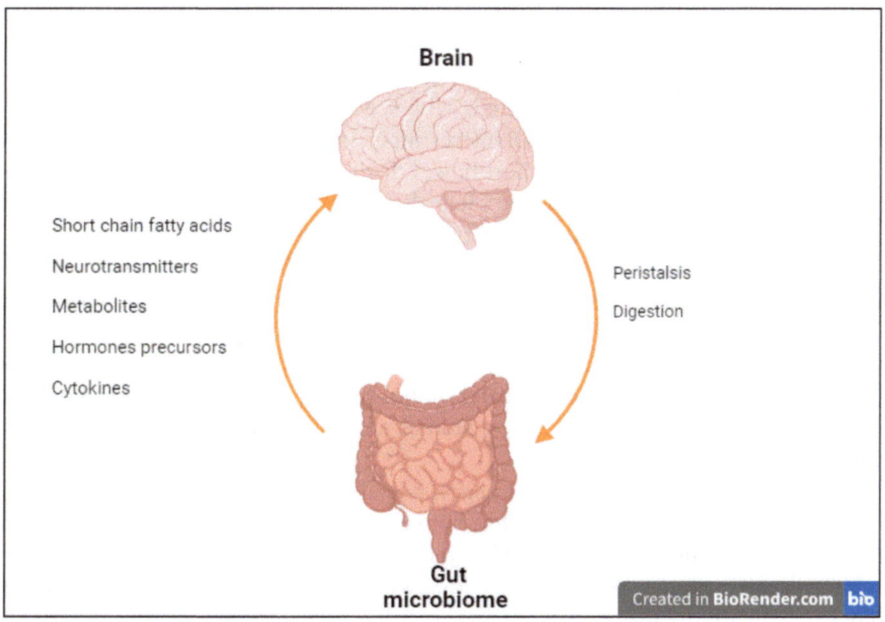

Fig. 7.2 Gut Brain axis (GBA) highlighting the key interactions between the gut and the brain

studies correlating positive symptoms of mental illness such as schizophrenia and depression to elevation of pro-inflammatory cytokines and decreased levels of anti-inflammatory cytokines (Kokai et al. 2002, Na et al. 2014, Parlindungan et al. 2023, R Caso et al. 2016, Schirmer et al. 2016). Faecal microbiota transplantation from individuals with pre-existing mental illness into germ free mice have shown to induce the positive symptoms of the mental disorder (Chinna Meyyappan et al. 2020; Jang et al. 2021; Settanni et al. 2021, Zheng et al. 2019). These findings are the cornerstones to directly imply the role of the gut to the cognitive function of the brain.

A study by Shen et al. (2018) investigates the trend of gut microbiota in individuals with schizophrenia in comparison to healthy control. It was reported that a higher abundance of Proteobacteria was observed among schizophrenic individuals at the genus level (Shen et al. 2018). In contrast, another study focusing on individuals with chronic schizophrenia showed decrease in Proteobacteria (Nguyen et al. 2019a, b). However, both studies noted decreased relative abundance of the commensal bacteria Clostridium among individuals with schizophrenia. Alternatively, the relative abundance of Firmicutes, Actinobacteria, and Bacteroidetes were altered among patients with depression (Zheng et al. 2016). In a meta review, a total of nine investigations reported *Prevotellaceae* to be in lower abundances among individuals with depression (Gao et al. 2023). In addition, echoing same sentiments, another meta-review comparing the gut microbiome of individuals with depression reported a common trend of decreased *Corprococcus*, and *Faecalibacterium*,. Moreover, it was also reported that abundance of lactic acid producing bacteria such as Lactobacillus and Bifidobacterium among individuals with schizophrenia and major depressive disorder (Ernst et al. 2017, McGuinness et al. 2022). Increased faecal lactate have been widely studied among individuals with schizophrenia (Krzyściak et al. 2023, Pruett and Meador-Woodruff 2020). Interestingly, increase in faecal lactate also associated with GI diseases such as short bowel disease and ulcerative colitis (Mayeur et al. 2013). These studies reported clear disparity or trend of certain microbiota among those afflicted with the disease, but the results, however, remain inconclusive due to other external factors such as the age, diet and lifestyle choices (Gao et al. 2019, Ma et al. 2020, Maranduba et al. 2015, Nguyen et al. 2019a, b, Sanada et al. 2020, Shen et al. 2018, Winter et al. 2018). Collectively, these studies implicate the importance and significant association of gut microbiota to mental health disorders.

In Asia, there are scarcity of data and studies investigating the gut microbiota and relationship with mental illness. Most studies are from eastern Asian countries like China as summarized in Table 7.2.

Based on Table 7.2 (see the next page), it is obvious that there is significant lack in studies identifying the relationship between gut microbiota and mental illness in southeast Asian countries. In Malaysia, our research group have worked on schizophrenia patients specifically on trans-kingdom interaction within gut microbiota specifically focusing on fungi and parasitic infection such as *Blastocystis* sp. (Franklin et al. 2024). Nonetheless, more data from this region is needed before microbiota-targeted treatment can be implemented.

The available microbiota data, whereby more than 71%, comes mainly from highly developed countries (Abdill et al. 2022). Large scale data, representative of Asia or

Table 7.2 Association of gut microbiota with mental illnesses from epidemiological studies in Asian countries

Type of mental illness	Sample size	Gut microbiota alterations in mental illness	Country	References
Anxiety	14 adults with social exclusion experience; 25 healthy control	↓ *Firmicutes/Bacteroidetes* and *Faecalibacterium* sp.	South Korea	Kim et al. (2022)
Anxiety and depression	129 patients with ulcerative colitis (UC), 49 patients with depression and anxiety (non-UC), and 62 healthy controls	↓ *Prevotella_9*, *Lachnospira* ↑ *Lactobacillales*, *Sellimonas*, and *Streptococcus*	China	Yuan et al. (2021)
Anxiety	40 patients with anxiety disorder; 36 healthy controls	↑ *Fusobacteria*, *Bacteroidetes* spp. ↓ *Firmicutes* spp., *Lachnospira*, and *Butyricicoccus*	China	Jiang et al. (2018)
Depression	24 patients with current depressive episode; 16 healthy controls	↑ *Akkermansia*, *Clostridium_sensu_stricto_1*, *UBA1819* ↓ *Dialister*, *Fusicatenibacter*, and *Lachnospira*	China	Jiang et al. (2020)
Depression	60 patients; 60 healthy controls	↑ phylum Firmicutes, genera *Prevotella* and *Klebsiella* ↓ *Bacteroidetes*	China	Lin et al. (2017)
Depression	32 patients; 32 healthy controls	↑ *Rhodospirillaceae*, *Hungatella*, *Clostridium bolteae*, *Hungatella hathewayi*, and *Clostridium propionicum* ↓ *Ruminococcus genus*, *Ruminococcus callidus*, *Desulfovibrio piger*, *Coprococcus comes*	Thailand	Maes et al. (2023)
Depression	10 patients; 10 healthy controls	↑ Firmicutes, Actinobacteria ↓ Bacteroidetes, Proteobacteria	China	Chen et al. (2018)

(continued)

Table 7.2 (continued)

Type of mental illness	Sample size	Gut microbiota alterations in mental illness	Country	References
Schizophrenia	42 patients with acute schizophrenia; 40 patients with schizophrenia in remission; 44 healthy controls	↑ *Haemophilus* ↓ *Coprococcus*	China	Zhu et al. (2021)
Schizophrenia	38 schizophrenia patients; 38 normal controls	↑ *Veillonella* ↓ *Ruminococcus* and *Roseburia*	China	Li et al. (2021)
Schizophrenia	82 schizophrenia patients; 80 normal controls	↑ *Collinsella, Lactobacillus, and Succinivibrio* ↓ *Adlercreutzia, Anaerostipes, and Ruminococcus*	China	Li et al. (2020)

a specific region in it is unavailable. This paucity emphasizes the critical need to increase global representation of gut microbiota data from this region. Over the years, there has been increasing trend in the generation of microbiota data highlighting the importance of deciphering gut microbiota data specific to Asian population. The wealth of microbiota data may act as a major driver in the effort to target gut microbiota for mental health treatment.

7.3.2 Therapeutic Strategies for Gut Microbiota

Based on evidence above, it is clear that the consequences of gut microbiota on mental health is indispensable. In the past decade, research on maintaining a healthy gut microbiota has unveiled many therapeutic strategies that were supported by experimental evidence. With significant advancements in technological tools for microbial analysis, the following are some of the key strategies or approaches to restore the gut microbiota:

(a) **Prebiotics**: Prebiotics, generally refers to the indigestible complex carbohydrates such as fibres that are usually of plant-origin. They are known to selectively stimulate the proliferation and function of beneficial gut bacteria. Studies have demonstrated that incorporating prebiotics into the diet can boost the presence of beneficial or 'good' microbes such as Bifidobacteria and Lactobacilli, resulting in a healthier gut environment (de Paulo Farias et al. 2019, Quigley 2019). This certainly will influence the body's digestion, nutrient uptake, immune function, and defense against harmful pathogens, emphasizing their importance in promoting well-being.
(b) **Probiotics**: Probiotics are live microorganisms that confer health benefits to host when consumed in adequate amounts. Certain unique taxa of live bacteria that confers mental health benefits (psychotropic) are regarded as psychobiotics (Misra and Mohanty 2019). Past studies have demonstrated their ability to modulate gut microbiota composition, thus improving gut barrier function, and alleviate symptoms of various gastrointestinal disorders (Quigley 2019), including irritable bowel syndrome (IBS) and inflammatory bowel disease (IBD). On that note, *Faecalibacterium prausnitzii* has been identified as beneficial against inflammatory bowel disease in humans (Parsaei et al. 2021).
(c) **Postbiotics:** Postbiotics are metabolic by-products produced by probiotics during fermentation. Researchers have suggested that postbiotics, such as short-chain fatty acids (SCFAs) which include acetate, propionate, and butyrate, exhibit anti-inflammatory, immunomodulatory, and metabolic effects in the host, contributing to gut homeostasis and overall health (Mutalub et al. 2022).
(d) **Fecal Microbiota Transplantation (FMT)**: FMT involves transferring faecal matter from a healthy donor to a recipient with a dysbiosis gut microbiota, to restore microbial balance and treat various gastrointestinal disorders. Clinical trials and case studies have provided persuasive evidence for the efficacy of

FMT in conditions like recurrent *Clostridioides difficile* infection (Pomares Bascuñana et al. 2021), which have been positively correlated with ulcerative colitis, and Crohn's disease (Dalal and Allegretti 2021).

(e) **Bacteriophages**: Bacteriophages are viruses that infect and replicate (kill) in only specific bacterial cells. On that notion, the therapeutic potential of bacteriophages in selectively targeting pathogenic bacteria within the gut microbiota while preserving commensal bacteria, have been shown (Emencheta et al. 2023). In other words, phage therapy is suggestive for treating bacterial infections and restoring microbial balance in the human gut.

There are mounting interest for the treatment of mental health using microbiota-targeted therapies using dietary components such as probiotics, prebiotics, and dairy products. However, there are limitation in clinical trials and outcome have been contradicting. While some have recent trial from Asian countries have showed alterations in gut microbiota which could be beneficial (Ma et al. 2021; Tian et al. 2022; Zheng et al. 2021) but there are still trials that have showed otherwise, with no significant changes upon treated (Tarutani et al. 2022). More clinical trials with a more standardized prebiotics and probiotics unique for Asian population is required. Attention should be shifted to understanding the mechanism and function of prebiotics in reversing the mental illness especially in Asian populations. These studies will add confidence for the implementation of gut microbiota-targeted treatment and facilitate in establishing guideline for administration in healthcare setting.

7.3.3 Gut Microbiota as Treatment Target for Mental Illness: Past, Future and Challenges

In the past, animal models were used to demonstrate that altering gut microbiota composition through interventions like probiotics and their effect on behaviour and mood (Agranyoni et al. 2021). Such findings sparked interest in exploring similar effects in humans. Subsequent clinical studies involving human have provided preliminary evidence suggesting that modulating gut microbiota (via dietary changes, probiotics, prebiotics, or FMT) (Liu et al. 2019; Zhao et al. 2023) could have therapeutic potential for mental health conditions such as depression, anxiety, as well as autism spectrum disorders. Based on randomised clinical trials, probiotics reduce anxiety scores. Nevertheless, it is too preliminary to conclude with such statement, as to date, number of clinical trials pertaining to gut microbiota treatment for mental related illnesses are still scanty.

Looking to the future, the field (gut microbiota-targeted treatment) holds immense promise. As researchers delve deeper into understanding the complex interactions between gut microbiota and the brain using advanced scientific platforms, they hope to uncover novel therapeutic targets for mental illness. This may involve identifying specific microbial strains or their metabolites that play key roles in regulating mood and behaviour of individuals with various types of mental disorders.

Advancements in technology, such as high-throughput sequencing and metagenomics, are allowing scientists to analyse gut microbiota with extraordinary detail (Mathieu et al. 2023). This deeper understanding of microbial communities and their functions could lead to more precise interventions tailored to individual patients. In other words, personalized treatment options could be designed for various mental illnesses.

However, there are still many challenges to overcome. One of the key hurdles is the variability of gut microbiota across individuals and the lack of standardized protocols for microbial interventions. Additionally, the mechanisms by which gut microbiota influence mental health can be complex and multi-layered as other factors namely age, diet, race, and comorbidities may add on to the complexity of the gut microbiota-brain interactions. Although Asian diet vary within its continent, it is significantly different from that of other continents. Asian staples that is rich in carbohydrate, fibre, vitamins and antioxidants but low in fat and simple sugar are basically considered beneficial against metabolic diseases and inhibits dysbiosis of gut microbiota (Benno et al. 1989; Conteh and Huang 2020). This confers difference between Asian and western gut microbiota hence a baseline microbiota data specific to the local population is key in targeting microbiota. Population microbiota data is largely unachieved simply due to the lack in research fundings and prioritization for alternative treatment options.

Moreover, the awareness and knowledge about gut microbiota and its impact on human health is still low (Bendriss et al. 2020). Cultural beliefs contributing to the notion that new and modern treatments are harmful are still prevalent among Asian (Wibowo et al. 2022, Yoon et al. 2023). This may impact the compliance and disrupt adherence to the completion of microbiota-targeted treatment regime. Limitation in mental healthcare resources is one of the major factors contributing to increasing mental illness in Asia (Brian and Ben-Zeev 2014). Consequently, practice of microbiota-targeted treatment is regarded as a costly affair in this healthcare setting specifically during its inception.

Despite these challenges, the potential of gut microbiota-targeted interventions for mental illness remains a tantalising prospect. With continued research and clinical trials, we may see the development of innovative treatments that harness the power of the gut-brain axis to improve mental health outcomes in the future.

7.4 Future Directions

In the past, sporadic studies have targeted gut microbiota treating mental illness, specifically in a few Asian countries like China (Berding and Cryan 2021; Li et al. 2023). The implementation is by far has not been clearly described and discussed, particularly for Asian population. Generally, future of treatments that target the gut microbiota for mental illness involves several key steps and considerations:

(1) Research and Development

Before any treatment, it is important to understand the gut microbial distributions in a specific mental health condition such as depression, anxiety, and psychotic disorders. This involves sophisticated microbiome analysis techniques to correlate microbial composition with clinical outcomes. This aspect also includes preclinical studies in both animal models as well as human subjects. This will allow us to identify specific microbial strains, metabolites, or other factors within the gut microbiota that have significant effects on the type of mental illnesses (Andrioaie et al. 2022).

(2) Preparation of Interventions

Based on research findings, interventions such as probiotics, prebiotics, dietary modifications, or FMT may be developed. These interventions need to be carefully formulated to ensure consistency, efficacy, and safety.

(3) Clinical Trials

Rigorous clinical trials are essential to evaluate the effectiveness of gut microbiota-targeted treatments for mental illness. These trials should adhere to established protocols, including randomized controlled trials with appropriate placebo controls, and blinding to minimize bias.

(4) Regulatory Approval

Once clinical efficacy and safety are demonstrated, regulatory approval from local health authorities like the FDA (in the United States) or the EMA (in Europe) may be sought for the gut microbiota-targeted treatment. This involves submitting comprehensive data from preclinical and clinical studies to support the treatment's safety and effectiveness.

(5) Clinical Implementation

The treatment can be made available to healthcare providers for clinical use. Guidelines and protocols need to be established to ensure appropriate patient selection, dosing, and monitoring. During this process, the patient or individual need to be educated about the treatment, its potential benefits, risks, and alternatives. Informed consent should be obtained before initiating any gut microbiota-targeted intervention.

(6) Monitoring and Follow-Up

Patients undergoing gut microbiota-targeted treatment should be monitored closely for treatment response, adverse effects, and changes in microbial composition. Follow-up appointments may be scheduled to adjust treatment as needed and to assess long-term outcomes. Gut microbiota-targeted treatments should be integrated into comprehensive treatment plans for mental illness, which may include psychotherapy, medication, lifestyle modifications, and other interventions tailored to individual patient needs.

(7) Research and Continuous Improvement

Even after implementation, ongoing research is needed to further optimize gut microbiota-targeted treatments, understand long-term effects, and identify new therapeutic targets.

(8) Exploiting technological advancements

There has been increasing interest in identifying the hidden patterns from the hypervariable and complex dataset derived from gut microbiota. Continuous advancements in technology, reduces the cost of high-throughput sequencing, thus contributes to the increased availability of sequence data. Artificial intelligence algorithms and machine learning approaches are beginning to be implemented to extract valuable information from these datasets. Deeper analysis of gut microbiota profile, gut microbial interaction and bacterial derived metabolites will allow the prediction of a disease even before its onset.

Using the wealth of cohort-based gut microbiota data, a tool that can foretell the effects of gut microbiota on health and disease of the host can developed (Patil et al. 2024). Despite challenges from confounding factors and high dimensionality, the computational prediction of a disease will likely be the future for precision medicine and personalized therapies especially in early-detection and curbing increasing mental illness occurrence in Asia.

(9) Facilities and guidelines

There needs to be a gut microbiota facility with standardized equipment across the region and well-trained personnel. This should be followed with a stringent guideline. A preliminary guideline is as in Fig 7.3 (see the next page). It ensures safety, efficacy, and ethical standards. Some of the key areas covered will be patient eligibility and assessment, screening, and selection of donor microbiota before administration protocol.

7.4.1 Policy Implication from Microbiota-Targeted Treatment

Despite experimental evidence, there is no solid clinical trials that is population-specific on gut microbiota-targeted treatment for mental illness. Assessment of clinical and economic feasibility is complicated due to the absence of long-term epidemiological data. However, there are still broad policy implication due to the potential impact on public health, regulatory practices, and healthcare access. This includes the safety, efficacy, and quality control. Since there is no accepted classification of microbiota-targeted treatment into categories like drug, supplement or biologics by globally recognized bodies, regulation is generally weak and chance for misuse is relatively high. The United States Food and Drug Administration (FDA) lamented that, in the case of FMT, faecal microbiota should be considered as a biologic product and regulated as drug. A large-scale clinical trial will be required with high standards addressing dosing, delivery methods and efficacy.

The future of microbiota-targeted treatment will require a reassessment of ethical considerations. Informed consent from patients where clarity on potential risks and long-term impacts are needed. A robust regulation in data privacy and data ownership protecting patient-derived microbiota, its storage and use must be mooted to prevent

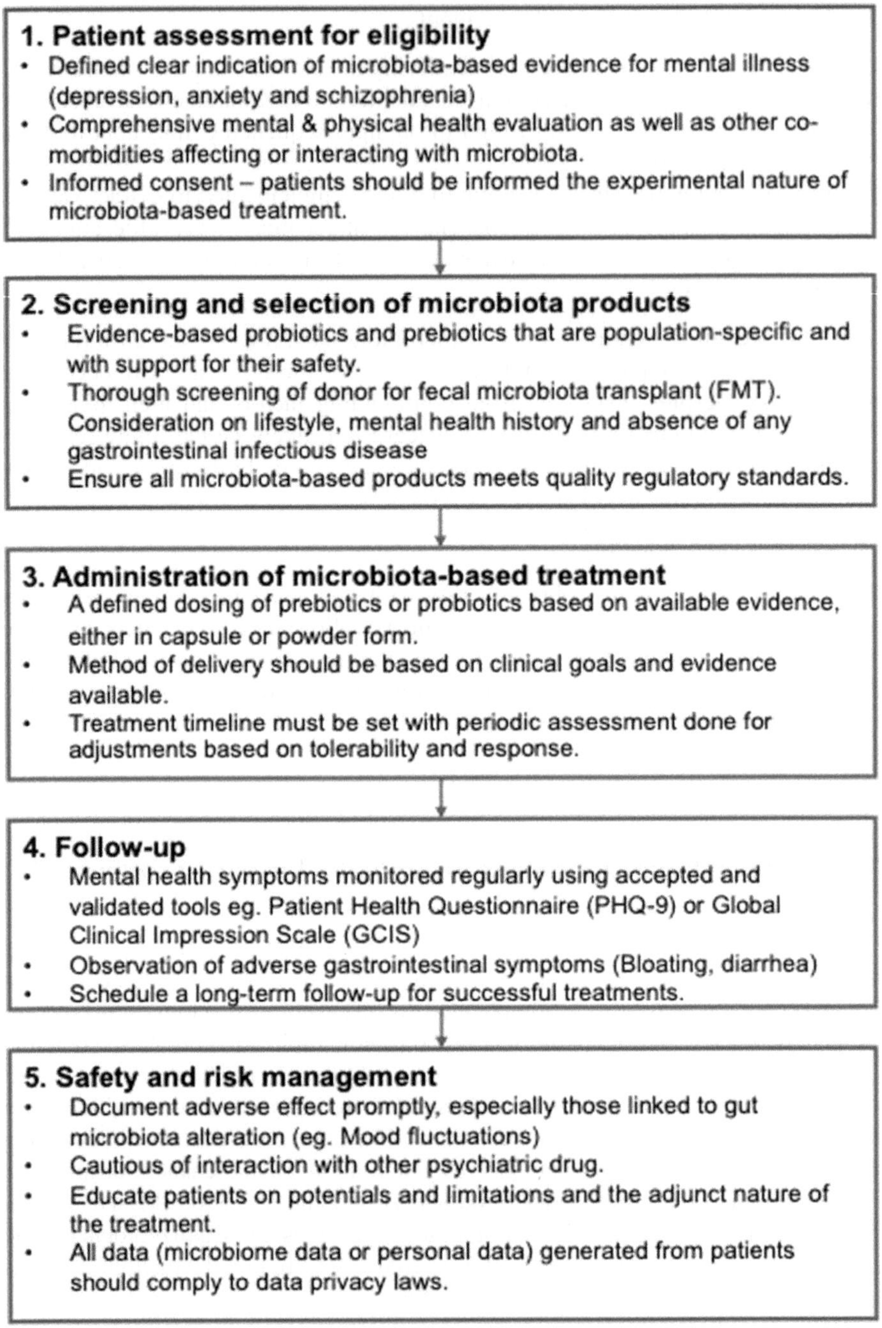

Fig. 7.3 Preliminary guidelines for microbiota-targeted treatment

misuse. Modifications in health insurance policies for coverage and reimbursement is another area of concern where coverage for FMT and other personalized microbiota-targeted therapy, which can be very costly, is essential for wider accessibility. This includes a clear billing standard for ease of reimbursements.

Lack of awareness and understanding is prevalent in Asian countries. Stakeholders from every government body should join forces in forming polices supporting research on microbiota and its dissemination and communication to the general public. Healthcare professionals should be trained on importance of microbiota and its restoration in human health. This will have a huge impact as every other treatment delivered will take into consideration of the health of microbiota.

In short, microbiota-targeted treatment implies policies addressing unique regulatory, ethical, and practical consideration. Striking a balance between innovative approach and patient safety, data privacy and equitable access will be paramount when microbiota-targeted research becomes mainstream.

7.5 Conclusion

In a nutshell, this review highlights the influence of gut microbiota on the physiology of mental health and sheds light on enormous potential of gut microbiota as a treatment target. However, adopting gut microbiota as treatment target will have to go through a thorough examination of confounders that can play a major role in impacting the gut microbiota and its associated co-morbidities. The implementation of gut microbiota-targeted treatments for mental illness, specifically in Asia, requires a multidisciplinary approach due to lack in clinical data and microbiota research in this region. Multipronged involvement from researchers, healthcare providers, regulators, and patients is critical to ensure safety, efficacy, and accessibility. With the involvement of government and other industrial stakeholders, a personalized and sustainable treatment can be implemented adjunct to existing conventional therapies to address mental health issues in this region.

Highlights

- Gut microbiota's role in mental illness should integrate advanced microbiome analysis, preclinical studies, and clinical research risks for microbiota-targeted intervention in Asia
- Incorporation of advancements of AI and machine learning will inform on the disease pattern and prediction of health
- Consideration on ethics, regulation, reliable guidelines are key for successful implementation microbiota-targeted treatments

References

Abdill RJ, Adamowicz EM, Blekhman R. Public human microbiome data are dominated by highly developed countries. PLoS Biol. 2022;20(2): e3001536.

Agranyoni O, Meninger-Mordechay S, Uzan A, Ziv O, Salmon-Divon M, Rodin D, et al. Gut microbiota determines the social behavior of mice and induces metabolic and inflammatory changes in their adipose tissue. NPJ Biofilms and Microbiomes. 2021;7(1):28.

Aizawa E, Tsuji H, Asahara T, Takahashi T, Teraishi T, Yoshida S, et al. Possible association of Bifidobacterium and Lactobacillus in the gut microbiota of patients with major depressive disorder. J Affect Disord. 2016;202:254–7.

Alibudbud R. Insights from an infodemiological study of mental health in Malaysia, the Philippines, Singapore, and Thailand from 2004 to 2021. Asian J Psychiatr. 2022;78: 103311.

Amoroso C, Perillo F, Strati F, Fantini M, Caprioli F, Facciotti FJC. The role of gut microbiota biomodulators on mucosal immunity and intestinal inflammation. 2020;9(5):1234

Andrioaie I-M, Duhaniuc A, Nastase EV, Iancu LS, Luncă C, Trofin F, et al. The role of the gut microbiome in psychiatric disorders. Microorganisms. 2022;10(12):2436.

Appleton JJIMACsJ. The gut-brain axis: influence of microbiota on mood and mental health. 2018;17(4):28.

Banerjee P, Adhikary K, Chatterjee A, Sarkar R, Bagchi D, Ghosh N, et al. Digestion and gut microbiome. Nutrition and Functional Foods in Boosting Digestion, Metabolism and Immune Health: Elsevier; 2022. p. 123–40.

Bendriss G, Al-Ali D, Shafiq A, Laswi I, Mhaimeed N, Salameh M, et al. Targeting the gut microbiome: A brief report on the awareness, practice, and readiness to engage in clinical interventions in Qatar. Qatar Med J. 2020;2020(3):47.

Benno Y, Endo K, Miyoshi H, Okuda T, Koishi H, Mitsuoka T. Effect of rice fiber on human fecal microflora. Microbiol Immunol. 1989;33(5):435–40.

Berding K, Cryan JF. Microbiota-targeted interventions for mental health. Curr Opin Psychiatry. 2021;35(1):3–9.

Bienenstock J, Kunze W, Forsythe PJNr. Microbiota and the gut–brain axis. 2015;73(suppl_1):28–31.

Bodeker G. Mental wellness in Asia. Asian Development Bank: Mandaluyong City, Metro Manila, Philippines. 2020.

Brian RM, Ben-Zeev D. Mobile health (mHealth) for mental health in Asia: objectives, strategies, and limitations. Asian J Psychiatr. 2014;10:96–100.

Chen Q, Huang S, Xu H, Peng J, Wang P, Li S, et al. The burden of mental disorders in Asian countries, 1990–2019: an analysis for the global burden of disease study 2019. Transl Psychiatry. 2024;14(1):167.

Chen Z, Li J, Gui S, Zhou C, Chen J, Yang C, et al. Comparative metaproteomics analysis shows altered fecal microbiota signatures in patients with major depressive disorder. NeuroReport. 2018;29(5):417–25.

Chinna Meyyappan A, Forth E, Wallace CJ, Milev RJBp. Effect of fecal microbiota transplant on symptoms of psychiatric disorders: a systematic review. 2020;20:1–19

Coello K, Hansen TH, Sørensen N, Munkholm K, Kessing LV, Pedersen O, et al. Gut microbiota composition in patients with newly diagnosed bipolar disorder and their unaffected first-degree relatives. Brain Behav Immun. 2019;75:112–8.

Conteh AR, Huang R. Targeting the gut microbiota by Asian and Western dietary constituents: a new avenue for diabetes. Toxicol Res. 2020;9(4):569–77.

Cuijpers P, Stringaris A, Wolpert M. Treatment outcomes for depression: Challenges and opportunities. The Lancet Psychiatry. 2020;7(11):925–7.

Dalal RS, Allegretti JR. Diagnosis and management of Clostridioides difficile infection in patients with inflammatory bowel disease. Curr Opin Gastroenterol. 2021;37(4):336–43.

de Paulo FD, de Araújo FF, Neri-Numa IA, Pastore GM. Prebiotics: Trends in food, health and technological applications. Trends Food Sci Technol. 2019;93:23–35.

Emencheta SC, Olovo CV, Eze OC, Kalu CF, Berebon DP, Onuigbo EB, et al. The role of bacteriophages in the gut microbiota: Implications for human health. Pharmaceutics. 2023;15(10):2416.

Ernst J, Hock A, Henning A, Seifritz E, Boeker H, Grimm SJMp. Increased pregenual anterior cingulate glucose and lactate concentrations in major depressive disorder. 2017;22(1):113-9

Exeter PB. Effective treatments for mental illness not being used, WHO says. BMJ. 2001;323.

Foster JA, Neufeld K-AMJTin. Gut–brain axis: How the microbiome influences anxiety and depression. 2013;36(5):305–12.

Franklin F, Rajamanikam A, Phang WK, Raju CS, Gill JS, Francis B, et al. Establishing associated risk factors, including fungal and parasitic infections among Malaysians living with schizophrenia. Scientific Reports. 2024;14(1).

Gao J-M, He Z-H, Xie Y-T, Hide G, Lai D-H, Lun Z-R. The association between Toxoplasma gondii infection and postpartum blues. J Affect Disord. 2019;250:404–9.

Gao M, Wang J, Liu P, Tu H, Zhang R, Zhang Y, et al. Gut microbiota composition in depressive disorder: a systematic review, meta-analysis, and meta-regression. 2023;13(1):379.

Halverson T, Alagiakrishnan KJAoM. Gut microbes in neurocognitive and mental health disorders. 2020;52(8):423-43

Hooper LV, Littman DR, Macpherson AJJs. Interactions between the microbiota and the immune system. 2012;336(6086):1268–73

Jang H-M, Kim J-K, Joo M-K, Shin Y-J, Lee CK, Kim H-J, et al. Transplantation of fecal microbiota from patients with inflammatory bowel disease and depression alters immune response and behavior in recipient mice. 2021;11(1):20406

Jiang H, Ling Z, Zhang Y, Mao H, Ma Z, Yin Y, et al. Altered fecal microbiota composition in patients with major depressive disorder. Brain Behav Immun. 2015;48:186–94.

Jiang H-y, Zhang X, Yu Z-h, Zhang Z, Deng M, Zhao J-H, et al. Altered gut microbiota profile in patients with generalized anxiety disorder. J Psychiatric Res. 2018;104:130–6.

Jiang Hy, Pan Ly, Zhang X, Zhang Z, Zhou Yy, Ruan B. Altered gut bacterial–fungal interkingdom networks in patients with current depressive episode. Brain Behav. 2020;10(8):e01677.

Kim C-S, Shin G-E, Cheong Y, Shin JH, Shin D-M, Chun WY. Experiencing social exclusion changes gut microbiota composition. Transl Psychiatry. 2022;12(1):254.

Kokai M, Kashiwamura S-I, Okamura H, Ohara K, Morita YJJoI. Plasma interleukin-18 levels in patients with psychiatric disorders. 2002;25:S68-S71.

Krzyściak W, Karcz P, Bystrowska B, Szwajca M, Bryll A, Śmierciak N, et al. The association of the oral microbiota with the effects of acid stress induced by an increase of brain lactate in schizophrenia patients. 2023;11(2):240

Lauber C, Rössler W. Stigma towards people with mental illness in developing countries in Asia. Int Rev Psychiatry. 2007;19(2):157–78.

Leichsenring F, Steinert C, Rabung S, Ioannidis JP. The efficacy of psychotherapies and pharmacotherapies for mental disorders in adults: an umbrella review and meta-analytic evaluation of recent meta-analyses. World Psychiatry. 2022;21(1):133–45.

Li B, Xu M, Wang Y, Feng L, Xing H, Zhang K. Gut microbiota: a new target for traditional Chinese medicine in the treatment of depression. J Ethnopharmacol. 2023;303: 116038.

Li S, Song J, Ke P, Kong L, Lei B, Zhou J, et al. The gut microbiome is associated with brain structure and function in schizophrenia. Sci Rep. 2021;11(1):9743.

Li S, Zhuo M, Huang X, Huang Y, Zhou J, Xiong D, et al. Altered gut microbiota associated with symptom severity in schizophrenia. PeerJ. 2020;8: e9574.

Lin P, Ding B, Feng C, Yin S, Zhang T, Qi X, et al. Prevotella and Klebsiella proportions in fecal microbial communities are potential characteristic parameters for patients with major depressive disorder. J Affect Disord. 2017;207:300–4.

Liu RT, Walsh RF, Sheehan AE. Prebiotics and probiotics for depression and anxiety: A systematic review and meta-analysis of controlled clinical trials. Neurosci Biobehav Rev. 2019;102:13–23.

Lloyd-Price J, Abu-Ali G, Huttenhower CJGm. The healthy human microbiome. 2016;8:1–11

Ma T, Jin H, Kwok L-Y, Sun Z, Liong M-T, Zhang H. Probiotic consumption relieved human stress and anxiety symptoms possibly via modulating the neuroactive potential of the gut microbiota. Neurobiol Stress. 2021;14: 100294.

Ma X, Asif H, Dai L, He Y, Zheng W, Wang D, et al. Alteration of the gut microbiome in first-episode drug-naïve and chronic medicated schizophrenia correlate with regional brain volumes. J Psychiatr Res. 2020;123:136–44.

Maes M, Vasupanrajit A, Jirakran K, Klomkliew P, Chanchaem P, Tunvirachaisakul C, et al. Exploration of the gut microbiome in Thai patients with major depressive disorder shows a specific bacterial profile with depletion of the ruminococcus genus as a putative biomarker. Cells. 2023;12(9):1240.

Maramis A, Van Tuan N, Minas H. Mental health in southeast Asia. The Lancet. 2011;377(9767):700–2.

Maranduba CMdC, De Castro SBR, Souza GTd, Rossato C, da Guia FC, Valente MAS, et al. Intestinal microbiota as modulators of the immune system and neuroimmune system: impact on the host health and homeostasis. 2015;2015.

Mathieu E, Léjard V, Ezzine C, Govindin P, Morat A, Giat M, et al. An insight into functional metagenomics: A high-throughput approach to decipher food–microbiota–host interactions in the human gut. Int J Mol Sci. 2023;24(24):17630.

Mayeur C, Gratadoux J-J, Bridonneau C, Chegdani F, Larroque B, Kapel N, et al. Faecal D/L lactate ratio is a metabolic signature of microbiota imbalance in patients with short bowel syndrome. 2013;8(1):e54335

McGuinness A, Davis JA, Dawson S, Loughman A, Collier F, O'hely M, et al. A systematic review of gut microbiota composition in observational studies of major depressive disorder, bipolar disorder and schizophrenia. 2022;27(4):1920–35.

Misra S, Mohanty D. Psychobiotics: a new approach for treating mental illness? Crit Rev Food Sci Nutr. 2019;59(8):1230–6.

Mutalub YB, Abdulwahab M, Mohammed A, Yahkub AM, Al-Mhanna SB, Yusof W, et al. Gut microbiota modulation as a novel therapeutic strategy in cardiometabolic diseases. Foods. 2022;11(17):2575.

Na K-S, Jung H-Y, Kim Y-KJPiN-P, Psychiatry B. The role of pro-inflammatory cytokines in the neuroinflammation and neurogenesis of schizophrenia. 2014;48:277–86

Naseribafrouei A, Hestad K, Avershina E, Sekelja M, Linløkken A, Wilson R, et al. Correlation between the human fecal microbiota and depression. Neurogastroenterol Motil. 2014;26(8):1155–62.

Nguyen TT, Kosciolek T, Maldonado Y, Daly RE, Martin AS, McDonald D, et al. Differences in gut microbiome composition between persons with chronic schizophrenia and healthy comparison subjects. J Schizophr Res. 2019a;204:23–9.

Nguyen TT, Kosciolek T, Maldonado Y, Daly RE, Martin AS, McDonald D, et al. Differences in gut microbiome composition between persons with chronic schizophrenia and healthy comparison subjects. 2019;204:23–9

O'Mahony L, McCarthy J, Kelly P, Hurley G, Luo F, Chen K, et al. Lactobacillus and bifidobacterium in irritable bowel syndrome: symptom responses and relationship to cytokine profiles. 2005;128(3):541–51

Parlindungan F, Hidayat R, Ariane A, Shatri HJCP, CP EiMH, EMH. Association between proinflammatory cytokines and anxiety and depression symptoms in rheumatoid arthritis patients: a cross-sectional study. 2023;19.

Parsaei M, Sarafraz N, Moaddab S, Leylabadlo HE. The importance of Faecalibacterium prausnitzii in human health and diseases. New Microbes New Infections. 2021;43.

Patil A, Singh N, Patwekar M, Patwekar F, Patil A, Gupta JK, et al. AI-driven insights into the microbiota: figuring out the mysterious world of the gut. Intelligent Pharmacy. 2024.

Pomares Bascuñana R, Veses V, Sheth C. Effectiveness of fecal microbiota transplant for the treatment of Clostridioides difficile diarrhea: a systematic review and meta-analysis. Lett Appl Microbiol. 2021;73(2):149–58.

Pruett BS, Meador-Woodruff JHJSr. Evidence for altered energy metabolism, increased lactate, and decreased pH in schizophrenia brain: a focused review and meta-analysis of human postmortem and magnetic resonance spectroscopy studies. 2020;223:29–42

Quigley EM. Prebiotics and probiotics in digestive health. Clin Gastroenterol Hepatol. 2019;17(2):333–44.

R Caso J, Balanza-Martinez V, Palomo T, Garcia-Bueno B. The microbiota and gut-brain axis: contributions to the immunopathogenesis of schizophrenia. J Curr Pharm Des. 2016;22(40):6122–33.

Rami FZ, Li L, Kang C, Han MA, Chung Y-C. Risk and protective factors for severe mental disorders in Asia. Neuroscience & Biobehavioral Reviews. 2024:105652.

Sanada K, Nakajima S, Kurokawa S, Barceló-Soler A, Ikuse D, Hirata A, et al. Gut microbiota and major depressive disorder: A systematic review and meta-analysis. 2020;266:1–13

Saxena S, Davidson L. The global mental health imperative and the role of the world health organization within the un 2030 agenda. The Routledge Handbook of International Development, Mental Health and Wellbeing: Routledge; 2019. p. 3–20.

Schirmer M, Smeekens SP, Vlamakis H, Jaeger M, Oosting M, Franzosa EA, et al. Linking the human gut microbiome to inflammatory cytokine production capacity. 2016;167(4):1125–36. e8.

Semahegn A, Torpey K, Manu A, Assefa N, Tesfaye G, Ankomah A. Psychotropic medication non-adherence and its associated factors among patients with major psychiatric disorders: a systematic review and meta-analysis. Systematic Reviews. 2020;9(1).

Settanni CR, Ianiro G, Bibbò S, Cammarota G, Gasbarrini AJPiN-P, Psychiatry B. Gut microbiota alteration and modulation in psychiatric disorders: Current evidence on fecal microbiota transplantation. 2021;109:110258

Shen Y, Xu J, Li Z, Huang Y, Yuan Y, Wang J, et al. Analysis of gut microbiota diversity and auxiliary diagnosis as a biomarker in patients with schizophrenia: a cross-sectional study. 2018;197:470–7

Silva YP, Bernardi A, Frozza RLJFie. The role of short-chain fatty acids from gut microbiota in gut-brain communication. 2020;11:508738

Sprouse ML, Bates NA, Felix KM, Wu HJJJI. Impact of gut microbiota on gut-distal autoimmunity: a focus on T cells. 2019;156(4):305–18

Strandwitz PJBr. Neurotransmitter modulation by the gut microbiota. 2018;1693:128–33.

Tarutani S, Omori M, Ido Y, Yano M, Komatsu T, Okamura T. Effects of 4G-beta-D-Galactosylsucrose in patients with depression: A randomized, double-blinded, placebo-controlled, parallel-group comparative study. J Psychiatr Res. 2022;148:110–20.

Tian P, Chen Y, Zhu H, Wang L, Qian X, Zou R, et al. Bifidobacterium breve CCFM1025 attenuates major depression disorder via regulating gut microbiome and tryptophan metabolism: A randomized clinical trial. Brain Behav Immun. 2022;100:233–41.

Ubeda C, Djukovic A, Isaac SJC, immunology t. Roles of the intestinal microbiota in pathogen protection. 2017;6(2):e128

Vigo D, Thornicroft G, Atun R. Estimating the true global burden of mental illness. Lancet Psychiatry. 2016;3:171–8.

Wibowo MINA, Yasin NM, Kristina SA, Prabandari YS. Exploring of determinants factors of anti-diabetic medication adherence in several regions of Asia–A systematic review. Patient preference and adherence. 2022:197–215.

Winter G, Hart RA, Charlesworth RP, Sharpley CFJRitN. Gut microbiome and depression: what we know and what we need to know. 2018;29(6):629–43.

World Health Organization. Mental health status of adolescents in South-East Asia: Evidence for action 2017.

Yoon S, Kwan YH, Yap WL, Lim ZY, Phang JK, Loo YX, et al. Factors influencing medication adherence in multi-ethnic Asian patients with chronic diseases in Singapore: a qualitative study. Front Pharmacol. 2023;14:1124297.

Yuan X, Chen B, Duan Z, Xia Z, Ding Y, Chen T, et al. Depression and anxiety in patients with active ulcerative colitis: Crosstalk of gut microbiota, metabolomics and proteomics. Gut Microbes. 2021;13(1):1987779.

Zhao Z, Xiao G, Xia J, Guo H, Yang X, Jiang Q, et al. Effectiveness of probiotic/prebiotic/synbiotic treatments on anxiety: A systematic review and meta-analysis of randomized controlled trials. J Affect Disorders. 2023.

Zheng P, Zeng B, Liu M, Chen J, Pan J, Han Y, et al. The gut microbiome from patients with schizophrenia modulates the glutamate-glutamine-GABA cycle and schizophrenia-relevant behaviors in mice. 2019;5(2):eaau8317.

Zheng P, Zeng B, Zhou C, Liu M, Fang Z, Xu X, et al. Gut microbiome remodeling induces depressive-like behaviors through a pathway mediated by the host's metabolism. 2016;21(6):786–96

Zheng Y, Yu Z, Zhang W, Sun T. Lactobacillus rhamnosus Probio-M9 improves the quality of life in stressed adults by gut microbiota. Foods. 2021;10(10):2384.

Zhu C, Zheng M, Ali U, Xia Q, Wang Z, Chenlong, et al. Association between abundance of Haemophilus in the gut microbiota and negative symptoms of schizophrenia. Frontiers in Psychiatry. 2021;12:685910.

Arutchelvan Rajamanikam is currently a senior lecturer and the principal investigator for the Microbiome Laboratory. His work primarily involves research on gut microbiota from healthy and disease with the association and interaction of intestinal protozoan parasite. He currently focuses on exploring gut microbiota profiles from schizophrenia and colorectal cancer patients as a predictive tool for preventive medicine. He also leads the Emerging Pathogen Laboratory 1, which explores various emerging parasites.

Chandramathi Samudi is serving as an Associate Professor working specifically on gut bacteria and its immunological implications. She also works with various mucosal immunity in the infection of gut bacteria and host-parasite interaction. Her research interest focuses on exploring natural products as antibiotic potentiator against multi-resistant bacterial strains.

Freddy Franklin is a postgraduate at the Faculty of Medicine, Universiti Malaya. His research focuses specifically on schizophrenia patients and associated parasitic and fungal infection. He also studies the immunophenotyping and gut microbiota of schizophrenia patients compared healthy control. He is currently interested on developing animal model to further study the gut-brain-axis in schizophrenia.

Suresh Kumar Govind is an honorary professor working closely on gut parasite and mental illness. He has worked extensively on gut parasites, namely Blastocystis sp., its pathobiology and interaction with bacteria. He has a strong mental health advocate and works closely with various organization on improving mental health conditions among local populations. He currently does various consultation works for research and development on gut microbiota-mental health research.

Chapter 8
Self-management Capacity of Persons Who Use Drugs (PWUDs) with Multimorbidity: Insights from Asian Healthcare Perspectives and Future Trends

Clusters of Future Studies: Post-normality and Complexity; Humanity at the Limen; Environmental Futures

Joseph Kristian B. Reyes, Neriza V. Bustillo, Danica P. Ydia, and Justine Bernard C. Fiestada

Abstract This study investigates the self-management capabilities of Persons Who Use Drugs (PWUDs) with multimorbidity in the Philippines, focusing on the Department of Health-Drug Treatment and Rehabilitation Center in Dagupan City, a 300-bed facility. It aims to enhance understanding of effective healthcare support strategies in this complex context. Specific objectives include exploring multimorbidity prevalence, evaluating self-management practices across seven domains, and exploring cultural factors and self-management practices among PWUDs. The study utilised a Retrospective Chart Review (RCR) to analyse existing patient data from the Department of Health-Drug Treatment and Rehabilitation Center in Dagupan, focusing on male PWUDs aged 18–59 with multimorbidity. Data collection included chart reviews and interviews with clinical professionals to establish study protocols. Descriptive statistics and generalized linear models were applied to analyse self-management across seven domains, offering insights into chronic care strategies for PWUDs. Descriptive statistics provide insights into self-management capacity across diverse healthcare settings, highlighting multimorbidity prevalence and proactive self-management strategies. The study concludes with recommendations for integrated interventions, collaborative approaches, and targeted strategies for PWUDs

J. K. B. Reyes (✉) · D. P. Ydia · J. B. C. Fiestada
DOH-Drug Treatment and Rehabilitation Center, Dagupan City, Philippines
e-mail: jkbreyes@gmail.com

N. V. Bustillo
University of Pangasinan, Dagupan City, Philippines
e-mail: nvbustillo.up@phinmaed.com

R. Xiang Ng et al. (eds.), *Future of Healthcare in Asia*,
https://doi.org/10.1007/978-3-032-03183-9_8

facing multimorbidity challenges, aligning with future healthcare trends and offering insights from an Asian country perspective.

Keywords Self-management · Multimorbidity · Substance use · Retrospective chart review (RCR) · Healthcare perspective · Healthcare

8.1 Introduction

Substance use disorders (SUDs) and multimorbidity present complex challenges in healthcare systems globally, particularly within Asian contexts. Individuals with SUDs often face compounded health issues, including the presence of multiple chronic conditions, which can significantly impact their overall well-being and treatment outcomes (Bahji 2024). Understanding the self-management capabilities of individuals with SUDs and multimorbidity is crucial for developing effective healthcare support strategies tailored to their needs. Recent studies highlight the necessity of integrated care approaches for managing these intertwined health conditions, emphasising the importance of self-management in improving health outcomes and quality of life (Volkow et al. 2021; Bhalla et al. 2022).

Recent research supports the study's focus on the interconnectedness of SUDs and multimorbidity, reinforcing the need for integrated care models. Boucher et al. (2022) highlight both the challenges and benefits of self-management strategies in managing chronic conditions among individuals with SUDs. Additionally, Franklin et al. (2023) emphasise that self-management is not merely an isolated individual endeavour but a social practice. This perspective underscores the importance of considering patient-professional interactions and embodied knowledge (habitus) when setting self-management goals. By recognising the social context, healthcare professionals can better support patients with SUDs in managing their conditions.

Furthermore, research on older adults by Abbasinia et al. (2020) highlights the benefits of self-management. By empowering older individuals with SUDs to actively manage their health, self-management programmes can enhance their quality of life and reduce reliance on the healthcare system.

This study explores the self-management capacity of individuals grappling with substance use and multimorbidity within the Drug Treatment and Rehabilitation Centre in the Philippines, offering valuable insights into Asian healthcare perspectives and future trends. A meticulous investigation conducted at the Department of Health-Drug Treatment and Rehabilitation Centre in Dagupan City aims to shed light on multimorbidity prevalence and self-management practices across seven domains.

Utilising a Retrospective Chart Review (RCR) methodology with existing patient-centred data, this research underscores the cost-effectiveness and efficiency of such an approach, particularly relevant for conditions characterised by prolonged latency periods. Focusing on persons with SUDs within a 300-bed facility, the study employs stringent inclusion criteria, ensuring a targeted analysis of individuals aged 18–59, admitted between January and December 2022, and diagnosed with multimorbidity.

Ethical considerations remain paramount, with approvals obtained and unique code assignments ensuring anonymity for all participants. Data collection instruments, including a standardised chart review instrument validated through initial interviews and chart analysis, are meticulously employed to gather comprehensive insights into self-management capacity.

Through descriptive statistics, this study provides a nuanced understanding of self-management capacity across diverse healthcare settings, elucidating demographic trends, multimorbidity prevalence, and proactive self-management strategies employed by individuals with SUDs and multimorbidity. The findings of this research are crucial in informing targeted interventions and collaborative approaches and hold significant implications for shaping future healthcare trends, particularly within the Asian context.

8.2 Literature Review

Substance use disorders (SUDs) are a significant public health concern globally, characterised by the harmful or hazardous use of psychoactive substances, including alcohol and illicit drugs. Multimorbidity, defined as the coexistence of two or more chronic conditions, is increasingly recognised among individuals with SUDs. This phenomenon complicates treatment and management, necessitating a comprehensive understanding of the interplay between different types of substance use and associated health conditions.

Multimorbidity is highly prevalent among older adults, with rates ranging from 30 to 87% in older populations (Roomaney et al. 2022). It is also increasingly common in low- and middle-income countries, with a pooled prevalence of 36.4% (Asogwa et al. 2022). The prevalence of multimorbidity increases with age and is higher among females, urban dwellers, and those who are well-off (Roomaney et al. 2022). Hypertension, diabetes, and cardiovascular diseases are frequently observed in multimorbidity clusters (Roomaney et al. 2022; Asogwa et al. 2022). The risk of multimorbidity is higher in those with unhealthy lifestyles and socioeconomic disadvantage (Paukkonen et al. 2022). Chronic conditions such as cardiovascular diseases, chronic respiratory diseases, cancer, and diabetes contribute to seven out of ten deaths in the Philippines (World Health Organization 2019a).

Among persons who use drugs (PWUDs) in the Philippines, several significant co-morbidities have been identified. Notably, mental health disorders, such as severe mental illness, frequently co-occur with substance use disorders. These mental health conditions, including schizophrenia and bipolar disorder, often lead to other physical health issues like cardiovascular disease and metabolic syndrome, exacerbating the overall health burden on individuals (Fiorillo and Sartorius 2021).

HIV/AIDS is another prevalent co-morbidity among PWUD, particularly those who inject drugs. The integration of HIV services with other health services has been highlighted as a critical strategy to improve health outcomes for this population. This integration aims to address the high incidence of co-morbidities related to HIV, such as tuberculosis and other opportunistic infections, which are common among PWUD (Bulstra et al. 2021a, b).

Additionally, infectious diseases like hepatitis C are prevalent among PWUD due to practices such as needle sharing. The co-occurrence of these infectious diseases with SUDs presents significant challenges for healthcare delivery and requires comprehensive, integrated treatment approaches to be effectively managed (Zerfu et al. 2023).

The Philippines has a notably high prevalence of drug use, particularly involving methamphetamine (shabu). Studies indicate that methamphetamine remains the most commonly used illicit drug in the country, with reports from the United Nations Office on Drugs and Crime (UNODC) and the Philippine Dangerous Drugs Board (DDB) highlighting its widespread availability and use (UNODC 2022; DDB 2021).

The government's approach to drug control has been heavily focused on punitive law enforcement strategies, most notably through President Rodrigo Duterte's "War on Drugs," launched in 2016. This campaign has drawn international criticism due to its alleged human rights violations, including extrajudicial killings of suspected drug offenders. Reports from Human Rights Watch (2021), Amnesty International (2020), and the United Nations Human Rights Council (UNHRC 2020) document widespread allegations of summary executions and abuses committed under the anti-drug campaign.

The heavy emphasis on law enforcement and punitive measures has largely overshadowed the need for comprehensive healthcare services for PWUD. Public health experts argue that criminalisation and stigmatisation discourage individuals from seeking treatment, exacerbating existing health inequities (Bulstra et al. 2021a, b). There remains a significant gap in access to integrated health services, particularly in addressing co-occurring mental health disorders, infectious diseases such as HIV/AIDS and tuberculosis, and harm reduction interventions (Global Commission on Drug Policy 2022).

Despite these challenges, the Philippines has taken a step towards improving healthcare access with the passage of the Universal Health Care (UHC) Act. This legislation aims to expand equitable access to quality and affordable healthcare services, aligning with the World Health Organization's (WHO) self-care framework—which emphasises health promotion, disease prevention, and the ability to manage one's own health with or without healthcare provider support (Jaminola 2023; WHO 2019b).

Meanwhile, self-management, which involves managing symptoms, treatment, consequences, and lifestyle changes, is crucial for people with multimorbidity. However, it can be very challenging and burdensome for those with multiple conditions, requiring coordination of care between providers. Patient activation, or having the knowledge, skills, and confidence to manage one's health, is key to effective self-management. Studies show that more activated patients are more likely to engage in positive health behaviours such as exercise, a healthy diet, medication adherence, and condition monitoring (Yang et al. 2021).

Self-management support interventions, tailored to individual patient needs, including education, cognitive-behavioural change, skills training, and goal-setting, can help empower patients to self-manage more effectively. However, most current programmes do not specifically target medication self-management capacity in those with multimorbidity (Yang et al. 2021).

Patient activation is essential for effective self-management. Studies show that patients who are more activated are more likely to engage in health-promoting behaviours, such as regular physical activity, a nutritious diet, adherence to medication regimens, and self-monitoring of conditions (Paukkonen et al. 2022).

Qualitative research on patient self-management demonstrates that it comprises different domains, including skills, abilities, and attitudes that patients use to address the challenges of living with a chronic condition (Riegel et al. 2021). A review of related studies by Wong et al. (2021) reveals gaps that can be addressed through further research. There remains a lack of sufficient studies on the coexistence of SUDs and chronic diseases (multimorbidity). Healthcare professionals face obstacles in engaging with self-management programmes, primarily due to uncertainty about their benefits and limited local evidence concerning their impact on patients' self-management abilities. Standardised referral processes across health and community settings are needed to improve coordination and accessibility to relevant programmes.

8.3 Methodology

The study employed a Retrospective Chart Review (RCR) approach to analyse pre-existing patient data for clinical research. This method served several key purposes, including auditing historical data, testing hypotheses on risk factors and outcomes, determining sample sizes for future studies, and investigating rare events impractical for prospective studies. RCR was chosen for its cost-effectiveness, access to existing data, and ability to generate hypotheses for prospective testing, facilitating a deeper understanding of the research subject.

Conducted at the Department of Health-Drug Treatment and Rehabilitation Centre in Dagupan, a 300-bed facility in Northern Luzon, the study focused on male individuals dealing with substance abuse from Northern Luzon and neighbouring provinces. Participants were recruited by reviewing medical records of PWUDs admitted to the residential programme between 2014 and 2022, aged 18–59 years, and diagnosed with multimorbidity. Exclusion criteria included incomplete programme participation or medical records.

Data collection involved initial interviews with clinical professionals to establish charting protocols, develop standardised documentation, and define study variables based on sample chart reviews. A checklist was created to classify participants based on sociodemographic factors and disease status. Ethical standards were maintained through approvals, data anonymisation, and secure storage protocols.

Descriptive statistics were used to analyse PWUD profiles with chronic conditions, focusing on self-management across seven domains. Generalised linear models accounted for repeated measures, adjusting for appointment types and specific self-management domains. This approach enabled a comprehensive analysis of self-management capacities among PWUDs with multimorbidity, informing clinical practices and future research directions.

8.4 Findings

8.4.1 Profile of PWUDs (Persons Who Use Drugs) with Chronic Conditions Based on Their Socio-Demographics

Based on the empirical findings derived from a sample of 216 participants, several key demographics and behavioural trends emerged among individuals with substance use disorders and multimorbidity. Table 8.1 (see the next page) summarises their characteristics about age, marital status, educational attainment, employment status, financial capability, lifestyle, and substance abuse.

A financially capacitated, *C1* patient whose monthly income per capita is above 180% but not 220%, *C2* patient whose monthly income per capita is above 140% but not 180%, *C3* patient whose monthly income per capita is below 140% of the latest PCPT for the region, *D* indigent/poor.

In terms of age distribution, most participants fell within the 18–29 age group (29.17%) and 30–39 age group (44.44%), indicating that a significant proportion of younger adults are grappling with these complex health issues. Marital status revealed a predominance of single individuals (65.28%), followed by those who were married (24.54%), with smaller proportions being widowed (2.78%), separated (3.24%), or in a live-in relationship (4.17%).

Educational attainment varied among participants, with a considerable proportion having attained a high school education (65.77% for graduates, 50.48% for undergraduates), followed by those with college qualifications (9.91% for graduates, 28.57% for undergraduates) or vocational qualifications (9.01% for graduates, 2.86% for undergraduates). Employment status showed that nearly half of the participants were employed (47.69%), while significant proportions were either unemployed (26.39%) or self-employed (25.93%).

Regarding financial capacity, most participants fell within the C3 category (88.89%), including semi-skilled and unskilled manual workers, as well as those who are unemployed. Smaller proportions were in categories A (higher managerial, administrative, and professional occupations), C1 (supervisory, clerical, and junior managerial, administrative, or professional occupations), C2 (skilled manual workers), and D (lowest income group, often encompassing semi-skilled and unskilled workers). Lifestyle factors indicated a high prevalence of smoking habits (70.83%) among participants. In terms of substance abuse, methamphetamine

Table 8.1 Profile of PWUDs (persons who use drugs) with chronic conditions based on their socio-demographics

Particulars	Frequency (N = 216)	Percentage
Age group		
18–29 y/o	63	29.17
30–39 y/o	96	44.44
40–49 y/o	42	19.44
50–59 y/o	13	6.02
60 and above	2	0.93
Marital status		
Single	141	65.28
Married	53	24.54
Widowed	6	2.78
Separated	7	3.24
Live-in	9	4.17
Educational attainment		
Graduate		
Elementary	17	15.32
Highschool	73	65.77
College	11	9.91
Vocational	10	9.01
Post-graduate	0	0.00
Under-graduate		
Elementary	19	18.10
Highschool	53	50.48
College	30	28.57
Vocational	3	2.86
Post-graduate	0	0.00
Employment status		
Employed	103	47.69
Unemployed	57	26.39
Self employed	56	25.93
Financial capacity		
A	3	1.39
C1	8	3.70
C2	5	2.31
C3	192	88.89
D	8	3.70
Lifestyle		
Smoker	153	70.83
Non-smoker	63	29.17

(continued)

Table 8.1 (continued)

Particulars	Frequency (N = 216)	Percentage
Substance of abuse		
Methamphetamine	174	80.56
Cannabis	27	12.50
Poly-users	14	6.48
Ethanol	1	0.46

(shabu) emerged as the primary substance of abuse (80.56%), followed by cannabis (12.50%), poly-drug use (6.48%), and ethanol (0.46%).

8.4.2 *Prevalence of Multimorbidity Among PWUDs and the Five Chronic Disease Groups*

The prevalence of multimorbidity among Persons Who Use Drugs (PWUDs) highlights the complex health challenges faced by this population. Table 8.2 provides a detailed breakdown of the frequency and prevalence of five chronic disease groups commonly observed among PWUDs, emphasizing the high occurrence of acute respiratory disease, skin disease, and musculoskeletal disorders.

Among the participants, food allergies were observed in 17.59% of cases, indicating a notable proportion affected by this allergic condition. Diabetes was diagnosed in 11.11% of individuals, highlighting the coexistence of metabolic disorders within this population. Acute respiratory diseases were highly prevalent, affecting 94.91% of participants, underscoring the susceptibility of individuals with substance

Table 8.2 Prevalence of multimorbidity among PWUDs (persons who use drugs) and the five chronic disease groups

	Frequency	Prevalence
Food allergies	38	17.59%
without food allergies	178	82.41%
Diabetes	24	11.11%
without diabetes	192	88.89%
Acute respiratory disease	205	94.91%
Without acute respiratory disease	11	5.09%
Skin disease	203	93.98%
without skin disease	11	6.02
Cardiovascular disease	35	16.02%
without cardiovascular disease	181	83.80%
Musculo skeletal disease	124	57.41%
Without musculo skeletal disease	92	42.59%

use disorders to respiratory health issues. Similarly, skin diseases were prevalent in 93.98% of cases, suggesting a high burden of dermatological conditions among this population.

Furthermore, cardiovascular diseases were identified in 16.02% of individuals, emphasising the increased risk of cardiovascular complications associated with substance use disorders and multimorbidity. Musculoskeletal diseases also exhibited a considerable prevalence, affecting 57.41% of participants, indicating the significant impact of these conditions on physical health and mobility. The prevalence rates of these co-occurring health conditions highlight the complex and multifaceted nature of multimorbidity among individuals with substance use disorders, necessitating comprehensive and integrated healthcare approaches to address their diverse medical needs.

8.4.3 Self-management Capacity of PWUDs with Multimorbidity Based on Seven Domains

The self-management capacity of Persons Who Use Drugs (PWUDs) with multimorbidity is critical to their overall health outcomes, particularly when addressing multiple chronic conditions. Table 8.3 outlines their engagement across seven self-management domains, including medical, emotional, behavioral, and resource management, providing insight into the extent of their health-related activities and support systems.

Table 8.3 Self-management capacity of PWUDs (persons who use drugs) with multimorbidity based on the seven domains

Self-management domain	Study findings	Average number of activities/engagements
Medical management	Regular healthcare appointments and diagnostic tests	7 appointments 10 diagnostic tests
Role management	Adapting daily routines to incorporate health practices	2 routine exercise/diet compliance activities
Emotional management	Activities to manage anxiety, depression, and stress	3 mental health management activities
Behavioral management	Lifestyle modifications for health improvement	2 routine exercise/diet compliance activities
Resource management	Use of nutrition assessments and counseling	3 nutrition assessments/ counseling sessions
Social interaction	Family visits to support health management	4 family visits
Information seeking	Engaging in health-related programs and self-reflection	4 engagements in health programs 3 self-awareness activities

Medical Management. The study indicates a high level of engagement with healthcare services, with participants averaging seven appointments and ten diagnostic tests, highlighting the importance of regular medical check-ups and adherence to treatment regimens.

Role Management. Participants reported integrating routine exercise or diet compliance activities into their daily lives, averaging two such activities, demonstrating the adaptation of daily roles to include health practices.

Emotional Management. The findings emphasise the importance of addressing mental health, with participants engaging in an average of three activities aimed at managing anxiety, depression, and stress.

Behavioural Management. Similar to role management, the focus on lifestyle modifications, such as exercise and diet, is crucial for overall health, with an average of two activities reported.

Resource Management. Nutrition assessments and counselling sessions (averaging three per participant) are essential for managing dietary needs and ensuring nutritional adequacy.

Social Interaction. Family visits play a pivotal role in the self-management framework, with an average of four visits per participant, indicating the positive impact of social support on health outcomes.

Information Seeking. Participants engaged in an average of four activities aimed at promoting participation in health-related programmes and three self-awareness activities, underscoring the significance of education and self-reflection in health management.

8.4.4 Cultural Factors and Self-management Practices Among PWUDs (Persons Who Use Drugs)

Cultural factors play a significant role in shaping the self-management practices of Persons Who Use Drugs (PWUDs), influencing how they navigate recovery and chronic illness management. Table 8.4 (see the next page) explores the impact of family structure, religious beliefs, stigma, and collectivist values on self-management behaviors, supported by statistical evidence demonstrating strong associations and variations across cultural dimensions.

Family Structure. The statistically significant Pearson correlation coefficient ($r = 0.68$, $p < 0.001$) suggests a strong positive relationship between family support and self-management practices among PWUDs, highlighting the crucial role of family networks in addiction recovery efforts.

Table 8.4 Cultural factors and self-management practices among PWUDs (persons who use drugs)

Cultural factor	Self-management practice	Statistical computation	Interpretation
Family structure	Greater reliance on family support networks for recovery	Pearson correlation coefficient ($r = 0.68$, $p < 0.001$)	Highlights the strong positive relationship between family support and self-management practices among PWUDs
Religious beliefs	Incorporation of spiritual practices in addiction recovery	Chi-square test ($\chi^2 = 15.21$, $df = 1$, $p < 0.001$)	Indicates a significant association between religious beliefs and adoption of spiritual practices in self-management strategies
Stigma and shame	Concealment of drug use due to fear of social stigma	Logistic regression ($OR = 2.34$, 95% CI [1.68–3.25], $p < 0.01$)	Shows higher odds of concealing drug use among PWUDs experiencing stigma, impacting self-management behaviors
Collectivist values	Preference for group-based therapy and communal support systems	ANOVA ($F = 12.78$, $df = 2$, $p < 0.001$)	Reflects significant differences in self-management practices based on the degree of adherence to collectivist values

OR odds ratio, *CI* confidence interval, *ANOVA analysis of variance, F* f-statistic, *df degrees of freedom, p* p-value.

Religious Beliefs. The significant Chi-square test result ($\chi^2 = 15.21$, $df = 1$, $p < 0.001$) indicates that religious beliefs are associated with the adoption of spiritual practices in self-management strategies among PWUDs, emphasising the influence of religious factors in treatment approaches..

Stigma and Shame. The logistic regression analysis ($OR = 2.34$, 95% CI [1.68–3.25], $p < 0.01$) reveals higher odds of concealing drug use among PWUDs experiencing stigma, underscoring the negative impact of stigma on self-management behaviours and treatment outcomes.

Collectivist Values. The ANOVA result ($F = 12.78$, $df = 2$, $p < 0.001$) demonstrates significant differences in self-management practices based on adherence to collectivist values, highlighting the preference for group-based interventions and communal support systems among PWUDs in the Philippines.

8.5 Discussion of Implications

8.5.1 Profile of PWUDs (Persons Who Use Drugs) with Chronic Conditions Based on Their Socio-Demographics

The findings regarding demographic profiles, socio-economic statuses, and behavioural patterns of individuals with substance use disorders and multimorbidity reveal several important implications for healthcare practice and intervention strategies. Firstly, the age distribution of participants, with a significant portion falling within the younger age groups of 18–29 years and 30–39 years, suggests that these complex health issues are prevalent among younger adults. This highlights the need for early intervention and prevention efforts targeting this demographic to mitigate the long-term health impacts associated with substance use and multimorbidity.

Moreover, the predominance of single individuals, along with high rates of unemployment and lower educational attainment, underscores the socio-economic vulnerabilities faced by this population. Addressing social determinants of health, such as access to education, stable employment, and social support systems, is crucial in tackling the underlying factors contributing to substance use disorders and multimorbidity.

Additionally, the high prevalence of smoking habits and substance abuse, particularly methamphetamine use, underscores the interconnectedness of lifestyle factors and health behaviours in shaping overall health outcomes. Comprehensive interventions addressing both substances use disorders and associated lifestyle factors, such as smoking cessation programmes and harm reduction initiatives, are essential in promoting health and well-being among individuals with multimorbidity.

Furthermore, the distribution of participants across different financial categories highlights disparities in socio-economic status within this population. Tailoring healthcare interventions to accommodate varying levels of financial resources and access to healthcare services is imperative in ensuring equitable healthcare delivery and reducing health disparities among individuals with substance use disorders and multimorbidity.

8.5.2 Prevalence of Multimorbidity Among PWUDs (Persons Who Use Drugs) and the Five Chronic Disease Groups

The findings regarding the prevalence of co-occurring health conditions among individuals with substance use disorders and multimorbidity carry significant implications for healthcare practice and policy development. The high prevalence rates of food allergies, diabetes, acute respiratory diseases, skin diseases, cardiovascular diseases, and musculoskeletal diseases underscore the complexity of health issues faced by this population. These findings emphasise the importance of comprehensive screening and assessment protocols within substance use disorder treatment settings to facilitate early detection and intervention.

Healthcare providers must be vigilant in identifying and addressing co-occurring health conditions to ensure holistic and integrated care delivery. Additionally, the prevalence of respiratory and dermatological conditions highlights the need for strategies to enhance respiratory and skin health among individuals with substance use disorders, including addressing vulnerabilities related to substance abuse and promoting preventive measures. Furthermore, the high prevalence of musculoskeletal diseases underscores the importance of multidisciplinary approaches to effectively address chronic pain and mobility issues. Collaborative efforts between addiction treatment providers, primary care physicians, and specialists are essential to comprehensively address the diverse medical needs of this population.

8.5.3 Self-management Capacity of PWUDs (Persons Who Use Drugs) with Multimorbidity Based on Seven Domains

The findings from the study highlight a comprehensive approach to health management across multiple dimensions. Participants demonstrated robust engagement with healthcare services, averaging numerous appointments and diagnostic tests. This underscores the importance of regular medical check-ups and adherence to treatment plans in maintaining optimal physical health. Moreover, the integration of health-related activities into daily routines, such as exercise and dietary compliance, reflects a proactive approach to role management. By prioritising these behaviours, individuals not only fulfil their daily responsibilities but also promote long-term health and well-being.

Addressing mental health emerged as another critical aspect, with participants actively engaging in activities aimed at managing anxiety, depression, and stress. This focus on emotional management suggests a recognition of the interconnectedness between mental and physical health, contributing to overall resilience and improved coping strategies. Similarly, the emphasis on lifestyle modifications underscores the significance of behavioural management in health outcomes. By incorporating regular exercise and nutritious diets into their lifestyles, participants promote disease prevention and better overall health.

Resource management, particularly through nutrition assessments and counselling sessions, highlights a proactive approach to addressing dietary needs and ensuring nutritional adequacy. This aspect supports optimal physical health by addressing specific nutritional requirements and promoting informed dietary choices. Social interaction, particularly through frequent family visits, plays a pivotal role in providing emotional support and fostering a sense of community, which is crucial for maintaining motivation and adherence to health goals.

Furthermore, the active engagement in seeking health-related information and participating in self-awareness activities underscores the importance of education and empowerment in health management. By actively seeking knowledge and understanding their health needs, individuals can make informed decisions and adopt effective self-care practices.

8.5.4 Cultural Factors and Self-management Practices Among PWUDs (Persons Who Use Drugs)

The findings underscore the importance of tailoring interventions to leverage strong familial support networks and integrate spiritual practices, which are deeply rooted in Filipino culture. Addressing stigma associated with drug use emerges as a critical priority, as concealment due to social stigma can hinder treatment adherence and access to healthcare services. Healthcare providers need to cultivate cultural competence to deliver effective, respectful care that aligns with collectivist values, emphasising group-based therapy and communal support systems preferred by PWUDs.

Moreover, policymakers and programme developers should integrate these cultural insights into addiction treatment policies, fostering community partnerships and family involvement to support holistic recovery approaches. Further research is essential to deepen understanding of how cultural factors influence self-management practices, guiding education and healthcare practices towards more equitable and effective treatment outcomes for PWUDs in the Philippines.

8.6 Future Directions

Longitudinal studies could be pursued to assess the long-term efficacy and sustainability of integrated care models identified in this research. By tracking health outcomes over time, researchers can better understand the durability of improvements in health status and healthcare utilisation patterns among PWUDs managing multiple chronic conditions.

There is a need to explore the scalability of effective self-care strategies identified in the study. This involves adapting and implementing these strategies across diverse healthcare settings and cultural contexts within Asia, considering regional variations in healthcare infrastructure, socio-economic factors, and access to resources.

Additionally, future research could delve deeper into the integration of technology in self-management support for PWUDs. This includes exploring advancements in digital health tools, such as artificial intelligence, wearable devices, and mobile health applications, tailored to the unique needs of this population.

Furthermore, ongoing education and training for healthcare providers represents a critical area for future study. Enhancing provider competencies in managing substance use disorders alongside complex chronic conditions can lead to more effective patient-provider interactions, reduced stigma, and improved treatment outcomes.

Lastly, considering the evolving landscape of healthcare policies and funding mechanisms, future studies could examine policy implications and economic evaluations of integrated care models. Assessing the cost-effectiveness and return on investment of these models will be essential for advocating for sustainable funding strategies and policy reforms that support holistic care approaches for PWUDs with multimorbidity.

8.7 Conclusion

The demographic profile, socio-economic status, and behavioural patterns highlight significant vulnerabilities among younger adults, particularly those aged 18–39 years, emphasising the urgency for early intervention and prevention efforts. Addressing social determinants of health, such as unemployment and lower educational attainment, is crucial to mitigating substance use disorders and multimorbidity risks among this population.

The study underscores the interconnectedness of lifestyle factors and health behaviours, with high prevalence rates of smoking, substance abuse (particularly methamphetamine use), and multiple chronic conditions such as food allergies, diabetes, respiratory diseases, skin diseases, cardiovascular diseases, and musculoskeletal issues. These findings highlight the complexity of health issues faced by PWUDs with multimorbidity, necessitating comprehensive screening, early detection, and integrated care approaches within substance use treatment settings.

Furthermore, the study emphasises the robust self-management capacities observed across various domains among participants. Engagement with healthcare services, adherence to treatment regimens, and integration of health-promoting activities into daily routines such as exercise and dietary compliance demonstrate proactive role and behavioural management. Active participation in managing mental health, lifestyle modifications, resource management through nutrition counselling, social interactions, and continuous information seeking indicate a comprehensive approach to health management among PWUDs with multimorbidity.

Highlights

- Key insights emphasize the need for early interventions for younger PWUDs, integrated care for multimorbidity, and addressing socioeconomic and cultural factors.

- Promoting family support, reducing stigma, and leveraging collectivist values can enhance self-management.
- Holistic, culturally sensitive healthcare approaches are crucial to improving outcomes for PWUDs with multimorbidity in Asia.

References

Abbasinia M, Ahmadi F, Kazemnejad A. Patient advocacy in nursing: a concept analysis. Nurs Ethics. 2020;27(1):141–51 https://doi.org/10.1177/0969733019832950.

Amit AML, Pepito VCF, Dayrit MM. Advancing universal health coverage in the Philippines through self-care interventions. The Lancet Regional Health—Western Pacific. 2022;26:100579.

Amnesty International. "They just kill us": End impunity for extrajudicial executions in the Philippines. 2020. Available from: https://www.amnesty.org/en/documents/asa35/3085/2020/en/.

Asogwa OA, Boateng D, Marzà-Florensa A, Peters S. Multimorbidity of non-communicable diseases in low-income and middle-income countries: a systematic review and meta-analysis. BMJ Open. 2022;12(1):e049133. Available from: https://bmjopen.bmj.com/content/12/1/e049133.

Bahji A. Navigating the complex intersection of substance use and psychiatric disorders: a comprehensive review. J Clin Med. 2024;13(4):999.

Barbara R. Dunbar S, Fitzsimons D, Freedland K, Lee C, Middleton S, Stromberg A, Vellone E, Webber D, Jaarsma T. Self-care research: where are we now? Where are we going?, Int J Nurs Stud. 2021;116. ISSN 0020-7489. https://doi.org/10.1016/j.ijnurstu.2019.103402.

Bautista JC, Pascual ML, Sison MA. Self-management capacity of persons who use drugs (PWUDs) with multimorbidity: a cross-sectional study in the Philippines. Philipp J Public Health. 2019;49(1):1–10.

Bhalla IP, Stefanovics EA, Rosenheck RA. Mental health multimorbidity and poor quality of life in patients with substance use disorders: An analysis of electronic health records data. Drug Alcohol Depend. 2022;229: 109156. https://doi.org/10.1016/j.drugalcdep.2022.109156.

Boehmer KR, Kyriacou M, Behnken E, et al. Patient capacity for self-care in the medical record of patients with chronic conditions: a mixed-methods retrospective study. BMC Fam Pract. 2018;19:164. https://doi.org/10.1186/s12875-018-0852-0.

Boucher LM, Shoemaker ES, Liddy C, Leonard L, MacPherson P, Presseau J, Martin A, Pineau D, Lalonde C, Diliso N, Lafleche T, Fitzgerald M, Kendall CE. "The drug use unfortunately isn't all bad": chronic disease self-management complexity and strategy among marginalized people who use drugs. Qual Health Res. 2022;32(6):871–86. https://doi.org/10.1177/10497323221083353

Bulstra CA, Honings S, Wieczorek Ł, Dąbrowska K. Punitive drug policies in the Philippines and their impact on public health and human rights: a review of existing literature. Harm Reduct J. 2021a;18(1):1–10. https://doi.org/10.1186/s12954-021-00513-3.

Bulstra CA, Hontelez JAC, Otto M, Stepanova A, Lamontagne E, et al. Integrating HIV services and other health services: a systematic review and meta-analysis. PLOS Med. 2021b;18(11): e1003836. https://doi.org/10.1371/journal.pmed.1003836.

Burke JR, Onwueegbuzie AJ, Turner LA. Towards a definition of mixed methods research. J Mixed Methods Res. 2007;1(2):112–33. https://doi.org/10.1177/1558689806298224.

Creswell JW, Plano Clark V. Designing and conducting mixed methods research, 2nd ed. Thousand Oaks, CA: SAGE Publications;2011.

Creswell JW, Plano Clark VL. Designing and conducting mixed methods research. 2nd ed. Thousand Oaks (CA): SAGE Publications; 2011.
Crowe S, Cresswell K, et al. The case study approach. BMC Med Res Methodol. 2011;11:100.
Dangerous Drugs Board (DDB). Annual report on drug use in the Philippines. Manila (PH): Dangerous Drugs Board;2021. Available from: https://www.ddb.gov.ph
Fetters MD, Curry LA, Creswell JW. Achieving integration in mixed methods designs—principles and practices. Health Serv Res. 2013;48(6pt2):2134–156. https://doi.org/10.1111/1475-6773.12117.
Fiorillo A, Sartorius N. Mortality gap and physical comorbidity of people with severe mental disorders: the public health scandal. Ann Gen Psychiatry. 2021;20:52. https://doi.org/10.1186/s12991-021-00374-y.
Franklin M, Willis K, Lewis S, Smith L. Chronic condition self-management is a social practice. J Sociol. 2023;59(1):215–231. https://doi.org/10.1177/14407833211038059
Global Commission on Drug Policy. The war on drugs and its impact on public health in Southeast Asia. 2022. Available from: https://www.globalcommissionondrugs.org.
Harrison H, Birks M, et al. Case study research: foundations and methodological orientations. In Forum qualitative sozialforschung/forum: qualitative social research, vol. 18, No. 1, Art. 19;2017
Hess DR. Retrospective studies and chart reviews. Respir Care. 2004;49(10):1171–4.
Human Rights Watch. "Our Happy Family is Gone": impact of the "War on Drugs" on children in the Philippines. 2021. Available from: https://www.hrw.org/report/2021/05/27/our-happy-family-gone/impact-war-drugs-children-philippines.
Institute of Medicine (IOM). Best care at lower cost: the path to continuously learning health care in America. Washington, DC (US): The National Academies Press; 2013.
Jaminola LI. The implementation of universal health care (UHC) in the Philippines: challenges and opportunities. J Public Health Policy. 2023;44(2):203–19. https://doi.org/10.1057/s41271-023-00354-9.
Jaminola LI, Negre JM, Pepito VCF, Loreche AM, Dayrit MM. The policy environment of self-care: a case study of the Philippines. Health Policy Plan. 2023;38(2):205–17. https://doi.org/10.1093/heapol/czac095.
Johnson RB, Onwuegbuzie AJ, Turner LA. Toward a definition of mixed methods research. J Mix Methods Res. 2007;1(2):112–33. https://doi.org/10.1177/1558689806298224.
Paukkonen L, Oikarinen A, Kähkönen O, Kyngäs H. Adherence to self-management in patients with multimorbidity and associated factors. J Clin Nurs. 2022. Available from: https://doi.org/10.1111/jocn.16099
Powell J, Sweeting M. Retrospective studies. Eur J Vasc Endovasc Surg. 2015;50(5):675. https://doi.org/10.1016/j.ejvs.2015.07.005.
Roomaney RA, Van Wyk B, Cois A. Inequity in the distribution of non-communicable disease multimorbidity in adults in South Africa: an analysis of prevalence and patterns. Int J Public Health. 2022. Available from: https://doi.org/10.3389/ijph.2022.1605072/pdf.
Riegel B, Dunbar SB, Fitzsimons D, Freedland KE, Lee CS, Middleton S, Strömberg A, Vellone E, Webber DE, Jaarsma T. Self-care research: where are we? Where are we going? Int J Nurs Stud. 2021;116:103402. https://doi.org/10.1016/j.ijnurstu.2019.103402
United Nations Human Rights Council (UNHRC). Report on the human rights situation in the Philippines. Geneva (CH): UNHRC; 2020. Available from: https://www.ohchr.org/en/documents/country-reports/report-human-rights-situation-philippines.
United Nations Office on Drugs and Crime (UNODC). Patterns and trends of amphetamine-type stimulants and other drugs in Asia and the Pacific. Vienna (AT): UNODC; 2022. Available from: https://www.unodc.org/roseap/uploads/documents/Publications/2022/Booklet_2_12th_Jan_2022.pdf.
Volkow ND, Baler RD, Compton WM. The challenges of co-occurring substance use disorders and mental health conditions. Neuropsychopharmacology. 2021;46(1):177–92. https://doi.org/10.1038/s41386-020-0778-5.

VonKoss KH, Moore ML, Richardson L. Using health records as sources of data for research. J Nurs Meas. 1995;3(1):3–12.
Wong EC, Maher AR, Motala A, Ross R, Danz M, Akinniranye G, et al. Methods for identifying health research gaps, needs, and priorities: a scoping review. Santa Monica (CA): RAND Corporation; 2021. Available from: https://www.rand.org/pubs/research_reports/RRA119-1.html
World Health Organization (WHO). Noncommunicable diseases country profiles 2019: Philippines. Geneva (CH): WHO; 2019a. Available from: https://www.who.int/publications-detail/ncd-country-profiles-2019.
World Health Organization (WHO). WHO guideline on self-care interventions for health and well-being. Geneva (CH): WHO; 2019b. Available from: https://www.who.int/publications/i/item/9789241550550
Yang C, Hui Z, Zeng D, Zhu S, Wang X, Lee D, Chair S. A community-based nurse-led medication self-management intervention in the improvement of medication adherence in older patients with multimorbidity: Protocol for a randomized controlled trial. BMC Geriatr. 2021;21(152). https://doi.org/10.1186/s12877-021-02097-x
Zerfu B, Kassa T, Legesse M. Epidemiology, biology, pathogenesis, clinical manifestations, and diagnosis of dengue virus infection, and its trend in Ethiopia: a comprehensive literature review. Trop Med Health. 2023;51(11). https://doi.org/10.1186/s41182-023-00504-0

Joseph Kristian Reyes is a registered and practicing Medical Technologist with 14 years of experience both in the Philippines and abroad. He has completed a Master's degree in Public Health and holds certifications as a Biosafety and Biosecurity Officer, Occupational Safety and Health II Officer, HIV and STI Analyst, HIV and STI Counselor, and Drug Analyst. He is currently employed at the Department of Health's Drug Treatment and Rehabilitation Center in Dagupan, Philippines. Research writing is one of his passions, as it thrills him to acquire new knowledge and contribute to the advancement of human well-being.

Neriza V. Bustillo is a Certified Research Specialist (CRS) and an educator from Dagupan City, Philippines. Her research spans public infrastructure, technology-driven innovations, and educational frameworks. Notable works include "E-Purse Transit Pass: The Potential of Public Transport Smart Card System in the Philippines," which explores smart card technology for transit systems, and "An Architectural Model on Employment Opportunities as Aftercare Program," which focuses on the architectural model. She is recognized for her interdisciplinary expertise and commitment to advancing practical solutions in her studies.

Danica P. Ydia is a registered Nutritionist-Dietitian and a member of various organizations in the field of Nutrition and Dietetics. She has various trainings in Diabetes Management, Clinical Nutrition, Nutrition in Emergencies, Food Safety, Health Research, and others. She has published works on food allergies among Persons who use drugs (PWUDs). Currently taking her master's degree in Public Health at Urdaneta City University, she has eight years of experience working in a government drug rehabilitation facility.

Justin Bernard C. Fiestada is an experienced IT Specialist with wide knowledge in programming, hardware and software troubleshooting, and network and system administration. His capability to design various web-based systems is quite exceptional, demonstrating an ability to create innovative, efficient, and functional IT digital solutions. He is always contributing towards the enhancement of organizational efficiency and performance with his proficient expertise in maintaining secure and efficient infrastructure. This attitude towards the IT field reflects the commitment to excellence and a passion for technology-driven progress.

Chapter 9
Exploring the Eco-Friendly and Health Benefits of Exergaming to Improve Physical Activity Level in Urbanised East Asian Cities

Cluster of Future Studies: Technological Trends

Maziah Mat Rosly

Abstract Rapid urbanisation of populated East Asian cities led to sedentarism that poses significant health issues among urban dwellers. Major city residents in Asia face challenges accessing exercise facilities, transportation issues and commitment to work and family that precludes exercise. The emergence of exergames, a combination of exercise and gaming, can be used to promote physical activity in urban settings. This case study explored the benefits and challenges exergaming can offer within the Malaysian cultural context, which parallels urbanised East Asian cities such as Japan, Singapore, Taiwan and China. Two sedentary urban Asians trained using exergaming for 12-weeks at home, utilising virtual reality-based sports exergaming. Physiological measures such as heart rate, physical activity level, enjoyment score and cardiorespiratory fitness levels were taken pre and post training. At the end of the training, open ended interviews were conducted individually to collate their experiences on the training program. Home-based exergaming training improved physical activity level, cardiorespiratory fitness and resting heart rate when performed at dose-potent intensities. Exergaming can also overcome common barriers to exercise among East Asians, and is space-efficient. However, wide-scale implementation of exergames face challenges for being too expensive to initiate, while also wrought by hardware and software issues.

Keywords Virtual reality · Exergames · Urbanisation · Malaysia · Sedentary behaviour · Carbon footprint

M. M. Rosly (✉)
Department of Physiology, Faculty of Medicine, Universiti Malaya, Kuala Lumpur, Malaysia
e-mail: maziahmr@um.edu.my

R. Xiang Ng et al. (eds.), *Future of Healthcare in Asia*,
https://doi.org/10.1007/978-3-032-03183-9_9

9.1 Introduction

Over the past few decades, East Asian countries faced rapid development and urbanisation that had caused overcrowding and traffic congestion. Many urban dwellers of major cities in Asia face lowered physical activity levels (PAL) owing to the limited exercise facilities, transportation issues and commitment to work and family that precludes exercise. These factors had accelerated sedentary lifestyles, which contributed to increasing co-morbidities and lowered quality of life. Exergames, which combines exercising with video gaming can be introduced as an innovative approach to encourage more physical activity (PA). A case study was conducted to determine the effectiveness of exergaming within a home-based setting. Data from two participants who volunteered for the study was collected to provide personalised and explorative information that can help researchers understand exergaming's potential uses.

9.1.1 The Significant Health Issues of Low Physical Activity Level and Introducing Exergaming

The rapid urbanisation of densely populated cities in Asia has led to sedentarism that poses significant health issues among urban dwellers. Major cities in Asia face challenges of overcrowding, congested facilities and expensive fitness centres, causing lowered physical activity levels (PAL). Low PAL has repeatedly been linked to poorer cardiometabolic health such as obesity, hypercholesterolemia, diabetes, and psychological impacts including mental health disorders, confidence and motivation (World Health Organization 2022a). The World Health Organization (WHO) has reported concerning PAL estimates among Asians, where roughly 16.2–36.8% of adults were considered sedentary, averaging less than 11 metabolic equivalent of task (MET) hours/week (World Health Organization 2022a, b). Similarly, the average PAL for individuals with physical disability most often fall in the lower end of the spectrum (Roberton et al. 2011; Mat Rosly et al. 2018). While the PAL values merely described levels of activity or vice versa, exercise habits, which requires a person to commit at least 75–150 min of moderate–vigorous exercise plus muscle strengthening training, were reportedly even lower (Chaput et al. 2020).

Therefore, the prevalence of low PAL can be costly, not only to the individual's health, but also to their countries' coffers (World Health Organization 2022a). Physical inactivity may lead to physiological and psychological health diseases that contribute approximately ~ 50% to the rise of non-communicable diseases (NCDs) (World Health Organization 2022a). As a consequence, health-care expenditures related to these NCDs can take up to 70% of the allocated health budget of a high-income country's yearly expenditure (World Health Organization 2022a). Instead, higher PAL correlated with reduced risks of cancer development, better brain cognition and osteoporosis prevention (World Health Organization 2022a).

The rapid expansion of technological innovations that accentuate sedentary lifestyles, are major obstacles urban East Asians face in order to maintain PA according to the levels recommended by WHO. Hence, the efforts to improve PA among modern urbanites, should include fun and portable exercises that can be done on the go, without relying on bulky exercise equipment or distant facilities that are situated far from their homes (Mat Rosly et al. 2018; Mat Rosly and Mat Rosly 2023). Therefore, a specific form of gaming genre could be introduced, called exergaming, a combination of exercise with video gaming. This new direction of inculcating higher forms of PA, which infuse exercise with the rising trend of gaming culture among youths (Lobel et al. 2017).

9.1.2 Case Study Objectives: Home-Based Exergaming in Malaysia

To ascertain the feasibility and positive effects of exergaming in promoting higher physical activity levels (PAL), two cases were conducted within a Malaysian urban setting. Since Malaysia is located within the East Asian region, countries with similar population density, demographics and weather conditions such as parts of China, Japan, Taiwan, Hong Kong, Thailand, Singapore, Brunei, Indonesia, Vietnam and Philippines may be applied. For this case study, two Asian adults were selected to exercise using exergames strictly at home, and record the overall PAL outcomes. The case study sought to investigate whether PAL can be improved via home-based exergaming in otherwise sedentary adults, confined to their homes the majority of the time. The case study also sought to determine whether their overall cardiorespiratory fitness (CRF) levels, in terms of PA score and resting heart rate (HRrest), can be improved through exergaming at home. An analysis of their exergaming perceptions and experiences will then be collated into themes via qualitative interviews.

9.1.3 Ethical Reasons for the Home-Based Exergaming Case Study

An important aspect of this case study involves understanding in-depth and multi-faceted explorations of the complex issues that might arise when participants are given the opportunity to exergame at home, on their own. Personalised and explorative data from such real-life settings can help researchers understand the nature of how the research design can be implemented or improved upon. Case studies therefore become a key research study design aspect where usability issues, major limitations and protocol errors can be avoided with repeated testing on a smaller scale. This reduces the cost of research studies and allowed for efficient means of data collection without exhausting resources related to the work.

9.2 Literature Review

Barriers to exercise in most East Asian countries include limited exercise facilities, transportation issues and commitment to work and family that precludes exercise. Sedentarism was more prominent among urban residents with higher education and higher income. However, since video games have amassed a large following among East Asians, the unique attributes of introducing exergames to improve PA can be advantageous. Previous works had shown that exergames provided health benefits to various types of interventions, including rehabilitation and physical training.

9.2.1 Cultural Nuances and Barriers to Physical Activities in East Asian Cities

In East Asia, sedentarism is apparently more prominent among higher income countries, older adult age groups and the female population (Mat Rosly et al. 2018; Huang et al. 2022; Lee et al. 2022; Teng et al. 2022; Cheng et al. 2023). Barriers to performing regular exercise among East Asians, whether it be individuals with or without physical disabilities (Mat Rosly et al. 2018; Mat Rosly and Mat Rosly 2022; Mat Rosly 2023) can largely be relegated to cultural nuances comprising:

- personal characteristics: such as laziness, lack of motivation, perceived boring and monotonous exercises or no interest, ethnicity, perceiving exercise as difficult
- physical limitations: pain, participants over the age of 35 years, health concerns
- socio-environmental issues: inaccessible facilities, lack of support, family responsibilities
- economic policies: lack of transport, expensive exercise equipment, health illiteracy, distortion of health information

Particularly in Malaysia, barriers to exercise are often faced not by individuals with physical disabilities (IPD) alone, but also affected non-disabled individuals who experience extreme pressures to support their families and progress in their careers (Mat Rosly et al. 2018, 2020; Mat Rosly 2022, 2023). The myth that barriers to exercise only tend to affect IPD may be the key 'elephant in the room' issue that many health organisations tend to ignore. Non-participation in the recommended dose of exercise has been linked to groups with higher education, urban residents and high-income earners (Mat Rosly 2023). Seeing that high income earners are often synonymous with higher education; it becomes imperative to include this group of population in the promotion of PA. Therefore, directions to improve physical activity levels (PAL). should now include non-disabled urban dwellers, who represent the country's key productivity output.

9.2.2 East Asian Trends in Exergaming

As opposed to the western cultural gaming phenomenon, which focuses on household-based gaming, urbanised East Asian cities have a wider adoption of smartphones and mobile gaming devices that fuel its demographics (Spire Academy 2017; Shah 2023). This notion stems from the cultural distinction East Asians emphasise in certain household demographics, such as having smaller spaces for console gaming, reserving sections of the house for relaxing and guest visitations or putting more importance on communal congregations (Koutsogeorga 2015). Mobile games are considered more popular among East Asian urban dwellers due to its portability, allowing it to be played anytime, anywhere on the go. These games often leverage on such features by including interactive gameplay mechanics that are less obtrusive in nature (Chen 2023). The gameplay styles developed were made to parallel current demographics of densely populated East Asian urban cities, featuring gaming aspects that can be done over a diverse range of commuting activities (Chen 2023). For instance, mobile games were made to be short, repetitive, and easily achievable targets that can be resumed at any time to be in conjunction with the routine of daily commuters.

Furthermore, virtual reality (VR) exergaming has gained traction among niche game enthusiasts seeking more immersive and challenging routines (Farrow et al. 2019). This iteration of gaming, immerses players into a VR environment that display preselected settings depicting a sense of presence and excitement while engaging in PA. Therefore, such diverse landscapes for extreme sports are most often in far-away locations or almost non-existent in most East Asian cities. Two forms of reality technologies are commonly used for digital simulation of the environment. These include VR, where digital simulation completely replaces a real-life environment, and augmented reality (AR), where digital elements are added to real-world environments to enhance the surroundings. The evolution of VR, and to a certain degree AR, provide multiple different experiences without requiring on-site location or expansive logistics, while keeping them in a relatively safe environment (Tieri et al. 2018). Players can now engage in activities such as rock climbing, diving, kayaking or speed boat riding via VR/AR environment that allows them to engage in rigorous activities (Miyachi et al. 2010; Mat Rosly et al. 2017a; França et al. 2024), without the fear of having potential predator attacks, natural disasters or physical accidents.

9.2.3 The Unique Attributes of Introducing Exergames to Improve Physical Activity

Space sharing and limited spaces in rooms/households are often constraining limitations felt by urban East Asians, especially among students and young adults. Such confined spaces severely restrict movements, preventing individuals from owning an exercise equipment to exercise regularly. By comparison to countries in the West, the average household space in most urbanised East Asian cities such as Singapore, Taiwan, Japan, South Korea and Malaysia are much smaller (World Population Review 2024). This is attributed to lower land mass availability per population when compared to big countries such as Canada, USA, Australia and Europe. Moreover, common spaces in households such as the living room within Asian context is often reserved for relaxing, watching TV and hosting guests, making it generally prohibitive for either exercising routines or video gaming in general (Koutsogeorga 2015).

9.2.4 *Health Benefits of Exergames*

Multiple studies had shown that integrating exergames into rehabilitative therapies, exercise regiments and physical training can help improve health parameters significantly (Gerber et al. 2016; Knights et al. 2016; Mat Rosly et al. 2017b, c). This includes cardiometabolic health, but also improved mental health status, quality of life, as well as cognitive functioning (Rosenberg et al. 2010; Staiano et al. 2013; Begde et al. 2023). Performing PA through exergaming generates positive emotions and promotes well-being of the emotional regulation (Marques et al. 2023). Additionally, integrating VR with exercise can also improve affective experience of PA for inactive adults, making them more adherent and engaged to the training (Sauchelli and Brunstrom 2022).

9.3 Methodology

Two participants who volunteered for the case study trained using exergaming for 12-weeks at home. Physiological data taken pre and post training include HRrest, PAL, enjoyment score and CRF levels. Following the conclusion of the training program, open ended interviews were conducted individually to collate their experiences on exergaming.

9.3.1 *Protocol and Analyses*

Ethical clearance was obtained from the Universiti Malaya Research Ethics Committee (UM.TNC2/UMREC_189), approved on 01st June 2022. Two sedentary adults within the age range of 18–65 years old, living in urban areas, were selected via convenience sampling to determine the PAL and fitness outcomes seen following exergaming training as an alternative exercise platform. Both were deemed healthy and devoid of any cardiovascular risk precluding vigorous exercise participation or having any previous history of stroke/brain lesions/illnesses affecting their cognitive

Fig. 9.1 Examples of how a home-based exergaming set-up is arranged in the common space or living area. *Note* identifying information has been removed from the image

function (Balady et al. 1998). They were able to follow instructions clearly in their comprehended languages. The exergaming equipment and accessories were posted to their residence once informed written consent and non-disclosure agreement were provided. Figure 9.1 depicts examples of the exergaming set up at home.

Individual coaching sessions, which includes gameplay techniques, and assistance on setting up the exergaming unit were provided to each participant at the start of training. Following this, participants performed exergaming training sessions independently for 12 weeks, as evidence show that training outcomes will be effective if performed between 4 and 16 weeks (Heesterbeek et al. 2005; Bakkum et al. 2015; Mat Rosly and Mat Rosly 2023). They must achieve at least 75 min of vigorous or 150 min of moderate intensity gaming spread two to three times per week (Garber et al. 2011). Participants must also include biweekly doses of muscle strengthening exercises at two to three repetitive sets as part of their exercise prescription. No external encouragement or motivation was provided. A graphical summary of the case study protocol is depicted in Fig. 9.2 (see the next page).

Quantification of PAL and enjoyment used the Malaysian adapted PA Scale (Mat Rosly et al. 2020; Mat Rosly 2023) and the physical activity enjoyment score (PACES) (Mat Rosly and Mat Rosly 2023). The PACES score ranks five aspects of PA perception (fun, enjoyment, dislike, feeling good physically and frustration) based on a 7-tier Likert scale ranging from 5 to 35 (from 5 = really disliked to 35 = really

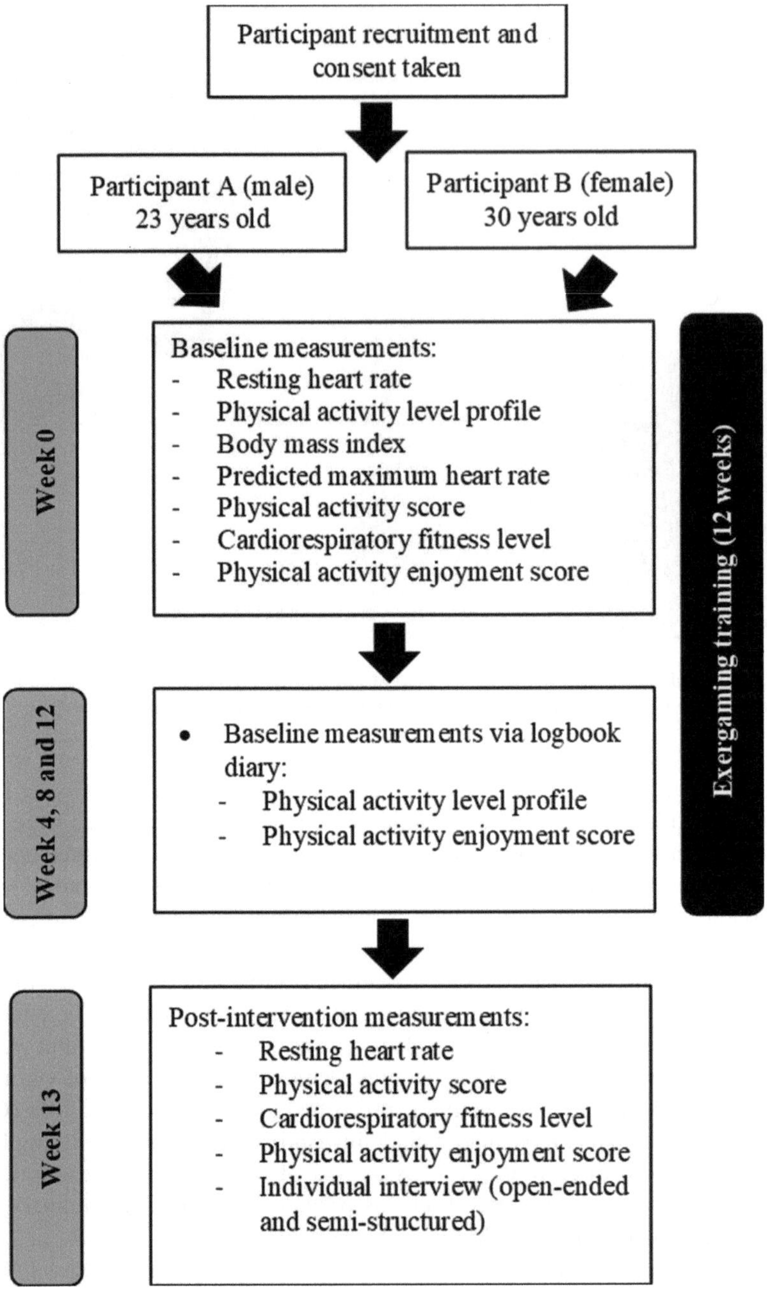

Fig. 9.2 Flow chart depicting the case study's 12-week intervention protocol

enjoyed) (Graves et al. 2010). Physical activities performed were re-categorised into 4 different factors as previously established (Mat Rosly 2023). These factors include factor 1: leisure activities (covering aspects related to leisure, light-to-vigorous exercises and endurance training), factor 2: home maintenance (home repair, lawn work and gardening), factor 3: household chores (light and heavy housework) and factor 4: career activities (occupation and volunteer activities). HRrest was taken using the HR monitor Polar® RS400. Participants were required to rest in supine position for 15 min and the average of three lowest HR value recorded was taken as their resting HR. Maximum HR (HRmax) was calculated using the Tanaka formula based on their age [208–(0.7 × age)] (Tanaka et al. 2001). The participants' estimated HRmax was then used to determine the range they should achieve (moderate to vigorous intensities) during their exergaming sessions (Garber et al. 2011). PA intensity estimations used rating of perceived exertion (RPE) (Borg 1982) or HR elevation (Garber et al. 2011). Cardiorespiratory fitness levels were estimated using predictive formulas established by Jurca et al (2005). Qualitative responses from post-intervention interviews were transcribed verbatim and were categorised into similar themes.

9.4 Findings

The basic demographics and limitations of each participant were described in Table 9.1. The results of the case study indicated that overall PAL increased by at least threefold for both participants following the exergaming intervention Table 9.2, Figs. 9.3 and 9.4. The participants' PACES also remained relatively high over the 12-week period, scoring at least 28 points and above over a 35-point total, as depicted in Fig. 9.5. Both participants showed improvements (14% and 0.82% respectively) in their overall CRF levels, however, only participant A displayed improvements that was deemed clinically significant according to guidelines (Jurca et al. 2005).

The individual feedbacks collected from the two participants' interviews, regarding their experiences with the home-based exergaming sessions were collated into two main themes: benefits and challenges. Six common themes were extracted from the participants' feedback, summarised in Table 9.3. Direct quotes and excerpts extracted were translated to English from the original Bahasa Malaysia format by the author. Some of the quotes were truncated and rearranged for brevity and higher clarity.

Table 9.1 Participants' gender, age, body mass index, locality profiles, type of exergames played and limitations to exercise

Attributes	Participant A	Participant B
Gender	Male	Female
Age (in years)	23	30
BMI	14.9	21.6
Locality	Semi-urban area	Densely populated urban area
Limitations to exercise	– No transport availability outside of home – Wheelchair bound the majority of the time – SCI level L4 (AIS D) with grade 3 lower limb motor power – Not under employment, education or training	– Works at home the majority of the time – Lives in a studio apartment with limited home space of only 298 square feet
E-sports exergames played	PS Move Gladiator Duel, Tennis and Boxing	PS Move Tennis and Badminton

Abbreviations: BMI: Body mass index; AIS D: American Spinal Injury Association Impairment Scale – D, SCI: Spinal cord injury; PS: PlayStation

Table 9.2 Participants' resting heart rate, physical activity level, physical activity score and cardiorespiratory fitness level pre and post-exergaming training intervention

Parameters	Pre-intervention (baseline)	Post-intervention (after 12 weeks of training)	Changes seen (in %)
HRrest (participant A)	86	82	−4.65%
HRrest (participant B)	71	68	−4.23%
PAL (participant A)	0.17	15.44	8982%
PAL (participant B) in MET/day	4.62	11.03	138.74%
PA score (participant A)	1	4	Positive improvements
PA score (participant B)	1	4	Positive improvements
CRF level in MET (participant A)	13.43	15.31	13.99%
CRF level in MET (participant B)	11.03	11.12	0.82%

Abbreviations: HRrest: Resting heart rate, PA: Physical activity, PAL: Physical activity level, MET: Metabolic equivalent of a task per day, CRF: Cardiorespiratory fitness

*PA score of 1 indicates inactive or little activity other than usual daily activities

PA score of 4 indicates participate in aerobic exercises such as brisk walking, jogging or running at a comfortable pace, or other activities requiring similar levels of exertion for 1–3 h per week. By (Jurca et al. 2005)

*Changes seen in percentage calculation $\frac{post\ intervention\ value - pre\ intervention\ value}{pre\ intervention\ value} \times 100$

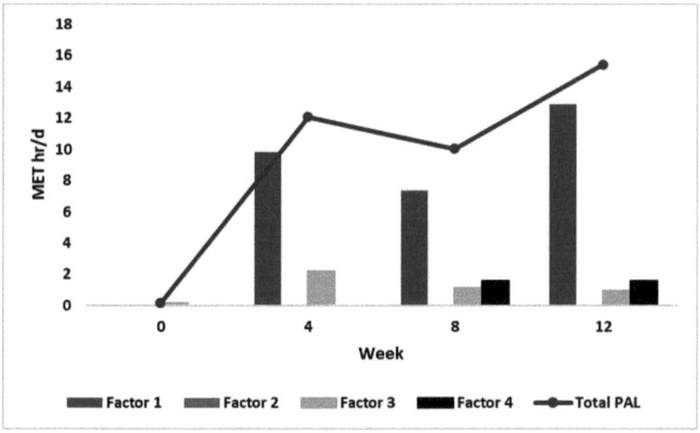

Fig. 9.3 Overall physical activity level performed by Participant A over the course of the 12-week exergaming training intervention in metabolic equivalent of a task (MET) in hours per day. *Note* MET hr/d: metabolic equivalent of a task in hours per day, PAL: physical activity level, Factor 1: leisure activities, Factor 2: home maintenance, Factor 3: household chores and Factor 4: career activities

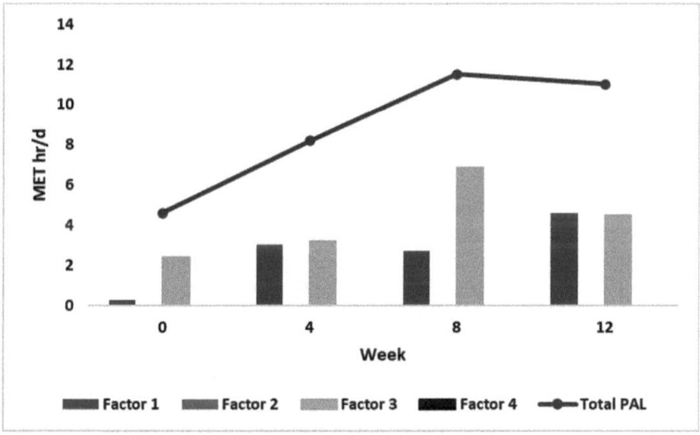

Fig. 9.4 Overall physical activity level performed by Participant B over the course of the 12-week exergaming training intervention in metabolic equivalent of a task (MET) in hours per day. *Note* MET hr/d: metabolic equivalent of a task in hours per day, PAL: physical activity level, Factor 1: leisure activities, factor 2: home maintenance, factor 3: household chores and factor 4: career activities

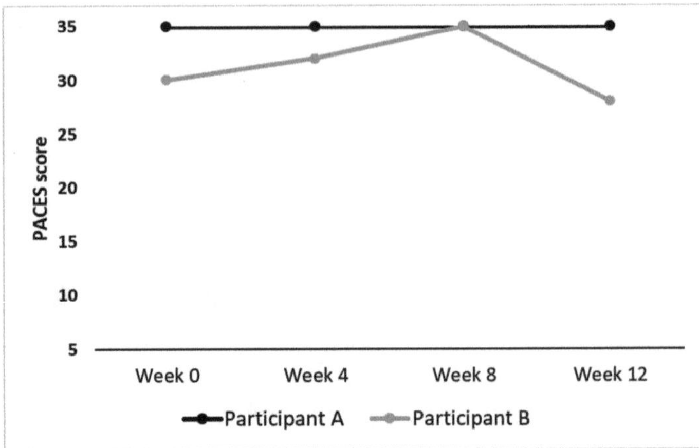

Fig. 9.5 Overall physical activity enjoyment score (PACES range 5–35) rated by participants while exergaming within 12 weeks

Table 9.3 The benefits and challenges described by participants' home-based exergaming training experiences

Themes	Descriptions
Benefits	
Exergaming transcends typical barriers to exercise	Notably, performing exercises require participants to either participate in group sports or having to travel to sports centres or gym facilities. This was difficult for them as transport and travel were key issues that prevented them from exercising. The amount of money they had to fork out for the trip, as well as to cover facility renting, simply to get a dose of exercise did not justify the expenditure. This was repeatedly highlighted by both participants as exergames allowed them to exercise at the comfort of their own home while 'working out a sweat'
	'I used to attend vocational school and there…they had dorms where I stay with other friends and we used to play sports together…and it was all free. Once I graduated (and returned home), the minute I stepped out of my home, everything required money…money I don't have and unwilling to spend just to participate in sports'
	'I do not have a car, and I work from home most of the time. I live in a small studio apartment which basically cannot fit an exercise equipment…Exercising means I have to pay for gym membership…no suitable parks nearby'

(continued)

Table 9.3 (continued)

Themes	Descriptions
Fun and interactive	Exergames are digitally-based recreations of its counterpart sports that are powered by artificially intelligent (AI) game mechanics. As such, players are able to interactively play sports alone against a computer or with another player. This AI feature becomes especially useful for individuals who live alone or do not have a regular partner to play with
	Additionally, the AI aspect allow for a more focused and higher quality training session, as players will not be bogged down by an exhausted partner, neither will they need a paid trainer for them to improve their skills
	'Playing against the computer (AI) is like playing with your friends. It feels the same. It's fun that I get to play (interactively)'
	'I don't feel so alone and bored anymore with this (exergame)'
	'It would be nice if I can play with my friends online with this (exergame platform)'
	'When I play (real life) sports, like Tennis for example, since I'm not good at it (yet), I tend to waste so much time chasing after the ball then actually hitting it (accurately). It's a waste of time, especially when renting the (Tennis) court or hiring a coach needs money'
Home-based feasible	Another beneficial factor that exergaming can offer is a home-based option for exercising that is also dose-potent for health benefits. Non-exercisers are most often too time-constrained or confined at home due to a variety of barriers to performing exercise. The fact that exergaming can be set up in small spaces at home and be interactive online or offline at the same time, is a huge advantage for the new demographic generation
	'…I tend to get self-conscious (anxiety) when I play in front of other people and I'm not good at it…so playing at home, in a safe environment (like this), gives me confidence to improve my skills, without burdening anyone with my poor skills'
	'I save time by exercising like this (exergaming) at home than having to go out'
	'Without the (exer)game, I wouldn't have anything much to do at home…now that I have (graduated) and returned home'
Compact, portable and flexible	In urban and semi-urban areas, space is a luxurious commodity that not many have. For instance, homes are often small-sized or cannot accommodate enough space for a typical bulky exercise equipment such as the treadmill or stationary bikes. However, the advent of exergames, which comes in small or compact modalities, is considered a handy exercise tool for urban dwellers. This is because the platform is interchangeable, doubling or tripling as an entertainment-work-exercise space that is more efficient than regular exercise equipment. The TV can be configured with different channels for different purposes, aimed at maximising small spaces for more uses

(continued)

Table 9.3 (continued)

Themes	Descriptions
	'Living in a small space like this (my apartment), I have to be mindful of my stuff. This exergame is perfect for my space. My work space is also my entertainment and exercising space'
	'If I were to travel (for work or others)...this will be my go-to platform. I love the fact that it is portable'
	'The entire (exer)game unit (can just) be posted. Everything fits in (the) small box and I can carry it around in (the) bag (if) I'm traveling. Yea, this is useful'
Challenges	
Expensive	The main challenge with having exergames as an exercise tool is the starting cost required to set-up the platform. The average price of the entire exergaming set needed, which does not include the TV, would be somewhere around RM1800-2200 (~USD400-550). Considering most East Asian countries are developing nations with low wage income, this can be a steep price tag to invest on for an exercise equipment. However, participants have noted that if assisted funding was provided, they would have preferred exergames over the regular exercise tool such as a treadmill or stationary bike
	'It's too expensive for me'
	'If I have to buy (it) myself, I can't (afford) it'
	'If you were to fund me RM1000 (~USD300) and ask me whether I want to buy a treadmill or this (exergame). I'd say I would go for the exergame. But I would be interested to know more about the 'Nintendo Switch features as well. So yes, it's the exergame'
	'With RM1000, yes -go for the exergame. I can play the games for entertainment and the exergames for exercise...worth it'
Software and hardware frustrations	The most apparent challenge with VR-based sports are issues related to hardware and software of the games. For instance, new iterations of games or consoles get released too often and are not backwards compatible. This creates a consistent need to update or upgrade the software and hardware, which burns through money most low- or middle-income earners are unwilling to fork out
	Another issue comes from the frustrations of unresponsive player input–output within the game. Avatar feedback from the players did not translate well with the game's haptics, creating outputs that were not as efficient or high quality as it was intended
	'The one I used, is a PlayStation (three). They don't sell this anymore in the market...and it gets more and more expensive with the next generation. I can't keep up...'
	'The control does not respond as intended at times. It becomes difficult to master (the movements) sometimes'

9.5 Discussion of Implications

The improvements seen in overall PAL, HRrest and CRF indicate exergaming, even at home, can provide health benefits for users with or without physical disability. Exergames have the advantage of providing a fun and interactive e-sports exercise that can be done within the comfort of one's home. Recommended training duration can range anywhere between 4 and 16 weeks when performed at the correct dose and intensity. However, since the sample size was small, the findings may not be applicable to the general population until more robust study designs have been replicated.

9.5.1 Benefits of Home-Based Exergaming

The results of the case study indicated that overall PAL in both participants increased at least threefold following the exergaming intervention (see Table 9.2, Figs. 9.3 and 9.4). Marked increases were seen in factors 1 (leisure activities), 3 (household chores) and 4 (career activities). These included activities such as light-to-vigorous exercises and leisure, doing more light or heavy housework and being more active in volunteering work (Figs. 9.3 and 9.4). This indicated that having performed leisure activities of adequate intensity and duration, as prescribed by international guidelines not only improve overall PAL, but also help motivate individuals to be self-dependent in performing activities of daily living. Such is seen in participant A, where much of the improvements were seen not only in his PA related to exergaming training, but also in performing more household-related chores, other PA associated with leisure, as well as carrying out volunteering work career-wise (Fig. 9.3). Multiple studies have shown that higher PAL correlates well with higher productivity output and better quality of life (World Health Organization 2022b).

Additionally, participants were also able to improve their CRF levels to better prognosis. This can be seen in higher PA scores, as well as lower HRrest following the 12-week intervention training (refer to Table 9.2). Both parameters had been associated with better health and fitness levels (Jurca et al. 2005; Gonzales et al. 2023), indicating potential benefits if larger cohort studies or more robust control trials were conducted in the future. The exergaming training in this case study, provides initial evidence of improved health among sedentary populations in urbanised Asian cities. However, readers must take these findings with caution since the results presented were based on a small sample size ($N = 2$) limited by a case study.

Another interesting factor to note is that their PACES towards the exergaming training sessions also remained relatively high (Fig. 9.5) (at least 28-points from the maximum score of 35) throughout their intervention. Decrease in enjoyment stems mostly from the frustrations they felt while playing higher level exergames that had harder opponents to beat. This ironically becomes a double-edge sword to achieving the recommended exercise dose-potency, as higher intensity equates to harder levels that cause frustrations to players who lose to the game. However, it is important to understand that winning or losing, or making frustrating mistakes are all part of exercise that players must go through in all forms of sports.

9.5.2 Comparisons with Other Case Studies in Asian Regions

A promising strategy to enhance everyday PA in communities with mobility challenges, isolation, or confinement is the introduction of home-based exergaming, particularly in East Asia. Performing 'dose-potent' amount of exergaming is key to achieving CRF and metabolic improvements (Garber et al. 2011; Mat Rosly and Mat Rosly 2023). The training duration can range between short intervals of 4 weeks to longer 16-week interventions (Heesterbeek et al. 2005; Tripette et al. 2014; Bakkum et al. 2015; Staiano et al. 2018; Mat Rosly and Mat Rosly 2023).

Unfortunately, current available studies in East Asia related to home-based exergaming were limited, and globally, had largely focused on introducing home-based exergaming for vulnerable and diseased populations such as individuals with spinal cord injury (Mat Rosly and Mat Rosly 2023), the elderly with Parkinson's disease (van der Kolk et al. 2019), stroke or cognitive impairment (Ghahfarrokhi et al. 2021; Begde et al. 2023; Lee 2023), children with Attention Deficit Hyperkinetic Disorder (Benzing and Schmidt 2019) or obesity (Staiano et al. 2018). In contrary, the notion of introducing home-based exergaming for otherwise healthy and non-disabled populations had been very limited and sparse. It is apparent that this may be because either exergaming is unpopular among the non-disabled or because of the larger options for sports participation available to them. As such, having the idea of combining video games with exercise may seem nonsensical.

However, the arrival of the Coronavirus disease (Covid-19) pandemic is one significant scenario where home-based exergaming can be helpful to moderate PA at home (Sultan et al. 2020; Mat Rosly and Mat Rosly 2023). Additionally, post-partum mothers confined at home for physiological recovery, individuals with physical disability and work from home workers, are also target populations who can benefit from home-based exercises such as exergaming. In a study by Tripette et al (2014) among post-partum mothers in Japan, home-based exergaming has been found to reduce body weight, body fat and overall body mass index. The introduction of exergames at home, not only helps increase their daily PA, but also helps manage their new maternal lifestyles while caring for a newborn, with interactive exercises they can do at home.

9.5.3 Context that Might Influence the Case Study's Findings

Although the use of exergaming in this case study may present promising findings, there are certain contexts to the research that should be addressed which may influence the feasibility within a wider scale approach. Firstly, since the study was opened to participants who volunteered, it may present an evident bias that those who willingly participated in the research may have had an inclination towards games in general, making them more motivated to train in the program. This modality of exercise training may not be as effective if participant selection were at random or they were not particularly keen on exercise or games in general. Data collected from larger pools of participants may show different adherence and efficacy rate towards the exergaming intervention. Additionally, in-depth understanding and passion in exergaming mechanics or hardware features from researchers during individual coaching sessions may also play a role on first impressions of the exercise on participants. These factors may encourage participants more and prevent drop outs when enduring the long study duration.

9.6 Future Directions

The existence of exergames as an exercise equipment that is weather-resistant, space-efficient and accessible online becomes a promising approach to greener global initiatives. Setting up of exergaming equipment can be costly to the lower-income group, where external financial support could prove useful. Collaborative efforts from countries from the North and South can pave new ventures in e-sports expansion, leveraging fair play among players that is independent of physical attributes. Although data in this case study point towards positive health benefits of exergaming to users, its social acceptance can be challenging, as video games in general carry certain negative connotations among public perception.

9.6.1 Potential Challenges or Limitations of Widespread Exergaming Adoption and How to Address Them

The three most significant challenges that could prevent widespread use of exergaming adoption include the high cost of setting up the exergaming platform, social isolation tendencies leading to disengagement, and serious concerns over gaming addiction. Larger proportions of exergamers can help reduce output costs, making it more affordable and attractive for the general population. However, the initial setup requiring expensive gaming consoles or subscriptions, still present a prohibitive challenge for lower income earners in urban areas. Government incentives or tax rebates may provide some relief for potential users to initiate exergaming.

Socially, video games have always been associated with negative characteristic undertones such as an 'addiction' or being 'problematic'. Prolonged use of video games can be a hazard especially to children and young adults, but the problem lies not on the tool itself, rather the user. Therefore, it is imperative that duration of gameplay is kept in check using automated timers and moderated by the respective administrations, commonly seen in all forms of media content.

9.6.2 Practical Recommendations for Exergame Use in Communities

The introduction of exergames is parallel to introducing any new exercise equipment found in the market. However, exergames become advantageous in view of its portability, flexible arrangements at home, as well as interchangeable uses. A key beneficial element of exergaming at home, is the larger global community network that can be achieved via online platforms. This multiplayer online availability becomes a futuristic approach to social interactions while exercising that can transcend international boundaries. Furthermore, the unnecessary need to be physically present or travel to a specific facility for exercise, indirectly reduces carbon footprint, by negating transportation or travelling needs. Reducing unnecessary travel frequencies will ultimately reduce road traffic, especially during peak hours. These home-based activities, popularised especially during the Covid-19 pandemic, becomes crucial elements in providing more efficient means of car usage in the future. Therefore, implementation of climate-conscious policies in PA promotion, helps reduce carbon emissions, by introducing home-based, online-interconnected series of e-sports exergaming instead.

In addition, exergaming also has the advantage of being an indoor-based exercise tool that is weather-resistant. Unfavourable weather conditions outside had no effect on exergaming feasibility, making it a robust form of sports engagement that is independent of weather changes. Being indoors, especially within the Malaysian demography, presents a multitude of advantage, owing to the hot and humid climate apparent in the country. This makes exergaming a type of exercise that is unaffected by hot or rainy days, common among East Asian countries, allowing greater adherence and continuity.

Moving forward, the promotion of exergames by integrating AR/VR in communities or health services should include wider reaches and innovative platforms. Recommendations by WHO (World Health Organization 2022b) highlights the importance of promoting PA in primary health care clinics, childcare centres, workplace initiatives, community-based or public open spaces and commercial settings.

9.6.3 Collaborative Efforts Between the Global North and Global South

With the emerging popularity of e-sports dominating the sports arena globally, the findings from this case study may be an important stepping stone towards collaborative efforts to fight against ongoing sedentarism within a digitally expanding world. Since online connectivity between players from the North and South is now possible, e-sports exergaming may mark its first venture by allowing players to compete against one another without the need to physically travel to a specific location. The online networking feature can help reduce carbon footprint that has contributed to climate change, and potentially expand connectivity within the comfort of one's own home.

Furthermore, with the advent of digital game avatars replacing players' physical attributes, the e-sports gaming arena can be leveraged to provide fair play among players with different body sizes—apparent among players of different ethnicity, gender or physical disability. This advantage can help foster more diversity and reduce the physical gaps between players, providing equal play and equal opportunities. E-sports exergaming can potentially be synonymous with the anthem of one world sports competition based solely on the player's skill.

9.6.4 Ethical Considerations for Widescale Implementation

The widescale integration of PA promotion using exergaming ensures all levels of residents are able to benefit from the training roll-out, instead of targeting only specific groups. However, health practitioners will first need to design specialised programs for populations who are deemed high risk for moderate-vigorous exercise engagement, such as those with multiple co-morbidities, life-threatening diseases and high-level paralysis. For this population, aspects related to safety while performing self-initiated sports on their own at home must come with emergency protocols outlined by qualified experts in place. Implementation of exergaming for vulnerable populations will need unbiased evidences from randomised clinical trial studies, as this can provide robust scientific evidence for its efficacy. Comparative studies against an exergaming's real-life sports counterpart may also be advantageous in comparing benefits and challenges of the two modalities. Additionally, specific forms of skill training based on using virtual-reality based sports such as exergaming may need to consider the real-life transferable skill acquisition it can offer.

9.7 Conclusion

This case study was designed to be of a small sample size to provide proof-of-concept evidence that home-based exergaming can be advantageous in densely populated East Asian cities. This is because the setting where the research participants were based in is urban areas with high population and limited spaces. Exergaming has shown potential in being an enjoyable, interactive and space-efficient solution to increase overall PAL, improve CRF and lower HRrest. These parameters contribute to the overall health quality and can help reduce morbidity. The integration of exergaming, benefits users by overcoming barriers to exercise, is fun and interactive, home-based feasible, compact, flexible and portable. However, wide-scale implementation of exergames, face challenges for being too expensive to initiate, while also wrought by hardware and software issues.

Highlights

- Home-based exergaming can improve overall PAL, CRF and HRrest when performed at dose-potent intensities.
- Exergames can overcome common barriers to exercise among East Asians, while being fun, interactive, compact and portable.
- Challenges in implementing exergaming include being too expensive to initiate and having multiple frustrating software/hardware issues.

Acknowledgements This case study was supported by Universiti Malaya Sustainable Development Centre (UMSDC) grant LL2023ECO015. The funding source had no influence on the outcome of this study.

References

Bakkum AJ, de Groot S, Stolwijk-Swüste JM, van Kuppevelt DJ, van der Woude LH, Janssen TW. Effects of hybrid cycling versus handcycling on wheelchair-specific fitness and physical activity in people with long-term spinal cord injury: a 16-week randomized controlled trial. Spinal Cord. 2015;53:395–401.

Balady GJ, Chaitman B, Driscoll D, Foster C, Froelicher E, Gordon N, et al. American heart association/American college of sports medicine joint position statement: recommendations for cardiovascular screening, staffing, and emergency policies at health/fitness facilities. Med Sci Sports Exerc. 1998;30:1009–18.

Begde A, Alqurafi A, Pain MTG, Blenkinsop G, Wilcockson TDW, Hogervorst E. The effectiveness of home-based exergames training on cognition and balance in older adults: a comparative quasi-randomized study of two exergame interventions. Innov Aging. 2023;7:igad102.

Benzing V, Schmidt M. The effect of exergaming on executive functions in children with attention-deficit hyperactivity disorder: a randomized clinical trial. Scand J Med Sci Sports. 2019;29:1243–53.

Borg GAV. Psychophysical bases of perceived exertion. Med Sci Sports Exerc. 1982;14:377–81.

Chaput JP, Willumsen J, Bull F, Chou R, Ekelund U, Firth J, et al. World Health Organization guidelines on physical activity and sedentary behaviour for children and adolescents aged 5–17 years: summary of the evidence. Int J Behav Nutr Phys Act. 2020;17:141.

Chen F. The mobile gaming craze in Asia: Unveiling its secrets. The Value Engineers. 2023. https://www.thevalueengineers.com/the-mobile-gaming-craze-in-asia-unveiling-its-secrets/2023. Accessed 10 May 2024.

Cheng HI, Hsu WT, Wang WH, Li RH, Tang FC, Huang SL. Health behavior models and regular exercise levels among workers in Taiwan. J Multidiscip Healthc. 2023;16:4149–59.

Farrow M, Lutteroth C, Rouse PC, Bilzon JLJ. Virtual-reality exergaming improves performance during high-intensity interval training. Eur J Sport Sci. 2019;19:719–27.

França C, Ashraf S, Santos F, Dionísio M, Ihle A, Marques A, et al. Estimated energy expenditure in youth while playing active video games: a systematic review. Sports. 2024;12:39.

Garber CE, Blissmer B, Deschenes MR, Franklin BA, Lamonte MJ, Lee IM, et al. American college of sports medicine position stand. Quantity and quality of exercise for developing and maintaining cardiorespiratory, musculoskeletal, and neuromotor fitness in apparently healthy adults: Guidance for prescribing exercise. Med Sci Sports Exerc. 2011;43:1334–59.

Gerber CN, Kunz B, van Hedel HJ. Preparing a neuropediatric upper limb exergame rehabilitation system for home-use: a feasibility study. J Neuroeng Rehabil. 2016;13:33.

Ghahfarrokhi MM, Banitalebi E, Negaresh R, Motl RW. Home-based exercise training in multiple sclerosis: a systematic review with implications for future research. Multiple Scler Related Disord. 2021;55: 103177.

Gonzales TI, Jeon JY, Lindsay T, Westgate K, Perez-Pozuelo I, Hollidge S, et al. Resting heart rate is a population-level biomarker of cardiorespiratory fitness: the Fenland study. PLoS ONE. 2023;18: e0285272.

Graves LE, Ridgers ND, Williams K, Stratton G, Atkinson G, Cable NT. The physiological cost and enjoyment of Wii Fit in adolescents, young adults, and older adults. J Phys Act Health. 2010;7:393–401.

Heesterbeek PJC, Berkelmans HWA, Thijssen DHJ, van Kuppevelt HJM, Hopman MTE, Duysens J. Increased physical fitness after 4-week training on a new hybrid FES-cycle in persons with spinal cord injury. Technol Disabil. 2005;17:103–10.

Huang Y, Ng OL, Ha ASC. A qualitative exploration of facilitators and barriers to physical activity participation among Chinese retired adults in Hong Kong. Int J Environ Res Public Health. 2022;19:3495.

Jurca R, Jackson AS, LaMonte MJ, Morrow JRJ, Blair SN, Wareham NJ, et al. Assessing cardiorespiratory fitness without performing exercise testing. Am J Prev Med. 2005;29:185–93.

Knights S, Graham N, Switzer L, Hernandez H, Ye Z, Findlay B, et al. An innovative cycling exergame to promote cardiovascular fitness in youth with cerebral palsy: a brief report. Dev Neurorehabil. 2016;19:135–40.

Koutsogeorga G. Contrasting ideals: comparing American and Japanese homes. ZenVita. 2015. https://www.zenvita.com/blog/contrasting-ideals-comparing-american-and-japanese-homes.html. Accessed 10 May 2024

Lee K. Home-based exergame program to improve physical function, fall efficacy, depression and quality of life in community-dwelling older adults: a randomized controlled trial. Healthcare. 2023;11:1109.

Lee GY, Kim KO, Ryu JH, Park SH, Chung HR, Butler M. Exploring perceived barriers to physical activity in Korean older patients with hypertension: photovoice inquiry. Int J Environ Res Public Health. 2022;19:14020.

Lobel A, Engels RC, Stone LL, Burk WJ, Granic I. Video gaming and children's psychosocial wellbeing: a longitudinal study. J Youth Adolesc. 2017;46:884–97.

Marques LM, Uchida PM, Barbosa SP. The impact of Exergames on emotional experience: a systematic review. Front Public Health. 2023;11:1209520.

Mat RM. Are Malaysians exercising? A psychometric analysis of their physical activity habits, physical literacy and exercise participation rates among adults with and without disability. Behav Sci. 2023;13:570.

Mat Rosly M, Mat RH. E-sports exergaming for youth with physical disabilities: towards equality in sports. In: Mohd Jaladin RA, Shanmuganathan T, editors. Empowering persons with disabilities: a multidisciplinary perspective. Kuala Lumpur: University of Malaya Press; 2022. p. 177–200.

Mat Rosly M, Mat RH. Home-based exergaming training effects for two individuals with spinal cord injury: a case report. Physiother Theory Pract. 2023;39:208–18.

Mat Rosly M, Halaki M, Mat Rosly H, Cuesta V, Hasnan N, Davis GM, et al. Exergaming for individuals with spinal cord injury: a pilot study. Games Health. 2017a;6:279–89.

Mat Rosly M, Mat Rosly H, Davis GM, Husain R, Hasnan H. Exergaming for individuals with neurological disability: a systematic review. Disabil Rehabil. 2017b;39:727–35.

Mat Rosly M, Mat Rosly H, Hasnan N, Davis GM, Husain R. Exergaming boxing versus heavy bag boxing: are these equipotent for individuals with spinal cord injury? Eur J Phys Rehabil Med. 2017c;53:527–34.

Mat Rosly M, Halaki M, Hasnan N, Mat Rosly H, Davis GM, Husain R. Leisure time physical activity participation in individuals with spinal cord injury in Malaysia: barriers to exercise. Spinal Cord. 2018;56:806–18.

Mat Rosly M, Halaki M, Mat Rosly H, Davis GM, Hasnan N, Husain R. Malaysian adaptation of the physical activity scale for individuals with physical disabilities in individuals with spinal cord injury. Disabil Rehabil. 2020;42:2067–75.

Mat Rosly M. Physical activity and exercise participation among Malaysian children (able-bodied vs. physical disability): a cross-sectional study. Children. 2022;9:704.

Miyachi M, Yamamoto K, Ohkawara K, Tanaka S. METs in adults while playing active video games: a metabolic chamber study. Med Sci Sports Exerc. 2010;42:1149–53.

Roberton T, Bucks RS, Skinner TC, Allison GT, Dunlop SA. Barriers to physical activity in individuals with spinal cord injury: a western Australian study. Aust J Rehabil Counselling. 2011;17:74–88.

Rosenberg D, Depp CA, Vahia IV, Reichstadt J, Palmer BW, Kerr J, et al. Exergames for subsyndromal depression in older adults: a pilot study of a novel intervention. Am J Geriatr Psychiatry. 2010;18:221–6.

Sauchelli S, Brunstrom JM. Virtual reality exergaming improves affect during physical activity and reduces subsequent food consumption in inactive adults. Appetite. 2022;175: 106058.

Shah K. Share of mobile gamers in Southeast Asia in 2021, by country and age. Business YouGov. 2023. https://business.yougov.com/content/48108-console-vs-mobile-gaming-understanding-trends-across-18-markets. Accessed 10 May 2023.

Spire Academy. Gaming in Asia—sizing opportunities in the region. 2017. https://www.spireresearch.com/gaming-in-asia-sizing-opportunities-in-the-region-3/. Accessed 10 May 2017.

Staiano AE, Abraham AA, Calvert SL. Adolescent exergame play for weight loss and psychosocial improvement: a controlled physical activity intervention. Obesity. 2013;21:598–601.

Staiano AE, Beyl RA, Guan W, Hendrick CA, Hsia DS, Newton RLJ. Home-based exergaming among children with overweight and obesity: a randomized clinical trial. Pediatr Obes. 2018;13:724–33.

Sultan N, Khushnood K, Awan MMA. Exergaming: An effective way to maintain physical and mental health at home during Coronavirus Disease-2019 (COVID-19) pandemic. J Coll Phys Surg Pakistan. 2020;30:143.

Tanaka H, Monahan KD, Seals DR. Age-predicted maximal heart rate revisited. J Am Coll Cardiol. 2001;37:153–6.

Teng B, Rosbergen ICM, Gomersall SR, Hatton A, Brauer SG. Barriers to and facilitators of adherence to prescribed home exercise in older adults at risk of falling in Singapore: a qualitative study. J Aging Phys Act. 2022;31:48–58.

Tieri G, Morone G, Paolucci S, Iosa M. Virtual reality in cognitive and motor rehabilitation: facts, fiction and fallacies. Expert Rev Med Devices. 2018;15:107–17.

Tripette J, Murakami H, Gando Y, Kawakami R, Sasaki A, Hanawa S, et al. Home-based active video games to promote weight loss during the postpartum period. Med Sci Sports Exerc. 2014;46:472–8.

van der Kolk NM, de Vries NM, Kessels RPC, Joosten H, Zwinderman AH, Post B, et al. Effectiveness of home-based and remotely supervised aerobic exercise in Parkinson's disease: a double-blind, randomised controlled trial. Lancet Neurol. 2019;18:998–1008.

World Health Organization. World Health Organization guidelines on physical activity and sedentary behaviour: At a glance. 2022b. https://www.who.int/publications/i/item/9789240014886. Accessed 31 Dec 2022.

World Health Organization. Physical inactivity a leading cause of disease and disability, warns World Health Organization. Geneva. 2022a. Accessed May 2022.

World Population Review. House size by country 2024. 2024. https://worldpopulationreview.com/country-rankings/house-size-by-country. Accessed 10 May 2024.

Maziah Mat Rosly is an expert on serious games and exergaming for active living. She received her MBBS degree from Manipal Academy of Higher Education in 2011 and obtained her Dual PhD in Exercise Physiology and Health Sciences from Universiti Malaya and The University of Sydney in 2019. She is affiliated with numerous serious games programs built for long-term sustainability, cost-efficient deployment and effective health management. Dr. Maziah is mainly responsible for overseeing the development, design and implementation of serious games for education and innovating virtual reality-based exergames to improve adaptability, reduce carbon footprint and, provide highly cost and space-efficient mechanisms for sports and exercise training. Her latest project involves designing a Virtual Reality cube e-sports exergame feasible for deployment in East Asian urban cities and for space exploration.

Chapter 10
Future of Personalised Medicine in Asia: Trends, Challenges, and Collaborative Pathways

Clusters of Future Studies: Past and Futures; Technological Trends

Sydney Tjandra, Muhammad Iqhrammullah, Shakira Amirah, Vindasya Almeira, and Don Eliseo III Lucero-Prisno

Abstract Personalised medicine has been prophesied to be the future of various medical facets, from epigenetics to treatment, serving Asia's large population and promising technological capabilities. This review chapter includes peer-reviewed journal articles from PubMed, Scopus, Google Scholar, and Web of Science, as well as legitimate web-based sources, entailing two main sections. Firstly, it aims to analyse past historical trends and challenges in personalised medicine in various aspects: research, industry, governance, as well as ethical and legal considerations. Secondly, alternative futures for personalised medicine are depicted with those aspects in mind. We categorised three country tiers: tier 1 countries with a head start in genomics, such as China, tier 2 countries with newly-made genomic facilities, such as Thailand and Indonesia, and tier 3 countries, which include all other nations that have not initiated independent genomic facilities. These three tiers entail their unique opportunities, low-hanging fruits, and sociopolitical nuances, hence projecting different futures as consequences. This chapter also highlights the potential for disparities in which these futures merge in the context of global collaborations with or within Asia (e.g. GenomeAsia), key opportunities in a preferred future, and recommendations for a multisectoral approach in personalised medicine.

Keywords Personalised medicine · Historical trends · Future · Disparities · Genomics

S. Tjandra (✉) · S. Amirah · V. Almeira
Faculty of Medicine, Universitas Indonesia, Jakarta, Indonesia
e-mail: sydney.tjandra@gmail.com

M. Iqhrammullah
Universitas Muhammadiyah Aceh, Banda Aceh, Indonesia

D.E.III Lucero-Prisno
London School of Hygiene and Tropical Medicine, Bloomsbury, UK

R. Xiang Ng et al. (eds.), *Future of Healthcare in Asia*,
https://doi.org/10.1007/978-3-032-03183-9_10

10.1 Introduction

The field of personalised medicine in Asian countries is complex and continually evolving, presenting both unique challenges and opportunities. The origins of individual-targeted medicine can be traced back thousands of years to ancient times, where various therapeutic approaches were first established. Despite these early advancements, conventional therapies today often fail to consider individual idiosyncrasies and genetic makeup, resulting in less effective treatments for some patients. The demand for more precise and effective treatment methods led to the emergence of the scientific field now known as "personalised medicine" (Visvikis-Siest et al. 2020).

Significant technological breakthroughs have positioned personalised medicine as the next generation of diagnosis and treatment. However, despite the increasing attention it has garnered in recent years, there remain several obstacles to its widespread application in clinical practice. These limitations have been further highlighted by the COVID-19 pandemic (Agyeman and Ofori-Asenso 2015).

The emergence of sophisticated biomedical assays, including deoxyribonucleic acid (DNA) sequencing (such as multiplex next-generation sequencing (m-NGS), proteomics, imaging protocols, and wireless monitoring devices, has revealed substantial inter-individual diversity in understanding disease processes. These technologies hold the promise of tailoring medical interventions to individual genetic profiles, lifestyle factors, and environmental influences. This shift towards personalised medicine is propelled by high-throughput, data-intensive methodologies, which illuminate a spectrum of inter-individual variability. Recognising the unique biochemical, physiological, environmental, and behavioural attributes of each individual, personalised medicine advocates for tailored interventions to optimise disease prevention, monitoring, and treatment strategies (Khoury 2016; Morganti et al. 2019) (Khoury 2016; Morganti et al. 2019). Personalised medicine is often interchangeably referred to as precision medicine due to its focus on tailoring medical treatment to the individual characteristics of each patient. The term "precision" highlights the goal of precisely matching treatments to the specific genetic makeup, lifestyle, and environmental factors of a patient (Goetz and Schork 2018).

Collaborations among academic institutions, healthcare providers, and technology stakeholders are essential for advancing personalised and precision medicine. These partnerships are designed to facilitate the translation of research into technology development, clinical validation, and healthcare applications (Dhawan 2016). They play a pivotal role in the identification, validation, and implementation of biomarkers crucial for individualized treatments (Landeck et al. 2016). Such collaborations are fundamental for establishing sustainable governance and organization of precision medicine, as well as addressing ethical, legal, and access-related issues (Stenzinger et al. 2023).

The use of personalised medicine is projected to continue growing over time. Data indicate that the market value of personalised medicine will reach USD 50.2 billion by 2028, up from USD 29.1 billion in 2023, with a compound annual growth rate (CAGR) of 11.5%. (Global Forecast 2023) As personalised medicine becomes increasingly prominent, it encounters a multitude of challenges encompassing regulatory approval, acceptance among stakeholders, and cost-effectiveness (Hayden 2016). Notably, promising therapies such as a personalised cancer treatment where a patient's T cells are genetically engineered (using chimeric antigen receptors, CARs) to recognize and attack cancer cells (autologous CAR-T cell transplants) and mutation-specific medicines offer targeted solutions for disease management but also raise concerns due to their high costs (Shah et al. 2018). Regulatory agencies face the arduous task of assessing the safety and efficacy of personalised therapies for standard clinical use (Hopkins 2023). Moreover, healthcare stakeholders, including physicians, administrators, insurers, and patients, grapple with issues related to accessibility, affordability, and the comparative effectiveness of personalised approaches compared to conventional treatment methods (Goetz and Schork 2018).

This chapter delves into the multifaceted landscape of personalised medicine, examining historical trends, current challenges, and alternative futures across different tiers of countries in Asia. This review provides a broader perspective, noting the significant role of Asian countries in the global pharmaceutical and medical device industry, and the potential for further growth and innovation in the field of personalised medicine.

10.2 Review Methodology

10.2.1 Search Terms and Databases

Literature search was performed on PubMed, Scopus, Google Scholar, and Web of Science databases for studies published as of April 1st, 2024. To identify relevant studies, the search strategies were comprised of the following search terms: "personalized medicine" OR "personalised medicine" OR "precision medicine". The search was applied to the title, abstract, and keywords of the indexed literature metadata. The search was also performed using forward- and backward-tracking reference lists. Web-based sources were also searched to retrieved information pertaining to the progress of personalised medicine, including the presence of related facilities (i.e. biobanks, genomic research centres, sequencing technologies).

10.2.2 Literature Selection Criteria

We included peer-reviewed journal articles and legitimate web-based sources into the present review. For journal articles, we prioritized cohort studies and randomized controlled trials. Only in cases where they were not available, we included the other observational or interventional studies such as case–control, cross-sectional, and non-randomized controlled studies. Case reports and case series were given the least priority. To support conceptual data in the present review, we included theoretical or conceptual studies, expert opinions, and proof-of-concept studies. We excluded low-quality studies appraised using Cochrane Risk of Bias Tool 2.0 for randomized controlled trials, STROBE (Strengthening the Reporting of Observational Studies in Epidemiology) for observational studies, and Joanna Briggs Institute Critical Appraisal tools for various study designs. For web-based sources, we prioritized government official websites. If none of such website available, the search was performed for articles written by respected experts in personalised medicine or published by reputable media. We excluded web-based articles written by unknown authors or by non-reputable media in Asia. The exclusion was made for journal or web-based articles that were not reporting the personalised media in Asian countries (though some articles were included for comparison purposes). No restriction on the publication date and the language. Articles written in languages other than English were translated using Google Translate. The selection was initially carried out by a single review author, and were then validated by all authors.

10.3 Findings

In this first half of the chapter, we hereby outline past trends and challenges in personalised medicine: research, translation into practical products, as well as ethical and legal considerations. Interesting examples in Asia are also discussed.

10.3.1 Research Trends

10.3.1.1 Acceleration of Genome Sequencing Through Novel Methods

The acceleration of genome sequencing technologies, particularly multiplex next-generation sequencing (m-NGS), has marked a transformative period in personalised medicine both in Asia and globally. (Park and Kim 2016) m-NGS allows for the rapid and comprehensive analysis of genomic data, crucial for the development of treatments tailored to individual genetic profiles (Chong et al. 2018).

Table 10.1 General overview of personalised medicine status in some countries in Asia and Southeast Asia

Country (Ref)	General overview	Biobank	Data privacy and protection
Tier 1 countries			
China42 (Regalado 2020)	China has a global leading genomic centre and participates in the Human Genome Project. The country is involved in international projects through genomic data sharing. The price of commercial genomic sequencing is 100 USD	Has state- and private-owned biobanks (i.e. China National GeneBank)	Full integration into the Chinese healthcare system Regulated under personal information protection law (pipl) 2021, cybersecurity law 2017, and genetic resources regulation
South Korea (Lee and Lee 2023; Kim et al. 2021)	South Korea spearheads numerous genome-wide initiatives (i.e. KoGES), alongside managing biobanks through institutions such as the Genome Research Foundation. The advancement of personalised drugs is facilitated by GIFT (global innovative products on the fast track)	Has state- and private-owned biobanks (i.e. KoGES, and Macrogen)	The bioethics and safety act (2005) is specified for genetic discrimination prevention Advanced regenerative medicine and advanced biological products policy review committee is responsible for the clinical research approvals Health information regulation agency (HIRA) regulates data protection
Japan (Illumina 2024, Muto et al. 2023)	Japan participates in the Human Genome Project and has well-established biobanks including the National Institute of Genetics founded in 1949	Has state- and private-owned biobanks	The personal data protection law which was amended in 2015 has introduced terms such as "individual identification code" and "special-care-required personal information"

(continued)

Table 10.1 (continued)

Country (Ref)	General overview	Biobank	Data privacy and protection
Singapore	Singapore, with institutions like the Genome Institute of Singapore (GIS) and National University of Singapore (NUS), has made significant contributions to genomic research in Southeast Asia	Has state- and private-owned biobanks (i.e. GIS)	Data protection is regulated under personal data protection act (PDPA) 2012 Data collection and storage are regulated under human biomedical research act (2015)
Tier 2 countries			
India	Personalised medicine in India is led by Institute of genomics and integrative biology (CSIR–IGIB). The country participates in the global consortium for genomics research (GCG)	Has state- and private-owned biobanks (i.e. CSIR-IGIB human bio-resource repository, medgenome, and mapmygenome)	Regulating bodies: Indian council of medical research (ICMR) and Indian genome variation consortium (IGVC)
Philippines (Phillipines House of Representative 2022a, 2022b)	Philippines Genome Centres (PGC) was established in 2009, with collaborative projects with other countries in ASEAN	Has state- and private-owned biobanks including that in the University of the Philippines—Philippine General Hospital	Data privacy act of 2012 or republic act no. 10173 regulate data privacy and protection Amendments to specify biometric and genetic data (House Bill No. 898 and House Bill No. 9651) are on pending status since their first proposal in 2022

(continued)

Table 10.1 (continued)

Country (Ref)	General overview	Biobank	Data privacy and protection
Thailand (Gunn 2023)	BIOTEC, a national center for genetic engineering and biotechnology, was established in Thailand since 1983. Thailand's Ministry of Public Health has issued a "Standard of practices for genomic medicine service" guideline for the standardization of personalised medicine providers	Has state- and private-owned biobanks (i.e. GIS)	Regulated under personal data protection act, B.E. 2563 (2021) The law categorizes genetic, health, and biometric data as sensitive data
Indonesia (BGSi 2023)	Personalised medicine research in this country was previously led by the Eijkman Institute, but now it is primarily carried out by BGSi. The country contributes to global research through Indonesia Human Genome Reference project. The price for genomic sequencing ranges from 300 to 600 USD	Has state- and private-owned biobanks (i.e. BGSi, MGI, and PT. Asa Nusantara)	• Registry Indonesia (RegINA) emphasising accessibility and security • Regulated under: Health Law No. 17/2023; Government regulation no. 21/2005 on biosafety of genetically engineered products; law no. 27/2022 on personal data protection
Malaysia (Amin et al. 2021; Jamal et al. 2021)	The research has been led by Malaysian Genomics Resource Centre (MGRC) since 2004, Malaysia Genome & Vaccine Institute (MGVI), and various institutions within universities	Has state- and private-owned biobanks (i.e. MGVI, MGRC, and AGTC)	• Infrastructure for data privacy and security is under development • Some related regulations include human tissue act 1974 and the personal data protection act 2010 (not specifically explain the bio-repository protection)

(continued)

Table 10.1 (continued)

Country (Ref)	General overview	Biobank	Data privacy and protection
Vietnam (Government of the Socialist Republic of Vietnam 2023; Ton 2023)	The Institute of Genome Research (IGR) in Vietnam was established in 2010 focusing on human and agricultural genomics research. The research centre collaborates with various international institutions. It collaborates with international partners through 1000 Genomes Project, SEAPharm, and E-Dengue projects. Some personalised drugs are approved and fully reimbursed (i.e. imatinib)	Has state- and private-owned biobanks (i.e. Vietnam Academy of Science and Technology, OUCRU, VinGroup's Institute of Big Data and OGCDC)	Regulation on data protection under Decree No. 13/2023/ND-CP of the Government on Personal Data Protection ("PDPD") Personalised medicine-related data protection is regulated per the standard of the provider
Tier 3 countries			
Cambodia (Cohen et al. 2023)	Despite being unspecific to personalised medicine, the genomic research in Cambodia is led by Cambodia Institute of Research and Development (CIRD). No information regarding international collaboration in this country	The government has partnered with private-owned biobanks. Some of which include Rodolphe Mérieux Laboratory of Phnom Penh and Institut Pasteur du Cambodge (IPC)	Legal framework for data privacy and protection in cambodia is still in its infancy Sub-Decree No. 252 is issued by the Royal Government of Cambodia to cater data privacy and protection The term 'biometric data' was mentioned but not defined
Myanmar 54 (Thean-ngarm and Oo 2023)	Myanmar Genome Research Center (MGRC) was established in 2016 with the support of South Korea. The research centre focuses on improving healthcare through human genomics	Does not have biobanks	The regulation is nearly absent, but are provisioned under the Constitution of the Republic of the Union of Myanmar 2008 and the Law Protecting the Privacy and Security of Citizens (Union Parliament Law 5/2017)

(continued)

Table 10.1 (continued)

Country (Ref)	General overview	Biobank	Data privacy and protection
Brunei (Katsanis and Katsanis 2013, Azmi et al. 2022)	Brunei does not currently have a dedicated genomic centre. However, genomic research is conducted within various institutions, like Universiti Brunei Darussalam and the Ministry of Health	A few information regarding biobanks in this country. But it has Malua BioBank that focuses on microorganisms in forests. It has a private-owned biobank Borneo Genomics Innovation Sdn. Bhd	Personal data protection is regulated under personal data protection order (PDPO) but not specified, even sensitive information is not distinguished from other data
Laos (Than et al. 2022; Kutanan et al. 2019, 2021)	There is limited genomic surveillance in Laos, but there are several collaborative international projects investigating the Lao population (i.e. Lao Genome Project in 2016). The country also has the Lao-Oxford-Mahosot Hospital-Wellcome Trust Research Unit (LOMWRU) conducting projects focusing on infectious diseases	Does not have biobanks	The law on electronic data protection enacted in 2017 can be used for the protection of data collection, processing and storage in electronic media, but not specified for health-related or genetic data

The investment in m-NGS has been particularly noteworthy in Asia, where countries like China, Japan, and Singapore are leading the charge (Global Data 2023). The availability of genomic sequencing technologies in these countries along with other Asian countries can be seen in Table 10.1. Reductions in sequencing costs and times, coupled with technological advancements, have brought personalised medicine to the forefront of healthcare innovation (Wong et al. 2023). Moreover, Asian countries' commitment to bolstering their biotechnological landscape through governmental policies further propels the growth and application of m-NGS (Manolio et al. 2015). Contributing to this acceleration are initiatives such as the GenomeAsia 100 K Project, which aims to sequence the genomes of individuals from diverse Asian populations to understand genetic variability and its implications for disease and treatment. This endeavour, along with other public–private partnerships, emphasises the importance of collaborative research efforts. These international consortia serve to enrich the databases of genetic material, which involves many physicians who collected patients' data from both public and private sectors, such as the National Health System (NHS) and Illumina (Illumina 2024) are expected to facilitating a deeper understanding of population-specific genomic markers and accelerating the advent of precision medicine. The expansion of genomic data has the potential to revolutionise healthcare, offering better diagnostic tools, more efficient drug development processes, and more effective treatment plans (Wall et al. 2019; MGI 2023).

Key factors such as government investment, policy support, and international collaborations are instrumental in developing and implementing m-NGS. In Japan, for instance, the government's backing of genomic initiatives has positioned it at the forefront of cancer genomics, proven by hospital-laboratory readiness to provide gene panel tests for sequencing. Japanese projects aimed at integrating genomics into clinical settings serve as examples of effective policy driving scientific innovation. Beyond technology and finance, the formation of robust frameworks addressing the ethical, legal, and social implications of genome sequencing ensures the responsible use of genetic information, which is essential for its clinical application (Takeda et al. 2019).

10.3.1.2 Global Collaboration

As seen in m-NGS, personalised medicine has provided precise individual profiles and distribution of common-type genomic characters in populations. Since its report on significance and efficacy, personalised medicine has started to be recognised worldwide. Whole genome sequencing, metagenomics, transcriptome, and other approaches of individualised and genomics medicine—a subset of personalised medicine—have been fully supported in all countries, tempted by its sophistication to treat and diagnose common highly-contributed-mortality diseases, such as malignancies (Venner et al. 2022). Health research and technology majors in universities, genomic-based companies, as well as health stakeholders worldwide show boundless interest as many are committed to getting involved in maximising its potential through investment on national and international projects, such as Biomedical and

Science Initiative (BGSi) initiated by Indonesia's Ministry of Health, which involving Faculty of Medicine University of Indonesia and Cipto-Mangunkusumo Hospital. Unfortunately, resources and capabilities to evolve personalised medicine present challenges, especially in developing countries lacking the technology, resources, and experience. Indisputably, technology advancement is quite important to establish appropriate personalised medicine. However, addressing the importance of this approach to government stakeholders has encountered many obstacles since they would possibly have different values from the mission of the Ministry of Health and their designated agenda. Considering the hefty cost in developing those tools and inadequacy of supported legal policies, personalised medicine seems to be hampered for indefinite periods. Nevertheless, as patient characteristics data and biorepositories are urgently needed to combat mortality escalation from deadly disease and the need to broaden research population coverage of developing countries are in demand, indeed global collaboration with more developed countries would be the best solution. Personalised medicine is already well-established in developed countries and could be positively impacted if opportunities were constantly presented for developing countries to collaborate and learn to make their own frameworks. Collaboration can involve academia, health and technology stakeholders, and many healthcare providers in diagnosing, treating, and preventing disease which could be used as a reference for treatment guidelines and further research topics.

10.3.2 Industrial Outlook of Personalised Medicine

10.3.2.1 Translation of Research into Practical Products

The process of translating research findings into practical applications within the realm of personalised medicine has been characterised by a dynamic interplay of historical trends and challenges. An exemplary instance of this translational endeavour lies in the conversion of genomic research insights into clinical implementations of targeted therapies. The discernment of specific genetic mutations associated with particular cancers has precipitated the development of targeted therapeutic agents such as trastuzumab, a monoclonal antibody targeting the HER2 receptor, has markedly enhanced treatment outcomes for patients with HER2-positive breast cancer after the identification of HER2 as a pivotal oncogene in this context. Similarly, imatinib, a paradigmatic cancer therapy targeting a specific molecular aberration, primarily addresses chronic myeloid leukaemia by antagonising the BCR-ABL tyrosine kinase, the product of a chromosomal translocation event. Such transition has significantly enriched the spectrum of personalised treatment modalities available to patients (Zakaria et al. 2023).

Another significant domain where research findings have been effectively translated into clinical practice is pharmacogenomics. This interdisciplinary field investigates the influence of genetic variations on individual responses to pharmacological agents. For instance, medications like clopidogrel utilised as anticoagulants are now prescribed with consideration to genetic testing results. Genetic variation of loss-of-function (LOF) alleles in CYP2C19 gene has been proven to reduce the oxidative conversion of clopidogrel prodrug into its active metabolites (Pereira et al. 2019). This personalised approach aids in dosage determination and contributes to the mitigation of adverse drug reactions (Karczewski et al. 2012). Furthermore, pharmacogenomic assessment for warfarin dosage management incorporates considerations of genetic variations affecting drug metabolism and sensitivity. The bioavailability and efficacy of warfarin active metabolites were affected by polymorphic genes of CYP2C9, VKORC1 and CYP4F2.CYPF2 (Asiimwe and Pirmohamed 2022). Thus, genotyping is suggested to improve the treatment regimens in minimising the risks of bleeding or thrombosis (Beigh 2016; Bayer and Galea 2015; Katsanis and Katsanis 2013).

10.3.2.2 Global Collaboration in Personalised Medicine Products

Personalised medicine products have evolved and are effectively used to treat noncommunicable diseases (NCDs) and rare diseases with the high-mortality risk. Its massive number of successful rates, 50–86% complete remissions, Graham et al. (2018) has attracted public health attention and raised research interest for these approaches. However, the development of almost all genomic products has been quite difficult and costs a lot of money, hindering many countries from leveraging the effectiveness of these products. For example, there are many types of cellular products targeting exact genomes for treatment in the oncological department, such as the chimeric antigen receptor of T-cell (CAR-T) to treat multiple myeloma which has been proven by the Food and Drug Administration (FDA) for its effectiveness. CAR-T cell itself requires $150.000–$1.000.000 USD (Cliff et al. 2023) to be developed, making its affordability was getting questionable.

Nonetheless, FDA reported other global challenges to establish this product, such as manufacturing challenges due to lack of technologies, clinical development timelines, and different global regulatory requirements. In their summary reports, FDA explained their concern and recommended collaboration for all countries to provide needs of advanced laboratory machines and technologies and create protocols for manufacturing processes in order to achieve agreements and adjust regulatory requirements towards all states. FDA also initiated the global gene therapy collaborations, Collaboration on Gene Therapies Global Pilot (CoGenT), under the Centres of Biologic and Research Evaluation. This collaboration has already established certain products, including CAR-T and accelerated gene therapy accessibility for the country members. Collaboration towards private companies. Aneizi and Chokeir (2019) such as Pfizer and Cellectis collaboration on Allogeneic CAR-T (Mastrojohn and Valachs 2014) and Turnstone Biologics and Takeda Pharmaceuticals Company collaboration to provides RIVAL-0, recent immunotherapy which utilize inherent activity from SKV Vaccinia to serve oncolytic function, also shows how collaborative products have opened up opportunities to increase product accessibility, solve research-funding problems, and reduce regulatory barriers for the countries involved.

10.3.3 Governance: Rise of National Genomic Facilities

Genetic materials on a large scale are stored in biobanks. Biobanks may be either population-based, where the main interest is to oversee the progression of diseases over time, or disease-based, where specific disease data are concentrated to develop personalised treatment approaches. Such databases allow research to identify the biomarkers for various uses from susceptibility, diagnostics, cancer metastasis, to treatment responsiveness. The large data repositories are crucial for genotyping validation, a process that cannot be done within smaller samples (Policiuc et al. 2018).

Addressing such importance, numerous countries have initiated their own national genomic facilities and biobanks. The global largest is China's BGI, said to be leading with larger populations providing larger datasets of diseases (Cyranoski 2016). The country already leads several strategic areas, including metagenomic next-generation sequencing research, (He et al. 2023) gene expression research (Wang et al. 2022) and single-cell sequencing technology (Wang et al. 2021). More recently, Indonesia has newly minted their Biomedical and Genome Science Initiative (BGSi) in 2022. The country plans to sequence at least 10.000 genomes for its first database by 2024 (Arlinta 2024) a leap from 140 genomes sequenced in 2020 (Kementerian Kesehatan Indonesia 2022). This step is similar to Singapore in their first phase of establishing the National Precision Medicine program in 2017, before recruiting over 100,000 healthy samples and 50,000 patients in its second phase (Kong et al. 2022) The overview of genomic research status and facilities among Asian countries has been summarised and presented in Table 10.1 (see page number 195).

10.3.4 Ethical and Legal Considerations Associated with Personalised Medicine

10.3.4.1 Legal and Regulatory Frameworks

Establishing regulations is crucial to securing funding for personalised medicine research, offering guidance for clinical practices, shaping healthcare insurance policies, and protecting patient privacy and personal data. Many countries in Asia, particularly those with lower- and middle-income statuses, unfortunately, lack robust regulatory frameworks for genetics. Compounded by weak enforcement of regulations, this further underscores the underdevelopment of personalised medicine. Enabling the practice of personalised medicine in such countries, therefore, requires extra efforts in policymaking and policy enforcement. For instance, many Southeast Asian countries continue to struggle with formulating policies aimed at reducing maternal mortality rates, largely stemming from home births. (As et al. 2020; Laksono et al. 2023) Conversely, personalised medicine necessitates access to advanced medical technologies and services, such as genotyping and cord blood collection for stem cell therapy. Given these circumstances, Southeast Asian nations may be less inclined to prioritise policies that support personalised medicine initiatives.

For instance, giving birth at home, often with the help of traditional birth attendants, is common (Kutanan et al. 2019). In some regions of Indonesia, a partnership was even built between traditional birth attendants and midwives to reduce the maternal mortality rate (MMR) (As et al. 2020). Similarly, other countries in Southeast Asia are still more focused on safe pregnancy and delivery. Whilst countries are pursuing MMR reduction, promoting the collection of cord blood for stem cell therapy could be far-reaching.

Regulatory framework readiness for personalised medicine also includes a pharmacovigilance system. In low-middle-income countries, the selling of prescription-only medicines (PoMs) is loosely controlled. (Alenezi et al. 2023) Non-prescription sales of antibiotics which contribute to the substantial increase of drug-resistant pathogens persist as the main public health threat. (Ferdiana et al. 2021; Belachew et al. 2021) Community pharmacies mostly are unable to provide sufficient medication advice and diagnostic testing. (Alenezi et al. 2023) In Vietnam, however, the treatment recommendation for tuberculosis by the community pharmacy is significant in managing and preventing tuberculosis transmission, where 46% of the total cases received treatment recommendations from pharmacies. The lack of robust pharmacovigilance among community pharmacies, which serve as the primary point of contact for many individuals, poses a fundamental challenge to the adoption of pharmacogenomics. (Duong et al. 2011).

In most Southeast Asian countries, their healthcare systems have already suffered from inadequate health financing policies and regulations. Malaysia and Indonesia have observed an influx of healthcare costs associated with population growth, increased health-seeking awareness, and rapid progress in medical technology and drugs. (Lakha et al. 2020; Sharma et al. 2020) At the same time, Southeast Asian countries are struggling to find funding sources. Most countries have little allocation of the total national spending to the healthcare system. Ineffective taxing methods, reliance on external funding, and constraints to collect premiums contribute to unhealthy financing (Dartanto et al. 2020; Tangcharoensathien et al. 2011; Sumriddetchkajorn et al. 2019). Moreover, insurance-based discrimination is experienced by patients in Thailand, Vietnam, the Philippines, and Indonesia, where insured patients received lower service quality compared to those who were non-insured (Lim et al. 2023). Corruptive behaviour, including falsifying diagnoses to increase the payment claim, is another deep-rooted challenge that has made healthcare insurance programmes inefficient (Naher et al. 2020). Personalised medicine presents a clear challenge to healthcare costs which further substantiates a strong regulatory framework for health financing.

The unpreparedness of developing countries in Asia to receive personalised medicine technology is also observable in the perception and enforcement of personal data and privacy protection acts. Cases of data leakage by government agencies, including by the health agency, have occurred with no parties being held accountable. Intention by politicians regarding personal data and privacy protection is low (Protergo 2021; Raj 2024). Legal frameworks to regulate the distribution and access of personal data should be established especially in the context of personalised medicine which comprises genetic and behavioural big data.

10.3.4.2 Data Privacy, Legal, and Regulatory Frameworks

Personalised medicine relies heavily on individual health data to tailor treatments and interventions to specific patients, hence the name precision medicine. This data can include genetic information, medical history, lifestyle factors, and even environmental exposures. Medical history and health-related information themselves are already sensitive data, and unauthorised access to this data can lead to privacy breaches and potentially harmful consequences. (Santaló and Berdasco 2022) Other than through data breaching, re-identification of sample donors is feasible for data stored in public databases, especially epigenomic data with a high-throughput nature. Data obtained from whole-genome bisulfite sequencing (WGBS) is possible for re-identification because it provides detailed DNA sequence at a base-pair resolution by assessing genome-wide CpG methylation (Gürsoy et al. 2022). Privacy is defined as individuals' right to control the acquisition, use, or disclosure of their identifiable health-related information. Meanwhile, confidentiality is defined as the obligation of the data recipients or keepers to uphold the privacy rights of the information giver (Rothstein et al. 2009).

Stigmatization, or even worse, discrimination based on an individual's specific genetic predispositions, may manifest in various settings such as workplaces, healthcare facilities, educational institutions, and even in the formation of romantic relationships. Individuals with genetic disposition could be rejected by employers because of their higher risk of developing a specific medical condition. Similarly, insurers may deny coverage, charge higher premiums, or impose coverage limitations on the aforementioned individuals. The two instances are among many other harmful consequences which could occur because of breaching data privacy and security measures. Such potential consequences urge risk evaluation prior to adopting epigenetic-based screenings in real-world settings, especially among marginalised groups (based on socioeconomic status, race and ethnicity, sexual orientation, or others).

The legal framework for data privacy protection related to personalised medicine in the Asian region is in its infancy compared to that of the United States. The prevention of genetic information-based discrimination in the United States covers two main areas namely health insurance and employment, carried out through The Genetic Information Nondiscrimination Act (GINA, signed in 2008 and became effective in 2009) (Joly et al. 2020). In India, genetic discrimination is only addressed by the Insurance Regulatory Development Authority of India, but only applies to the insurance sector (Kim et al. 2021). Guideline from the Indian Council of Medical Research and Indian Genome Variation Consortium (IGVC), respectively, provided specific data privacy protection but is non-legally binding (Indian Council Medical Research 2024). In the case of Southeast Asian countries, specific laws regulating the consequence of genetic information are not yet specifically formulated (Kim et al. 2021). The protection of genetic privacy and genetic discrimination in Singapore is largely addressed by Ministry of Health guidelines. Singapore, Malaysia, and Thailand have already established Personal Data Protection Acts which may encompass genetic information, but not in specific detail. In South Korea, under their constitution, discrimination based on disability or medical history is prohibited (National Human Rights Commission of Korea Act of 2001 2016). In 2005, South Korea passed The Bioethics and Safety Act (2005) which specifically addresses genetic discrimination. There were multiple efforts in the Philippines to amend their Data Privacy Act of 2012 (or Republic Act No. 10173) to specifically include to the terms biometric and genetic data, but the discussion is on hold (Indian Council of Medical Research 2024). The absence of specific regulation for genetic information might implicate the extent of legal protection. The current regulation relies heavily on the individual institutions conducting genetic testing, leading to inconsistencies and a lack of standardised protection. This variability introduces uncertainty that could have serious consequences for patients whose data is utilised in personalised therapy or during the research.

10.4 Future Directions

Considering the various challenges and opportunities in history and the present, we characterise countries into three tiers as they will have different futures. Tier 1 countries are highly developed countries that have long established their genomic centres and are currently leading the field, for instance, China. Meanwhile, tier 2 countries have started establishing their centres in the past decade or the past few years, but infrastructure gaps crucial to progress remain, such as in Thailand and Indonesia. Furthermore, the remaining tier 3 countries are still yet to have their genomic initiatives. The list of countries in each tier has been previously presented in Table 10.1 (see page number 195).

10.4.1 Tier 1 Asian Countries

10.4.1.1 Opportunities: Increased Disease Prevention and Treatment Efficacy

The advancement of personalised medicine in Tier 1 countries like China has the potential to considerably impact the accessibility and affordability of healthcare, although the journey has been complex. Precision medicine applications can lead to cost savings and improved healthcare outcomes when treatments are more effective and adverse reactions are minimised. However, the initial costs associated with personalised medicine can be high, both in terms of the development of new treatments and the required genetic testing (Gao et al 2020).

The implementation of personalised medicine is mostly impacted by its accessibility and affordability. Advances in personalised medicine can offer more effective disease management and potential cures, which over time may decrease healthcare costs by reducing hospitalisations and avoiding ineffective treatments. However, the up-front cost of genetic testing and tailoring treatments can be significant and may not be covered by all insurance programs, presenting financial challenges for patients. In other instances, the capacity to produce large amounts of genomic data has outpaced the ability to interpret it in a clinically meaningful way, which can lead to additional costs without immediate benefits. Access to personalised medicine may also be unevenly distributed, where those living in urban areas with access to specialised healthcare facilities may benefit more than those in rural areas. (Gao et al. 2020; Fernald et al. 2011; Phillips et al. 2014).

When it comes to Tier 1 countries, the strides made in personalised medicine have notably impacted healthcare accessibility and affordability, particularly in nations like China. This progress is greatly augmented by the integration of artificial intelligence (AI) and advanced data processing technologies. The utilisation of AI and data processing has allowed for the interpretation of large volumes of patient data, enabling healthcare providers to develop tailored treatment plans and interventions. This personalised approach has not only improved the efficacy of medical treatments but has also contributed to cost savings by minimising unnecessary procedures and reducing the risk of adverse reactions (Kong et al. 2022).

The successful implementation of precision medicine programs in Tier 1 countries, hinges on balancing the high cost of personalised therapies and diagnostics with their potential long-term benefits in disease prevention and treatment efficacy. Ongoing initiatives striving towards universal health coverage urge the incorporation of cost-effective, personalised interventions that ultimately aim to shield individuals from undue financial hardship while enhancing healthcare delivery (Chong et al. 2018).

10.4.1.2 Boosters: Artificial Intelligence and 5G Technology

Tier 1 countries are also well-positioned to incorporate boosters to personalised medicine: artificial intelligence (AI) and 5G technology. At the molecular level, AI can sequence protein configurations and execute complex analyses, in addition to planning therapies at the clinical level with existing patient characteristics and comorbidities (Johnson et al. 2021). Meanwhile, 5G technology, with its 1000-fold bandwidth and <1 ms latency, (Li 2019) serves reliable connectivity for AI and its enormous load of data, boosting precision medicine (Kang et al. 2023). For instance, 5G can be applicable in nano-oncology, where large datasets can now be sent in real-time, allowing precise insights, to suggest personalised cancer therapy and doses (Moglia et al. 2022). 5G networks can also support AI mobile nursing systems (Guo and Li 2022). Although limitations, such as unexpected delays, data safety, and biases, exist, (Johnson et al. 2021) we believe Tier 1 countries are likely to mitigate these challenges before other countries do, further leapfrogging the personalised medicine race.

10.4.1.3 Urgency and Potential Benefits of Global Collaborations

Many international collaborations have been carried out by various countries in terms of regional conference agreements or private-company expansion, such as the International Consortium for Personalised Medicine Project to foster collaboration between China and the European Union (CORDIS 2023), the Indonesian Human Genome Project for Indonesia and China collaboration, (Mochtar 2023) and Australian Genomics Therapeutics Centre (AGTC) as Australian-private company which expands business in Malaysia. (AGTC Genomics 2023) This collaboration is shown to be beneficial for Tier 3 countries (commonly from the Global South) from partnerships with Tier 2 and Tier 1 countries (more commonly from the Global North), especially for availability in technology and references from patient databases/biobanks. Researchers would acquire advantages from biobanks for continuously learning and discovering genomic patterns for certain populations and exploring valuable propositions for possible treatments that would be fit to treat multiple diseases across all states.

Information from biobanks and previous research would be valuable resources for healthcare providers to manage high-risk clinical signs and for policymakers to decide on certain medical policies. Private genomics companies could also expand their business, collect more data from various populations, and gain valuable profits through partnership programs across states. Then, further collaborations are expected to enhance possibilities to produce new approaches in genomic-based treatment, especially for NCDs along with the development of personalised medicine and significantly reduce mortality rates.

10.4.2 Tier 2 Asian Countries

10.4.2.1 Low-Hanging Fruits

Providing the essential infrastructure, including the regulatory frameworks and sequencing facilities needed, is crucial before undertaking ambitious projects. To set biobanks up for future success, adaptive governance capable of storing high-quality biospecimens is crucial (Olson et al. 2014). As data privacy is a frequently raised public concern towards biobanks, effective data management systems should be established with an oversight committee. Instead of relying on existing data privacy laws, it would be greatly beneficial if specific laws on biosafety, including data exchange with international stakeholders, were enacted to mitigate unwanted leaks. Governments should proactively take part because of the centralised power needed to regulate and the funding required to ensure its sustainability (Dumancas et al. 2023).

After comprehensive infrastructure has been established, it is then prudent to start data collection, initially on health subjects, then a second phase on over 100,000 healthy and diseased individuals. This has been done in Singapore, and a similar approach is currently being undertaken in Indonesia (Kong et al. 2022). Genotyping arrays of single genes have previously been common, but whole genome sequencing is increasingly common as it has become more cost-effective, (Sergouniotis et al. 2023) allowing databases for genetic variations that are accessible to international research communities and industries (Xue et al. 2021). Combined with health-related data from electronic health records and wearable devices, pharmacogenomics and precision medicine can be further progressed, especially with the numerous polymorphisms, insertions, and deletions available in numerous databases (Alzu'bi et al. 2019).

10.4.2.2 Context-Specific Potential Opportunities

Recent advances in high-throughput technologies are paving the way for personalised medicine, revolutionising healthcare practices and disease management. When these technologies become increasingly available and affordable in the future, there is an opportunity to tailor treatment and prevention strategies to individual patients based on their genetic predispositions and specific disease manifestations, such as through risk factor and treatment target capacity identification (see Table 10.2 on the next page). As personalised medicine continues to evolve, healthcare professionals and policymakers in Tier 2 countries need to explore how these advancements can be integrated into their healthcare systems to improve overall public health in their contexts that may differ from developed countries, such as maternal health and infectious diseases (Brothers and Rothstein 2015; Ormond and Cho 2014).

To achieve such promise, indeed some challenges, most notably the accessibility and affordability of high-throughput technologies in Tier 2 countries, need to be carefully considered. Implementing personalised medicine requires sophisticated infrastructure, including genetic testing facilities and specialised medical expertise, which may not be readily available in Tier 2 countries. This could result in unequal access to advanced healthcare interventions, widening the gap between those who can afford personalised treatments and those who cannot. Furthermore, the ethical implications of personalised medicine, including infectious disease management, cannot be overlooked. The use of genetic and molecular information raises concerns about patient privacy, data security, and potential discrimination based on genetic predispositions. In addition, there is a risk of over-reliance on genetic testing as one of the implication of low education and awareness level, which may overshadow other important factors such as environmental influences and socio-economic determinants of (Geller et al. 2014; Mentis et al. 2018).

Table 10.2 Treatment target capacity and risk factor identification through genotyping research in Asia

Ref	Country	Research institution	Intl. collaborator(s)	Funding institution	Population and sample size	Gene (SNP code)	Genotyping technique	Condition of interest	Main findings
Azman et al. (2024)	Malaysia	Universiti Sains Malaysia Hospital	NA	Ministry of Higher Education Malaysia	104 patients with anaemia	BMP2 (rs235756)	ARMS-PCR	Iron deficiency anaemia	↓ Total iron binding capacity in AA genotype
Lee et al. (2024)	Malaysia	Universiti Putra Malaysia	NA	Universiti Putra Malaysia	217 mother-neonate dyads	VDR (rs2228570)	PCR–RFLP	Vitamin D deficiency	↓ serum vitamin D in maternal GG genotype
Saif-Ali et al. (2024)	Malaysia	University of Malaya	Yemen and Saudi Arabia	University of Malaya	397 versus 285 Indian patients	PTPRD (rs649891 and rs17584499) SRR (rs4523957, rs391300, and rs8081273)	PCR-Taqman genotype assay	T2DM	↑ risk of T2DM in CC genotype of PTPRD ↓ risk of T2DM in TT genotype of SRR
Hardiany et al. (2024)	Indonesia	Universitas Indonesia	NA	Universitas Indonesia	92 elderly	FOXO3 (rs2802292)	Sanger sequencing	Longetivity	↑ longevity in GG genotype
Mardhiyah et al. (2024)	Indonesia	Universitas Sumatera Utara	NA	No external funding	194 Bataknese individuals with schizophrenia	IL-10 -1082G/A (rs1800896)	PCR–RFLP	Schizophrenia	↑ Risk of schizophrenia in GG genotype of G allele

(continued)

Table 10.2 (continued)

Ref	Country	Research institution	Intl. collaborator(s)	Funding institution	Population and sample size	Gene (SNP code)	Genotyping technique	Condition of interest	Main findings
Guo et al. (2020)	Southern China and Singapore	Guangdong key laboratory of nasopharyngeal carcinoma diagnosis; therapy National cancer centre Singapore	China	Multiple donors from China and Singapore	5553 Chinese and Singaporean patients with nasopharyngeal carcinoma	RPA1 (rs1131636-T)	PCR-Taqman genotype assay	Nasopharyngeal carcinoma	↓ Overall survival in TT genotype
Cataldo et al. (2021)	Singapore	Nanyang Technological University	Italy	Nanyang Technological University Singapore and Singapore Ministry of Education	313 Italian and Singaporean young adults	OXTR (rs53576)	Fragment Analysis (ABI 3730xl Genetic Analyzer)	Avoidance	↑ avoidance in GG or AA genotype of Singaporeans ↓ avoidance in GA genotype of Singaporeans ↓ avoidance in GG genotype of Singaporeans

(continued)

Table 10.2 (continued)

Ref	Country	Research institution	Intl. collaborator(s)	Funding institution	Population and sample size	Gene (SNP code)	Genotyping technique	Condition of interest	Main findings
Phuthong et al. (2020)	Thailand	Khon Kaen University	Japan	Khon Kaen Universit	204 women	Fok1 (rs2228570), Apa1 (rs7975232), and Taq1 (rs731236)	PCR-Taqman genotype assay	Cervical cancer	↑ risk of cervical cancer in TT genotype and T allele of Fok1 ↑ risk of cervical cancer TC genotype, CC genotype, C allele of Taq1
Kamdee et al. (2021)	Thailand	Walailak University	NA	Walailak University	200 patients with ischemic stroke	IL-6-174G/C (rs1800795) TNF-α-308G/A (rs1800629)	PCR-Taqman genotype assay	Ischemic stroke	↑ risk of ischemic stroke in GA or AA genotype and A allele of TNF-α-308

(↑), increase significantly; (↓), decrease significantly

10.4.2.3 Nuances of Sociopolitical Barriers

In the sociopolitical context, the development of personalised medicine is driven by the fact that most Southeast Asian countries are home to diverse ethnicities. For example, a study has identified different gut microbiota profiles across ethnicities living in Malaysia which implicates to prevalence of certain diseases and the efficacy of medical treatment (Dwiyanto et al. 2021) Addressing different ethnicities in a country also implicates genomic research as the individuals share different ancestries (Flanagin et al. 2021). Despite the urge to strengthen the root of personalised medicine in Southeast Asian countries, political willingness and public perception remain the contributing factors hindering the creation of related public policies. The evolution of policies has been reported to heavily rely on socio-technical imaginary, which is a concept describing the collective expectations and visions of desirable technological advancements in a specific society (Meunier and Herzog 2023).

The emergence of socio-technical imaginary requires strong public awareness regarding personalised treatment. Such phenomena might be the reason of slow progress of personalised medicine among Southeast Asian countries, except in a few countries such as Thailand and Singapore. Educational programs on pharmacogenomics testing have been widely implemented in Thailand in the form of books, reports, and even mobile apps targeting patients as well as the general public (Chong et al. 2018). Information related to targeted oncology therapy for breast cancer and HER2 in breast cancer is made available through online education provided by the National University Hospital Singapore and the Indonesian Breast Cancer Association. Similar campaigns to raise awareness about breast cancer are also conducted in Malaysia. As noticeable, the public information on personalised medicine is mostly limited to oncology. An assessment conducted in Brunei Darussalam found that the overall knowledge score on genetics was 69.4 ± 15.31 (out of 100). The score suggests knowledge adequacy, yet it was centralised only in individuals with high educational backgrounds (Jaya et al. 2023).

Given the low public awareness of personalised medicine, other driving factors are required to grow political support. Researchers and clinical practitioners are the frontline to remind policymakers of the importance of such technologies. Improving knowledge and awareness can be carried out, not only by professionals but also by medical students who are well familiar with genetic knowledge (Swandayani et al. 2021). The sociopolitical aspect is as essential as the technology itself because the basis of personalised medicine is not only about effective treatment but also an inclusive system. Further, appropriate policies are required to respond to issues that might arise from the introduction of the technology including the legality of the treatment, professional liability, as well as personal data protection.

10.4.3 Tier 3 Asian Countries

Although Tier 3 Asian countries still do not have genomic centres, several countries, such as Vietnam and Laos, have published their works in genomic projects. Nevertheless, biobanks mainly serve agricultural projects as opposed to human-based genomic research. There is also still a lack of access to reports and news about their progress, both from universities/institutes or government organisations. This gap is bothersome, and it needs to be addressed quickly, considering that Tier 3 countries are the highest contributors to NCD prevalences and mortalities. According to WHO, 77% of deaths due to NCDs came from low-middle income countries (LMIC), most of which are Tier 3 countries (World Health Organization 2023). As shown through research in higher-income countries, NCDs can also benefit from genome-based treatment targeting specific genes. Leaving Tier 3 countries behind will not only potentially increase the barrier of accessing personalised treatment products in the future and perpetuate the inequality between countries, but also waste the population data essential to optimise these products' efficacy towards their largest market.

In advancing the development of genome-based medicine in Tier 3 countries, collaboration with tier 1 and 2 countries is the key to opening up opportunities for building genomic centres centres and establishing some genome projects. Before initiating their major genomic facilities, Indonesia Human Genome (Policiuc et al. 2018) and Malaysia Genome and Vaccine (International Trade Administration 2023) were proofs of how collaboration is an impressive solution to accelerate their growth through shared know-how in executing and developing personalised medicine. Many seasoned scientists, including those of developed countries, are open to collaborate, particularly to establish biorepositories and databases. Such universal participation will eventually provide an enormous amount of efficacy and safety data improving genomic-based treatments worldwide (Institute of Medicine 2012) Moreover, as these countries become Tier 2 and Tier 1, they will be eligible markets of future personalised medical industries.

10.5 Conclusion

Latest research interests in medical treatment have already shifted to offering precise treatment, most often through genomic data, to depict patient characteristics and narrow potential errors on treating diseases; advancements were implemented in adjunct with technology stakeholders. Continued international cooperation and investments will be key in addressing ethical, legal, and regulatory challenges in personalised medicine. While Tier 1 and some Tier 2 countries, like Thailand and Singapore, are advancing in genomics, limited infrastructure in other Tier 2 and most Tier 3 countries risks widening healthcare disparities. Personalised medicine must continue to be developed in all countries to provide holistic solutions on preventing risk factors of severe diseases and combat life-threatening conditions. In conclusion, advancing personalised medicine in Asia demands comprehensive investment in infrastructure, public awareness, and strategic partnerships, ensuring equitable access to genomic advancements and contributing to the global evolution of personalised healthcare.

Highlights

- Advanced nations will integrate AI, 5G, and rapid genome sequencing, driving innovation.
- Emerging countries must prioritize biobank infrastructure, regulation, and global collaboration for equitable access.
- Stronger governance, public education, and policy reforms will address ethical concerns, regulatory gaps, and high costs, ensuring sustainable progress.

References

AGTC Genomics. AGTC Genomics and Universiti Malaya ink MoU to collaborate on genomics research. 2023. https://agtcgenomics.com/2023/01/12/agtc-genomics-sdn-bhd-and-universiti-malaya-ink-mou-to-collaborate-on-genomics-research/. Accessed 23 Jun 2024.

Agyeman A, Ofori-Asenso R. Perspective: does personalized medicine hold the future for medicine? J Pharm Bioallied Sci. 2015;7(3):239.

Alenezi S, Alanazi M, Aljazaeri R, Almuzaini M, Alrasheidi S, Shamlan W Bin, et al Community pharmacies in the Asian countries of developing health system: formation, regulation, and implication. Pharmacy. 2023;11(4):127.

Alzu'bi AA, Zhou L, Watzlaf VJM. Genetic variations and precision medicine. Perspect Health Inf Manag. 2019;16(Spring):1a.

Amin L, Olesen A, Mahadi Z, Ibrahim M. Current status and future challenges of biobank research in Malaysia. Asian Bioeth Rev. 2021;13(3):297–315.

Aneizi A, Chokeir H. Turnstone biologics announces global collaboration and license agreement with Takeda to develop novel viral immunotherapies. 2019. https://www.businesswire.com/news/home/20191219005187/en/Turnstone-Biologics-Announces-Global-Collaboration-and-License-Agreement-with-Takeda-to-Develop-Novel-Viral-Immunotherapies. Accessed 13 Apr 2019.

Arlinta D. Sebanyak 10.000 Sampel Genom Masyarakat Indonesia Akan Diteliti. Kompas. 2024.

As A, Mahsyar A, Malik I. Implementasi kebijakan kesehatan masyarakat dalam upaya menurunkan angka kematian ibu dan bayi (studi kasus di kabupaten bulukumba dan takalar). JPPM: J Publ Policy Manag. 2020;1(1):22–8.

Asiimwe IG, Pirmohamed M. Ethnic diversity and warfarin pharmacogenomics. Front Pharmacol. 2022;13: 866058.

Azman NA, Zulkafli Z, Bakar NS, Assyuhada MGSN, Mohammad SNNA. Association of single nucleotide polymorphism at BMP2 gene with iron deficiency status among anaemic patients in Hospital Universiti Sains Malaysia. Egypt J Med Hum Genet. 2024;25(1):46.

Azmi M, Chen L, Idris A, Abdullah MS, Lu ZH. Genomic insights of Bruneian Malays. 2022. Available from: https://doi.org/10.21203/rs.3.rs-1985778/v1

Bayer R, Galea S. Public health in the precision-medicine Era. N Engl J Med. 2015;373(6):499–501.
Beigh M. Next-generation sequencing: the translational medicine approach from "bench to bedside to population." Medicines. 2016;3(2):14.
Belachew SA, Hall L, Erku DA, Selvey LA. No prescription? No problem: drivers of non-prescribed sale of antibiotics among community drug retail outlets in low and middle income countries: a systematic review of qualitative studies. BMC Public Health. 2021;21(1):1056.
BGSi. BGSi official site. 2023. https://bgsi.kemkes.go.id/. Accessed 1 Apr 2023.
Brothers KB, Rothstein MA. Ethical, legal and social implications of incorporating personalized medicine into healthcare. Per Med. 2015;12(1):43–51.
Cataldo I, Bonassi A, Lepri B, Foo JN, Setoh P, Esposito G. Recalled parental bonding interacts with oxytocin receptor gene polymorphism in modulating anxiety and avoidance in adult relationships. Brain Sci. 2021;11(4):496.
Chong HY, Allotey PA, Chaiyakunapruk N. Current landscape of personalized medicine adoption and implementation in Southeast Asia. BMC Med Genom. 2018;11(1):94.
Cliff ERS, Kelkar AH, Russler-Germain DA, Tessema FA, Raymakers AJN, Feldman WB, et al. High cost of chimeric antigen receptor T-cells: challenges and solutions. Am Soc Clin Oncol Educ Book 2023; (43).
Cohen J, Vannavuth S, Ing C. Cambodia–data protection overview. 2023. https://www.dataguidance.com/notes/cambodia-data-protection-overview. Accessed 1 Apr 2024.
Cyranoski D. China embraces precision medicine on a massive scale. Nature. 2016;529(7584):9–10.
Dartanto T, Halimatussadiah A, Rezki JF, Nurhasana R, Siregar CH, Bintara H, et al. Why do informal sector workers not pay the premium regularly? Evidence from the national health insurance system in Indonesia. Appl Health Econ Health Policy. 2020;18(1):81–96.
Dhawan AP. Collaborative paradigm of preventive, personalized, and precision medicine with point-of-care technologies. IEEE J Transl Eng Health Med. 2016;4:1–8.
Dumancas GG, Smith KFK, Bugayong-Janagap AM, Zamora PRFC, Ferriols VMEN, Liwag AG, et al. Challenges and opportunities in establishing a regional biobank center in a developing country: A case from the Philippines. In: Health policy and technology. Elsevier B.V.; 2023.
Duong D, Le T, Binns CW. Diarrhoea management by pharmacy staff in retail pharmacies in Hanoi, Vietnam. Int J Pharm Pract. 2011;5(2):97–100.
Dwiyanto J, Hussain MH, Reidpath D, Ong KS, Qasim A, Lee SWH, et al. Ethnicity influences the gut microbiota of individuals sharing a geographical location: a cross-sectional study from a middle-income country. Sci Rep. 2021;11(1):2618.
Ferdiana A, Liverani M, Khan M, Wulandari LPL, Mashuri YA, Batura N, et al. Community pharmacies, drug stores, and antibiotic dispensing in Indonesia: a qualitative study. BMC Public Health. 2021;21(1):1800.
Fernald GH, Capriotti E, Daneshjou R, Karczewski KJ, Altman RB. Bioinformatics challenges for personalized medicine. Bioinformatics. 2011;27(13):1741–8.
Flanagin A, Frey T, Christiansen SL. Updated guidance on the reporting of race and ethnicity in medical and science journals. JAMA. 2021;326(7):621.
Gao XR, Cebulla CM, Ohr MP. Advancing to precision medicine through big data and artificial intelligence. In: Genetics and genomics of eye disease. Elsevier; 2020. p. 337–49.
Geller G, Dvoskin R, Thio CL, Duggal P, Lewis MH, Bailey TC, et al. Genomics and infectious disease: a call to identify the ethical, legal and social implications for public health and clinical practice. Genome Med. 2014;6(11):106.
Global Forecast. Precision medicine market by type (inhibitors, monoclonal antibodies, cell & gene therapy, antivirals, antiretroviral), indication (oncology, rare diseases, hematology, infectious), end user (hospitals & clinics, home care) & region to 2018. 2023.
GlobalData. Reducing cost of next-generation sequencing: unlock personalised medicine [Internet]. 2023. https://www.medicaldevice-network.com/analyst-comment/ngs-reduced-cost-personalised-medicine/?cf-view. Accessed 31 Mar 2024.
Goetz LH, Schork NJ. Personalized medicine: motivation, challenges, and progress. Fertil Steril. 2018;109(6):952–63.

Government of the Socialist Republic of Vietnam. Decree No. 13/2023/NĐ-CP on Personal Data Protection. 2023.
Graham C, Hewitson R, Pagliuca A, Benjamin R. Cancer immunotherapy with CAR-T cells—behold the future. Clin Med. 2018;18(4):324–8.
Gunn S. World of genomics: Thailand. 2023. https://frontlinegenomics.com/world-of-genomics-thailand/. Accessed 7 Apr 2024.
Guo C, Li H. Application of 5G network combined with AI robots in personalized nursing in China: A literature review. Front Public Health. 2022; 10.
Guo Y, Chen J, Feng Y, Chua MLK, Zeng Y, Hui EP, et al. Germline polymorphisms and length of survival of nasopharyngeal carcinoma: an exome-wide association study in multiple cohorts. Adv Sci. 2020; 7(10).
Gürsoy G, Li T, Liu S, Ni E, Brannon CM, Gerstein MB. Functional genomics data: privacy risk assessment and technological mitigation. Nat Rev Genet. 2022;23(4):245–58.
Hardiany NS, Nurfiyana W, Iswanti FC. Polymorphism of the Forkhead box-O3 (FOXO3) longevity gene rs2802292 and senescence-associated secretory phenotype (SASP) in Indonesian elderly population. Nutr Healthy Aging. 2024;9(1):47–54.
Hayden EC. Promising gene therapies pose million-dollar conundrum. Nature. 2016;534(7607):305–6.
He S, Wei J, Feng J, Liu D, Wang N, Chen L, et al. The application of metagenomic next-generation sequencing in pathogen diagnosis: a bibliometric analysis based on Web of Science. Front Cell Infect Microbiol. 2023; 13.
Hopkins CE. Emerging technologies in personalized medicine. Mol Aspects Med. 2023;91: 101182.
Illumina. Genomics consortia. 2024. https://www.illumina.com/science/consortia.html. Accessed 16 Jun 2024.
Indian Council of Medical Research. Indian Council of Medical Research National Ethical Guidelines. 2024. https://ethics.ncdirindia.org/ICMR_Ethical_Guidelines.aspx. Accessed 4 Apr 2024.
Institute of Medicine. Genome-based therapeutics. Washington, D.C.: National Academies Press; 2012.
International Trade Administration. Malaysia precision medicine initiative. 2023. https://www.trade.gov/market-intelligence/malaysia-precision-medicine-initiative. Accessed 16 Jul 2024.
Jamal ARA. Precision medicine: making it happen for Malaysia. Malaysian J Med Sci 2021; 28(3):1–4.
Jaya H, Idayu Matusin SN, Mustapa A, Abdullah MS, Haji Abdul Hamid MRW. Public knowledge of and attitudes toward genetics and genetic testing in Brunei Darussalam. Front Genet. 2023; 14.
Johnson KB, Wei W, Weeraratne D, Frisse ME, Misulis K, Rhee K, et al. Precision medicine, AI, and the future of personalized health care. Clin Transl Sci. 2021;14(1):86–93.
Joly Y, Dupras C, Pinkesz M, Tovino SA, Rothstein MA. Looking beyond GINA: policy approaches to address genetic discrimination. Annu Rev Genomics Hum Genet. 2020;21(1):491–507.
Kamdee K, Panadsako N, Mueangson O, Nuinoon M, Janwan P, Poonsawat W, et al. Promoter polymorphism of *TNFα* (rs1800629) is associated with ischemic stroke susceptibility in a southern Thai population. Biomed Rep. 2021;15(3):78.
Kang CC, Lee TY, Lim WF, Yeo WWY. Opportunities and challenges of 5G network technology toward precision medicine. Clin Transl Sci. 2023;16(11):2078–94.
Karczewski KJ, Daneshjou R, Altman RB. Chapter 7: Pharmacogenomics. PLoS Comput Biol. 2012; 8(12):e1002817.
Katsanis SH, Katsanis N. Molecular genetic testing and the future of clinical genomics. Nat Rev Genet. 2013;14(6):415–26.
Kementerian Kesehatan. Indonesia pimpin pengembangan genomik level Asia Tenggara. 2022. https://www.kemkes.go.id/id/rilis-kesehatan/indonesia-pimpin-pengembangan-genomik-level-asia-tenggara. Accessed 31 Mar 2024.

Khoury MJ. The shift from personalized medicine to precision medicine and precision public health: words matter! Center for Disease Control and Prevention. 2016.

Kim H, Ho CWL, Ho CH, Athira PS, Kato K, De Castro L, et al. Genetic discrimination: introducing the Asian perspective to the debate. NPJ Genom Med. 2021;6(1):54.

Kong D, Yu H, Sim X, White K, Tai ES, Wenk M, et al. Multidisciplinary effort to drive precision-medicine for the future. Front Digit Health. 2022; 4.

Kutanan W, Kampuansai J, Srikummool M, Brunelli A, Ghirotto S, Arias L, et al. Contrasting paternal and maternal genetic histories of Thai and Lao populations. Mol Biol Evol. 2019;36(7):1490–506.

Kutanan W, Liu D, Kampuansai J, Srikummool M, Srithawong S, Shoocongdej R, et al. Reconstructing the human genetic history of Mainland Southeast Asia: insights from genome-wide data from Thailand and Laos. Mol Biol Evol. 2021;38(8):3459–77.

Lakha F, Suriyawongpaisul P, Sangrajrang S, Leerapan B, Coker R. Breast cancer in Thailand: policy and health system challenges to universal healthcare. Health Policy Plan. 2020;35(9):1159–67.

Laksono AD, Wulandari RD, Matahari R, Rohmah N. The choice of delivery place in Indonesia: does home residential status matter? Heliyon. 2023;9(4): e15289.

Landeck L, Lessl M, Reischl J, Busch A, Carrigan P, Gottwald M, et al. Collaboration for success: the value of strategic collaborations for precision medicine and biomarker discovery. Adv Precis Med. 2016; 1(1).

Lee SS, Ling KH, Subramaniam R, Tusimin M, Rahim KF, Loh SP. Genetic variants and non-genetic factors associated with a high prevalence of vitamin D deficiency in full-term neonates in Malaysia. Hum Nutri Metab. 2024;36: 200259.

Lee S, Lee JH. Cell and gene therapy regulatory, pricing, and reimbursement framework: with a focus on South Korea and the EU. Front Public Health. 2023; 11.

Li D. 5G and intelligence medicine—how the next generation of wireless technology will reconstruct healthcare? Precis Clin Med. 2019;2(4):205–8.

Lim MY, Kamaruzaman HF, Wu O, Geue C. Health financing challenges in Southeast Asian countries for universal health coverage: a systematic review. Arch Public Health. 2023;81(1):148.

Manolio TA, Abramowicz M, Al-Mulla F, Anderson W, Balling R, Berger AC, et al. Global implementation of genomic medicine: we are not alone. Sci Transl Med. 2015; 7(290).

Mardhiyah SA, Effendy E, Nasution NM. IL-10 (−1082 G/A) polymorphism in Bataknese with schizophrenia. J Taibah Univ Med Sci. 2024;19(1):64–9.

Mastrojohn D, Valachs P. Pfizer and cellectis enter into global strategic cancer immunotherapy collaboration. Pfizer. 2014. https://www.pfizer.com/und/news/press-release/press-release-detail/pfizer_and_cellectis_enter_into_global_strategic_cancer_immunotherapy_collaboration. Accessed 9 Feb 2024.

Mentis AFA, Pantelidi K, Dardiotis E, Hadjigeorgiou GM, Petinaki E. Precision medicine and global health: the good, the bad, and the ugly. Front Med (Lausanne). 2018; 5.

Meunier R, Herzog C. Omics and AI in precision medicine: Maintaining socio-technical imaginaries by transforming technological assemblages. TATuP - Zeitschrift Für Technikfolgenabschätzung in Theorie und Praxis. 2023;32(3):48–53.

MGI. MGI partners with national cancer centre Singapore to undergo comprehensive genomic profiling of asian-prevalent cancers [Internet]. News-Medical. 2023. https://www.news-medical.net/news/20230918/MGI-Partners-with-National-Cancer-Centre-Singapore-to-Undergo-Comprehensive-Genomic-Profiling-of-Asian-Prevalent-Cancers.aspx. Accessed 31 Mar 2024.

Mochtar I. The Indonesian Human Genome Project [Internet]. Kompas.id. 2023. https://www.kompas.id/baca/english/2023/07/05/en-proyek-genom-manusia-indonesia. Accessed 9 Sep 2024

Moglia A, Georgiou K, Marinov B, Georgiou E, Berchiolli RN, Satava RM, et al. 5G in healthcare: from COVID-19 to future challenges. iEEE J Biomed Health Inform. 2022;26(8):4187–96.

Morganti S, Tarantino P, Ferraro E, D'Amico P, Duso BA, Curigliano G. Next generation sequencing (NGS): a revolutionary technology in pharmacogenomics and personalized medicine in cancer. In: Ruiz-Garcia E, Astudillo-de la Vega H, editors. Translational research and onco-omics applications in the era of cancer personal genomics. Edinburgh: Springer, Cham; 2019. p. 9–30.

Muto K, Nagai A, Ri I, Takashima K, Yoshida S. Is legislation to prevent genetic discrimination necessary in Japan? An overview of the current policies and public attitudes. J Hum Genet. 2023;68(9):579–85.

Naher N, Hoque R, Hassan MS, Balabanova D, Adams AM, Ahmed SM. The influence of corruption and governance in the delivery of frontline health care services in the public sector: a scoping review of current and future prospects in low and middle-income countries of south and south-east Asia. BMC Public Health. 2020;20(1):880.

National Human Rights Commission of Korea Act of 2001. South Korea; 2016.

Olson JE, Bielinski SJ, Ryu E, Winkler EM, Takahashi PY, Pathak J, et al. Biobanks and personalized medicine. Clin Genet. 2014;86(1):50–5.

Ormond KE, Cho MK. Translating personalized medicine using new genetic technologies in clinical practice: the ethical issues. Per Med. 2014;11(2):211–22.

Park ST, Kim J. Trends in next-generation sequencing and a new era for whole genome sequencing. Int Neurourol J. 2016;20(Suppl 2):S76-83.

Pereira NL, Rihal C, So D, Rosenberg Y, Lennon RJ, Mathew V, et al. Clopidogrel pharmacogenetics: state of the art review and the TAILOR-PCI study. Circ Cardiovasc Interv. 2019;12(4): e007811.

Philippines House of Representatives. House Bill No. 892, 19th congress. An act amending republic act no. 10173, otherwise known as the "Data Privacy Act of 2012", to adopt a more systematic, comprehensive and effective protection program, to foster the imposition of all existing policies and guidelines, increasing its penalties and sanctions, and for other purposes. 2022. https://ldr.senate.gov.ph/bills/house-bill-no-892-19th-congress

Phillipines House of Representatives. House bill no. 898, 19th congress. An act instituting a national land use policy, providing the implementing mechanisms therefor, and for other purposes. 2022. https://ldr.senate.gov.ph/bills/house-bill-no-898-19th-congress.

Phillips KA, Ann Sakowski J, Trosman J, Douglas MP, Liang SY, Neumann P. The economic value of personalized medicine tests: what we know and what we need to know. Genet Med. 2014;16(3):251–7.

Phuthong S, Settheetham-Ishida W, Natphopsuk S, Ishida T. Genetic polymorphisms of vitamin D receptor gene are associated with cervical cancer risk in Northeastern Thailand. Asian Pac J Cancer Prev. 2020;21(10):2935–9.

Policiuc L, Buse M, Gulei D, Zimta AA, Braicu C, Irimie AI, et al. The foundation of personalized medicine is the establishment of biobanks and their standardization. J BUON. 2018;23(3):550–60.

Protergo. Indonesia's ministry of health calls data leak in old version of eHAC, saves important user information. 2021. https://protergo.id/indonesias-ministry-of-health-calls-data-leak-in-old-version-of-ehac-saves-important-user-information/. Accessed 3 Apr 2024.

Raj A. Malaysian telco provider has data breach—again. Techwire Asia. 2024.

Regalado A. China's BGI says it can sequence a genome for just $100. MIT Technol Rev. 2020.

Rothstein MA, Cai Y, Marchant GE. The ghost in our genes: legal and ethical implications of epigenetics. Health Matrix Clevel. 2009;19(1):1–62.

Saif-Ali R, Al-Hamodi Z, Salem SD, AL-Habori M, Al-Dubai SA, Ismail IS. Association of protein tyrosine phosphatase receptor type D and serine racemase genetic variants with type 2 diabetes in Malaysian Indians. Indian J Endocrinol Metab. 2024; 28(1):55–9.

Santaló J, Berdasco M. Ethical implications of epigenetics in the era of personalized medicine. Clin Epigenet. 2022;14(1):44.

Sergouniotis PI, Fitzgerald T, Birney E. From genetic variation to precision medicine. Cambridge Prisms: Precis Med. 2023;1: e7.

Shah GL, Majhail N, Khera N, Giralt S. Value-based care in hematopoietic cell transplantation and cellular therapy: challenges and opportunities. Curr Hematol Malig Rep. 2018;13(2):125–34.

Sharma M, Teerawattananon Y, Luz A, Li R, Rattanavipapong W, Dabak S. Institutionalizing evidence-informed priority setting for universal health coverage: lessons from Indonesia. INQUIRY: J. Health Care Organ. Provis. Finan. 2020; 57:004695802092492.

Stenzinger A, Moltzen EK, Winkler E, Molnar-Gabor F, Malek N, Costescu A, et al. Implementation of precision medicine in healthcare—a European perspective. J Intern Med. 2023;294(4):437–54.
Sumriddetchkajorn K, Shimazaki K, Ono T, Kusaba T, Sato K, Kobayashi N. Universal health coverage and primary care, Thailand. Bull World Health Organ. 2019;97(6):415–22.
Swandayani YM, Cayami FK, Winarni TI, Utari A. Familiarity and genetic literacy among medical students in Indonesia. BMC Med Educ. 2021;21(1):524.
Takeda M, Sakai K, Takahama T, Fukuoka K, Nakagawa K, Nishio K. New Era for next-generation sequencing in Japan. Cancers (Basel). 2019;11(6):742.
Tangcharoensathien V, Patcharanarumol W, Ir P, Aljunid SM, Mukti AG, Akkhavong K, et al. Health-financing reforms in southeast Asia: challenges in achieving universal coverage. The Lancet. 2011;377(9768):863–73.
Than KZ, Muisuk K, Woravatin W, Suwannapoom C, Srikummool M, Srithawong S, et al. Genetic structure and forensic utility of 23 autosomal STRs of the Ethnic Lao Groups From Laos and Thailand. Front Genet. 2022; 13.
Theanngarm Y, Oo N. Myanmar—data protection overview. 2023. https://www.dataguidance.com/notes/myanmar-data-protection-overview. Accessed 7 Apr 2024.
Ton VL. Vietnam—Data protection overview. 2023. https://www.dataguidance.com/notes/vietnam-data-protection-overview. Accessed 1 Apr 2024.
Venner E, Muzny D, Smith JD, Walker K, Neben CL, Lockwood CM, et al. Whole-genome sequencing as an investigational device for return of hereditary disease risk and pharmacogenomic results as part of the All of Us Research Program. Genome Med. 2022;14(1):34.
Visvikis-Siest S, Theodoridou D, Kontoe MS, Kumar S, Marschler M. Milestones in personalized medicine: from the ancient time to nowadays—the provocation of COVID-19. Front Genet. 2020; 11.
Wall JD, Stawiski EW, Ratan A, Kim HL, Kim C, Gupta R, et al. The GenomeAsia 100K project enables genetic discoveries across Asia. Nature. 2019;576(7785):106–11.
Wang S, Qian W, Chen S, Xian S, Jin M, Liu Y, et al. Bibliometric analysis of research on gene expression in spinal cord injury. Front Mol Neurosci. 2022; 15.
Wong E, Bertin N, Hebrard M, Tirado-Magallanes R, Bellis C, Lim WK, et al. The Singapore national precision medicine strategy. Nat Genet. 2023;55(2):178–86.
World Health Organization. Non communicable diseases [Internet]. 2023. https://www.who.int/news-room/fact-sheets/detail/noncommunicable-diseases. Accessed 16 Sep 2024
Xue Y, Bao Y, Zhang Z, Zhao W, Xiao J, He S, et al. Database resources of the national genomics data center, China National Center for Bioinformation in 2021. Nucleic Acids Res. 2021;49(D1):D18-28.
Wang Q, Yang KL, Zhang Z, Wang Z, Li C, Li L, et al. Characterization of global research trends and prospects on single-cell sequencing technology: bibliometric analysis. J Med Internet Res. 2021; 23(8):e25789.
Zakaria NH, Hashad D, Saied MH, Hegazy N, Elkayal A, Tayae E. Genetic mutations in HER2-positive breast cancer: possible association with response to trastuzumab therapy. Hum Genomics. 2023;17(1):43.

Sydney Tjandra is a medical student from Universitas Indonesia who has a vision to bridge health research and practice. She has been part of several research projects, both secondary and primary, including being a principal investigator of an implementation research on primary healthcare funded by TDR-WHO SEARO in 2021. Previously serving as Director of Global Health in AMSA International, Sydney not only had the opportunity to interact with medical students and health experts from 29 countries, but also initiated various collaborations within the network. She is passionate on various public and global health issues.

Muhammad Iqhrammullah a lecturer at the Postgraduate Program of Public Health, Universitas Muhammadiyah Aceh, Indonesia, was named among the World's Top 2% Scientists 2024 by Stanford University. His doctoral work focused on environmental monitoring via spectroscopy, spanning projects in herbal medicine, sensors, and pollution mitigation. Notably, he collaboratively addressed mercury pollution in Aceh's gold mining areas and received the Green Talent Award from the German Federal Ministry of Education and Research (BMBF). He also ranked second in the 30 under 30 list by the Opportunity Desk Impact Challenge in 2021. Currently, his research focuses on advancing environmental health in public health.

Shakira Amirah is a medical student from Universitas Indonesia. Her academic journey has been marked by a passion for research, resulting in the publication of 20 Scopus-indexed journals (H-index: 8) to date. With a commitment to global issues, Shakira previously served as Vice Overall Chairperson Internal in AMSA International. Her compassion for health challenges led her to create Isolasinfo, an application designed to address physical and mental issues during the COVID-19 pandemic. In collaboration with various stakeholders in Indonesia, Shakira strives to make a meaningful impact on public health and well-being, embodying a commitment to positive change.

Vindasya Almeira is a medical student at Universitas Indonesia, committed to medical knowledge development through research and passionate in public health major. Joining Journal of Asian Medical Student Association (JAMSA) as Capacity Building in Research Executives JAMSA 2023/2024 and Associate Editor of Conference Publication JAMSA 2024/2025 has broadened her perspective, strengthened her impact to share the same research opportunities to peers, and led her to various research conferences and publication opportunities. Enjoying community development and social projects, Vindasya was involved in the Public Health Committee in Centre for Indonesian Medical Student Activities Universitas Indonesia (CIMSA UI) and a policy document about Digital Transformation for Universal Health Coverage.Enjoying community development and social projects, Vindasya was involved in the Public Health Committee in Centre for Indonesian Medical Student Activities Universitas Indonesia (CIMSA UI) and a policy document about Digital Transformation for Universal Health Coverage.

Don Eliseo III Lucero-Prisno is a global health scientist, currently teaching Economics of Global Health Policy at the London School of Hygiene & Tropical Medicine (LSHTM). With over 400 publications, including 315 Scopus-indexed articles, his work focuses on global health equity and health research capacity building, particularly in developing countries. Founder of Global Health Focus, he has empowered over 20,000 emerging global health leaders. He has held academic roles across Asia, Europe, and Africa, consulted for WHO and other UN agencies, and received numerous awards, including the Global Health Promotion Practice Award.

Chapter 11
Ethical Stewardship in Asian Healthcare: The Imperative for Clinical Ethics Committees (CECs) in a Rapidly Evolving Landscape

Clusters of Future Studies: Technological Trends; Post-normality and Complexity; Humanity at the Limen

Nishakanthi Gopalan, Julia Patrick Engkasan, and Erwin Jiayuan Khoo

Abstract As healthcare in Asia modernises rapidly with technological advancements, the ethical ramifications of this evolving landscape have grown increasingly complex. This chapter employs a narrative review approach, synthesising findings from peer-reviewed literature, policy documents, and case studies to explore the role of Clinical Ethics Committees (CECs) in addressing these challenges in the Asian healthcare context. By integrating principles of ethical stewardship, cultural relativism, and operational dynamics, it offers insights into how CECs mediate between universal ethical standards and localised practices in diverse socio-cultural and economic contexts. CECs are presented as a critical mechanism for ensuring consistent ethical standards in patient care, safeguarding healthcare ethics amidst

N. Gopalan (✉)
Medical Humanities and Ethics Unit, Faculty of Medicine, Universiti Malaya, Kuala Lumpur, Malaysia
e-mail: nisha.gopalan@um.edu.my

J. P. Engkasan
Department of Rehabilitation Medicine, Faculty of Medicine, Universiti Malaya, Kuala Lumpur, Malaysia
e-mail: juliape@um.edu.my

E. J. Khoo
Department of Paediatrics, School of Medicine, International Medical University (IMU), Kuala Lumpur, Malaysia

Department of Global Health and Social Medicine, Harvard Medical School, Harvard University, Cambridge, USA

E. J. Khoo
e-mail: jiayuan_khoo@imu.edu.my; erwin_khoo@hms.harvard.edu

R. Xiang Ng et al. (eds.), *Future of Healthcare in Asia*,
https://doi.org/10.1007/978-3-032-03183-9_11

rapid innovation and population diversity. This chapter provides an overview and practical guide for establishing CECs within the Asian healthcare context, outlining actionable steps and operational guidelines. Additionally, it examines regional case studies that demonstrate how CECs have successfully navigated complex ethical dilemmas. By focusing on Asia, the chapter offers a timely resource for healthcare institutions, policymakers, and medical professionals striving to address the ethical complexities of modern healthcare while considering regional specificities.

Keywords Asian healthcare · Clinical ethics · Ethics committee · Ethical stewardship · Ethical oversight · Cultural competency

11.1 Introduction

In recent decades, Asia has emerged as a global leader in healthcare innovation, characterised by rapid development in vaccine development, regenerative therapy, genomic medicine, gene editing technology, precision medicine, telemedicine, and the integration of artificial intelligence (AI) into healthcare. These advancements, while promising unprecedented improvements in patient care, have introduced complex ethical dilemmas that challenge traditional moral frameworks and professional accountabilities (Mahara et al. 2023). The pace of technological change has outstripped the ability of healthcare systems to adapt ethically and operationally, leading to conflicts among stakeholders with differing values and expectations.

Against this backdrop, Clinical Ethics Committees (CECs) have become indispensable in Asia's healthcare landscape. Originating in the late 20th century as a response to growing concerns over patient rights and the ethical implications of high-stakes medical decisions, CECs have evolved from ad hoc panels to structured, multidisciplinary bodies that encompass experts in medicine, ethics, law, and community advocacy. These committees play a role in mediating between rapid medicine innovation and ethical healthcare delivery, ensuring that advancements in treatment are matched by an equally rigorous commitment to ethical practices.

By providing a forum for resolving ethical conflicts and deliberating on complex cases, CECs help to maintain the integrity of patient care and uphold moral principles within Asia's fast-evolving healthcare sector. Through their work, CECs foster a culture of ethical sensitivity and awareness, enhancing the overall quality of decision-making in healthcare institutions. This chapter employs a narrative review approach to critically analyse the challenges and advancements faced by CECs across diverse cultural and institutional contexts in Asia. By synthesising findings from peer-reviewed literature, policy documents, and case studies, the methodology integrates a multidisciplinary perspective to position CECs as critical mediators in addressing ethical dilemmas.

While research ethics committees guide health or clinical research, CECs focus on clinical practice and patient care. This narrative review contextualises the evolution of CECs within the broader socio-cultural and technological landscape of Asia, offering insights into their transformative potential to navigate complex ethical decision-making.

11.1.1 Overview of Clinical Ethics Committees (CECs)

Clinical Ethics Committees (CECs) actively engage in proposing solutions and framework to both real and hypothetical conflicts within the healthcare setting, playing a critical role in conflict resolution. They influence patient care protocols to enhance the quality of care, tailoring approaches that meet both medical and ethical standards (American Society for Bioethics and Humanities 2011).

CECs typically composed of a diverse group of members such as healthcare professionals (doctors, nurses, and other healthcare workers), ethicists, lawyers, social workers, religious leaders, and laypersons. This variety ensures a range of perspectives, which is critical given the cultural, social, ethical, and legal diversity found in clinical settings. Each member brings a unique set of skills and insights, enhancing the committee's ability to address complex ethical issues comprehensively (Rudd 2001). A key aspect of CECs is to enhance satisfaction among patients and their families by ensuring that healthcare decisions are made transparently and considerately. Furthermore, these committees are instrumental in providing ethical education to healthcare staff, equipping them with the necessary knowledge and skills to handle complex ethical situations. CECs also focus on developing strategies to prevent future ethical issues, thereby proactively safeguarding the interests of patients and healthcare providers.

Another significant role of CECs is to satisfy the perceived needs of the staff by providing moral support and guidance, which is essential for maintaining a positive work environment. They also have the authority to discontinue treatments that are deemed undesirable or 'futile', ensuring that patient care is effective and compassionate. Lastly, CECs are known to minimise the risk of legal liability for healthcare institutions. Through these diverse functions, CECs maintains a balance between the needs of the healthcare providers and the rights and well-being of the patients. However, it is essential to study CECs in local contexts to understand their unique challenges and adaptations. By examining regional differences, we can compare how Asian CECs operate relative to those in other parts of the world. This approach allows us to appreciate the influence of moral relativism in shaping ethical practices while also identifying universal ethical principles that apply globally. Such a comprehensive understanding will deepen our insights into the ethical dimensions of healthcare (American Society for Bioethics and Humanities 2011). By integrating both moral relativism and universal standards, we can better address the diverse needs of humanity and ensure that ethical considerations keep pace with technological advancements.

11.1.2 Navigating the Ethical Horizon in a Rapidly Evolving Asian Healthcare Landscape

In Asia, the healthcare sector is marked by a striking spectrum of innovation and disparity, shaped by distinct levels of socioeconomic development, cultural diversity, and geographical factors (Rudd 2001). Nations such as Japan, Singapore, and South Korea have become benchmarks in medical innovation, featuring advanced healthcare systems and pioneering technologies. Contrastingly, countries like Cambodia and Laos face fundamental challenges with basic healthcare access and infrastructure (Edwards and Ramachandran 2021; Chongsuvivatwong et al. 2011). These variations present unique global health challenges, from managing non-communicable diseases and ageing populations to integrating traditional and contemporary medical practices. The inequitable distribution of healthcare resources between regions, urban, and rural areas, compounded by the burgeoning sector of medical tourism and cross-border healthcare trade, adds layers of complexity to an already diverse landscape.

The integration of cutting-edge technologies like telemedicine and AI across Asia illustrates the region's commitment to providing state-of-the-art medical services to its diverse population, comparable to Western healthcare standards (Zuellig Pharma 2023). China, now the world's second-biggest pharmaceutical market, alongside Singapore's stature as a global biotech hub, exemplify this thrust (Jakovljevic et al. 2021). The widespread application of advanced technologies—from telemedicine services in remote communities to sophisticated genomic and regenerative medicine in metropolitan centres—showcases a commitment to enhancing healthcare accessibility and quality across diverse settings. Particularly notable are Japan's innovative advancements in pioneering pluripotent stem cell technology and South Korea's investments in stem cell research infrastructure, positioning them at the forefront of regenerative medicine in Asia. The expansion into 3D bioprinting technology, enabling the synthesis of living tissues and organs, marks a significant leap in medical capabilities, facilitating new possibilities in tissue regeneration and transplantation (Ramadan and Zourob 2021). Moreover, the advent of genomic medicine, driven by initiatives like the Human Genome Project, has revolutionised the approach to diagnosing and treating complex disease. This paradigm shift towards precision medicine allows for more personalised treatment plans, addressing individual genetic profiles to enhance treatment efficacy and reduce adverse effects (Jamal 2021; Carrasco-Ramiro et al. 2017).

However, these advancements bring forth significant moral challenges, particularly in the areas of genetic privacy, data security, and the ethical implications of genetic modifications. As the boundaries of medical innovation are pushed, it is essential to address the potential risks to individuals' genetic information and ensure robust protections against data breaches. Additionally, the ethical considerations surrounding genetic modifications raise questions about the long-term impacts on human genetics and the potential for unintended consequences. These issues are not just theoretical but have real implications for patient care at the bedside, where clinicians must navigate these complex ethical landscapes while maintaining trust and transparency with their patients.

Furthermore, the challenge of providing equitable access to these innovative treatments underscores ongoing issues of affordability and fairness. These advancements often come with high costs, making them inaccessible to many, particularly socioeconomically disadvantaged groups. Ensuring that all individuals, regardless of their economic status, can benefit from these medical breakthroughs is crucial to promoting health equity and justice. At the bedside, clinicians face the daily reality of these disparities, striving to provide compassionate care while advocating for broader systemic changes that ensure fairness and accessibility in healthcare.

The COVID-19 pandemic further accelerated the adoption of telemedicine, enabling healthcare providers to maintain continuity of care through remote consultations crucial for patients in isolated or underserved areas (Jha et al. 2021). Concurrently, the deployment of AI and machine learning in diagnostic imaging has transformed medical diagnostics, offering enhances precision and speed in identifying pathologies, which is particularly vital in cancer diagnosis (Bawa 2021; Hosny et al. 2018). Despite these technological strides, healthcare in Asia must navigate a multitude of ethical dilemmas—ranging from securing valid informed consent to ensuring timely and fair access to medical treatments for all, including vulnerable populations like children, pregnant women, and migrants. Recognising and addressing these ethical challenges necessitates a robust commitment to equity and justice within the healthcare system and governance structures, ensuring that all individuals receive high-quality care without discrimination (Pham et al. 2019; Yip 2019).

11.1.3 The Rise of Clinical Ethics Committees (CECs) in Asia

Historically, the establishment of CECs can be traced back to the efforts of senior clinicians, academics, and early government policies that recognised the need for ethical guidance in healthcare. Initially, these services were often informal, providing ad-hoc advice and support to healthcare professionals facing ethical dilemmas. These early efforts laid the groundwork for more structured and formalised approaches to clinical ethics.

As the complexity of medical treatments increased and global ethical standards evolved, there was a significant surge in the formation of CECs. This growth was also driven by public backlash against perceived ethical failures in healthcare, which highlighted the need for systematic ethical oversight. During this period, healthcare institutions began to integrate ethical deliberations more formally into clinical practice, ensuring that patient rights were protected, and professional integrity was maintained.

This shift marked a crucial development in the field of healthcare, as it underscored the importance of ethical considerations in medical decision-making. By embedding ethical consultations into the clinical environment, healthcare providers could better navigate the moral complexities of patient care.

This evolution is also particularly poignant in sensitive areas like palliative care, where healthcare professionals often face challenging decisions about life support, balancing the ethical principles of autonomy and beneficence with the patient and family wishes and the legal frameworks governing end-of-life care. Cultural preference for indirect communication may also impact the openness and effectiveness of discussions within CECs, sometimes complicating the resolution of ethical dilemmas and potentially leading to moral distress among healthcare professionals who struggle with the ethical justification of their actions (Khoo et al. 2019; Lantos et al. 2019).

CECs in Asia have adapted their structures, roles, and activities to meet unique cultural, legal, and healthcare contexts of each nation (Khoo et al. 2019; Nagao and Takimoto 2024). For example, in China, the focus on bioethics has intensified with rapid advancements in medical research and technology, prompting significant growth in the establishment of ethics committee. Similarly, India has experienced considerable evolution in bioethics, with an increasing emphasis on clinical ethics support systems to address ethical dilemmas in patient care and medical research. Regions like Taiwan have also made notable strides in integrating ethical considerations into healthcare by developing tailored ethics committees and guidelines (Yoo et al. 2023; Chen et al. 2022; Liu et al. 2013). These committees provide a valuable platform for interdisciplinary dialogue, education, and policy development in clinical ethics, enhancing the ethical quality of healthcare, fostering trust between patients and healthcare providers, and guiding institutions through the ethical challenges of contemporary medicine (Magelssen et al. 2017; McLean 2007). Their ongoing development and adaptation are crucial for ensuring that healthcare decisions across Asia are ethically sound and sensitive to the diverse cultural and societal contexts.

11.1.4 Case Studies: Leadership and Regional Practices

Across Asia, Clinical Ethics Committees (CECs) are relatively new compared to research ethics committees, which are well-established. These research ethics committees are often accredited by Strategic Initiative for Developing Capacity in Ethical Review (SIDCER) under the Forum for Ethical Review Committees in the Asian and Western Pacific Region (FERCAP) also known as SIDCER-FERCAP (SIDCER-FERCAP 2024), and in some instances, what is known as a Hospital Ethics Committee may oversee both research ethics and clinical practice. Some country in Asia has developed unique structures and leadership roles within their CECs to address their specific ethical challenges.

- **Singapore**: The Committee on Clinical Ethics supported by the Centre for Biomedical Ethics (CBmE) at the National University of Singapore (NUS) serves as a platform to enhance collaboration and mutual support among CECs in Singapore, providing training, sharing resources, and identifying areas for further development. This group holds regular meetings and contributes to building local capacity in clinical ethics consultation, aligning with the broader goals of the NUS bioethics initiatives. Recognising the increasing importance of ethics, NUS has expanded the CBmE, recruiting prominent international figures such as Julian Savulescu. Another notable leader in this field is Jacqueline Chin, a distinguished bioethicist with extensive experience. Dr. Chin has served on various ethics committees in Singapore, including the NUH Ethics Committee, significantly contributing to the development of ethical guidelines and the promotion of clinical ethics education (Elliott, Ho and Lim 2010) (p. xiv).

- **Japan**: The National Society for Clinical Ethics plays an essential role in promoting ethical standards across Japan's healthcare system. This organisation supports ethics committees in larger hospitals and assists in standardising ethical practices nationwide, ensuring consistent application of medical ethics considerations in various healthcare settings. A notable figure in this effort is Akira Akabayashi, the Founding Director of the Clinical Ethics Centre at the University of Tokyo Hospital. He has significantly shaped the role of clinical ethics in Japan. His contributions to developing ethics guidelines and his leadership within ethics committees and academic institutions have been crucial in enhancing the ethical framework of Japanese healthcare (Akabayashi 2009).
- **India**: All India Institute of Medical Sciences (AIIMS) Ethics Committee features a robust ethics committee that plays a critical role in addressing ethical dilemmas in both clinical care and research. This committee is essential in advancing ethical standards within one of the world's largest and most diverse healthcare systems. Arvind Kumar, a senior surgeon and ethicist at AIIMS, brings his extensive experience to the Ethics Committee. His efforts are focused on navigating the ethical challenges inherent in a resource-constrained environment, ensuring that ethical decision-making aligns with the clinical realities faced by the institution (Institute Ethics Committee 2023).
- **South Korea**: Seoul National University Hospital has an ethics committee that reviews ethical issues in clinical practice, research, and policy development. The hospital has a Centre for Palliative Care and Clinical Ethics that focuses on addressing the diverse ethical issues of cases referred to clinical ethics support after the new legislation on hospice palliative care and end-of-life decision-making implemented in Korea in 2018. This centre plays a crucial role in the hospital's approach to managing ethical dilemmas and ensuring that end-of-life care respects patient rights and aligns with ethical standards (Seoul National University Hospital 2024).
- **Malaysia**: Two national conferences in 2017 marked a pivotal step toward introducing formal CECs in Malaysia (Khoo et al. 2019; Khoo 2017). These conferences highlighted the potential benefits of CECs, such as improving the quality of care and reducing healthcare costs, while also addressing challenges like cultural barriers and limited resources. The urgency of ethical decision-making in healthcare became even more apparent in 2020, during the COVID-19 pandemic. In

response to the complex ethical dilemmas that arose, the Ministry of Health Malaysia recognised the importance of strengthening ethical frameworks in clinical practice and initiated the establishment of CECs in several hospitals. These committees are tasked with providing guidance on ethical issues in clinical settings, ensuring that healthcare decisions adhere to the highest ethical standards. The establishment of CECs demonstrates Malaysia's commitment to addressing ethical challenges in healthcare and fostering a more reflective, principled approach to patient care nationwide (Ministry of Health Malaysia 2023).

11.2 Methodology

This chapter employs a narrative review approach to integrate multidisciplinary perspectives and explore the role and impact of Clinical Ethics Committees (CECs) in the Asian healthcare context. By synthesising information from a broad range of sources, including peer reviewed articles, policy documents, and case studies, this review aims to provide a comprehensive understanding of the challenges and advancements in the field.

11.2.1 Conceptual Framework and Data Analysis

The narrative review is grounded in a theoretical framework that incorporates principles of ethical stewardship, cultural relativism, and the operational dynamics of CECs. Central to this framework are key ethical principles such as beneficence, justice, and respect for autonomy, which are adapted to reflect the socio-cultural and economic diversity that characterises the Asian healthcare landscape. Primary data sources include peer-reviewed journals, policy documents, case studies, and guidelines published by internationally related to CECs, which gathered by searching through databases such as Scopus, Web of Science, and google, using appropriate keywords such as clinical ethics committee, and healthcare ethics in Asia. Secondary sources encompass reports from healthcare institutions and narratives from regional experts, enabling a comprehensive review of existing practices and emerging trends.

To explore the key challenges and advancements in the establishment and functioning of CECs within the Asian context, a critical analysis was undertaken. This analysis sought to discern patterns in ethical dilemmas and trace the evolution of CECs across varied cultural and institutional landscapes. Using a combination of rigorous critical analysis and the PEEL framework (Point, Explain, Evidence, Link), the study synthesized findings to deliver actionable insights tailored for healthcare professionals and policymakers. The discussion is contextualized within the rapid technological advancements and the diverse cultural background of Asian countries.

By integrating insights from both high-income and low- and middle-income nations (LMICs), the chapter highlights disparities in healthcare access and ethical challenges while positioning CECs as critical mediators. These committees play a pivotal role in navigating the intersection between universal ethical standards and localised practices, effectively addressing the complexities of ethical decision-making in a culturally and economically diverse region.

11.3 Key Concepts and Issues

11.3.1 Concept of Clinical Ethics Committees Stewardship

The foundational concept behind CECs is to establish a structured platform for the rigorous discussion and analysis of ethical challenges, thereby facilitating informed and personalised decision-making. This framework respects patient rights, upholds cultural values, and adheres to the highest standards of medical care. CECs serves as crucial reviewers and advisors, meticulously analyse clinical-ethical dilemmas and grounding their guidance in relevant moral principles, values, and laws. This process fosters transparent and inclusive discussions among all stakeholders, proving essential not only for resolving individual cases but also for shaping broader healthcare policies.

As healthcare continues to evolve with technological advancements, CECs are increasingly tasked with understanding and integrating these developments into their ethical reviews. This involves balancing traditional ethical concerns with newer challenges brought about by innovations such as digital health records, telemedicine, and personalised medicine. Their role is to ensure that advancements are implemented practically and ethically, safeguarding both patient welfare and professional integrity.

CECs also functions as key disseminators of ethical awareness and enhancers of decision-making competence among healthcare professionals. As coordinators, they ensure seamless communication across a range of healthcare stakeholders, from clinicians to policymakers, effectively articulating ethical issues to inform organisational and policy decisions. Acting as an oversight committee and guardians of ethical standards, CECs advocate for the integrity of healthcare practices, playing a critical role in fostering an ethical culture within institutions, thus enhancing trust and confidence in healthcare systems. By serving as a bridge between clinical practice and policymaking, CECs translate ethical principles into actionable guidelines, that influence overall quality of healthcare. Furthermore, CECs are promoters of fair resource allocation, championing equitable access to healthcare resources. They address the ethical implications of resource distribution, ensuring justice and fairness in patient care. This role particularly significant in the context where resources are scarce or when new treatments emerge that are expensive and require careful allocation.

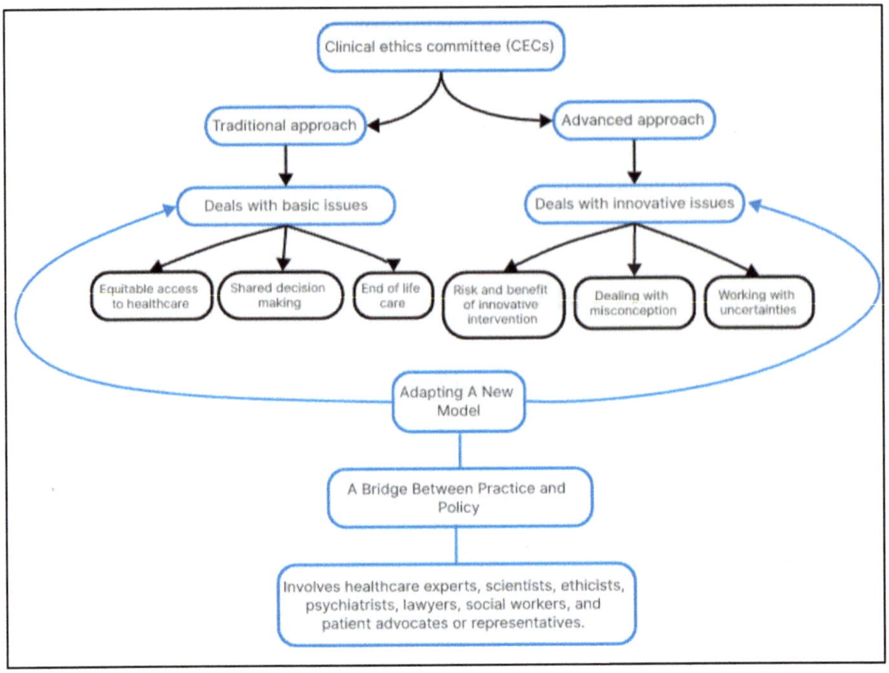

Fig. 11.1 Concept of Clinical Ethics Committees' (CECs) stewardship

The conceptual framework of CECs, depicted in Fig. 11.1, illustrates their comprehensive roles in addressing the ethical complexities of healthcare. It emphasised the need for CECs to be well-equipped with knowledge of ethics, law, and institutionalised systems of priority setting, highlighting the importance of a multidisciplinary approach to ethical decision-making in healthcare. This approach integrates insights from various fields to ensure that ethical deliberations are thorough and contextually relevant, enabling CECs to navigate the diverse and evolving ethical landscape effectively (McLean 2007).

11.3.2 Navigating Cultural Diversity in Asia's Evolving Healthcare Landscape

Asia's healthcare landscape is profoundly enriched by its cultural diversity, which plays a significant role in shaping the delivery and reception of healthcare. This diversity not only presents unique opportunities but also pose distinct challenges for CECs. Understanding the standpoint epistemology—a social theory framework that emphasises how individuals' unique perspectives and lived experiences, influence their understanding of the world—is crucial for CECs. Donna Haraway's (1988) concept of 'situated knowledge' highlights that knowledge and ethical understanding are always shaped by the context in which they are formed, a principle that is vital for the operation of CECs in culturally diverse settings (Haraway 1988).

Unique cultural practices significantly influence how families approach healthcare and interact with healthcare providers in Asia. Often, local customs and values take precedence over international norms, prompting CECs to adopt context-specific approaches to clinical decision-making. Recognising their own situatedness within specific institutional, cultural, and personal context, CECs actively seek to incorporate diverse viewpoints into their deliberations. This approach is crucial as it moves beyond the mere application of a universal ethical framework, fostering dialogue with patients, families, and other stakeholders to grasp their unique cultural perspectives. In many Asian cultures, family involvement in medical decisions may conflict with the patient's wishes or with standard medical practices advocating for patient autonomy and informed consent (Kröger et al. 2024; Louw 2016).

Moreover, religious beliefs significantly influence healthcare decisions across Asia, where spiritual considerations guide choices about medical interventions and end-of-life care. Additionally, language barriers can complicate the delivery of healthcare, potentially leading to misunderstandings about diagnoses, treatment plans, and consent. CECs play a critical role in advocating for culturally competent communication strategies ensuring that all patients, regardless of their linguistics background, fully understand their medical conditions and treatment options (Bandini et al. 2017). Furthermore, the integration of technological advancements in healthcare must be approached with cultural sensitivity to ensure that these innovations are accessible and beneficial to all segments of the population. This includes considering variations in technological literacy and access, which can differ widely across and within Asian countries. CECs are tasked with developing policies that promote equitable access to medical innovations, ensuring that these technologies do not exacerbate existing healthcare disparities but instead enhance the overall quality and accessibility of care.

Through these efforts, CECs not only navigate the complex interplay of cultural values and medical ethics but also enhance their capacity to make ethically sound decisions that respect the diverse cultural contexts of the patients and communities they serve. This nuanced approach to ethics in healthcare is essential for fostering trust and understanding between healthcare providers and patients, ultimately leading to better health outcomes and a more equitable healthcare system.

11.3.3 Ethical Ramifications of Technological Advancement

As Asia's healthcare systems evolve, they increasingly encounter numerous ethical challenges that needs comprehensive addressing, particularly as they adopt new technologies. One of the primary ethical concerns is the ethical implications of extensive research and clinical trials, which are resource-intensive and often pose significant challenges in low- and middle-income countries. This situation highlights disparities in access to cutting-edge treatments and raise concerns about equitable access and justice in healthcare delivery, reflecting a need for a historical perspective on how these disparities have developed and shifted over time (Dainow 2014). Informed consent remains a critical ethical issue, particularly salient in complex fields such as genomic and regenerative medicine. Ensuring that consent is genuinely informed is challenging, especially when patients have limited health literacy or when cultural factors influence their understanding and decision-making processes[19]. The evolution of consent practices, from early medical trials to contemporary complex interventions, illustrates the shifting ethical landscape in Asian healthcare.

Privacy and data security are paramount, especially with the increasing digitisation of health records and the use of AI in diagnostics. The historical development of data protection laws and the introduction of AI brings to light the continuous need to protect patient data from breaches and misuse, crucial for maintaining trust in healthcare systems. Additionally, AI's introduction raises questions about accuracy, accountability, and potential biases, which could lead to unjust outcomes if not properly managed (Kyung-min 2023). The dynamics of patient-provider relationships are affected by telemedicine and other remote care technologies. While these advancements improve access to care, they also introduce to concerns about the quality of care and the potential depersonalization of healthcare interactions. It is essential to ensure telemedicine services uphold the same ethical standards as in-person care, a challenge that CECs are uniquely positioned to navigate by developing guidelines that adapt as technological and societal norms evolve (Maroju et al. 2023; Bawa 2021). Cultural and religious diversity in Asia further complicates the ethics. Healthcare decisions are often influenced by cultural beliefs and religious practices, which can sometimes conflict with medical advice or patient autonomy. CECs must navigate these complexities by developing culturally sensitive guidelines that respect these diverse values while upholding universal ethical principles (Chattopadhyay and Simon 2008).

Resource allocation is another significant ethical concern, particularly in the context of introducing costly new treatments. Ensuring that resources are distributed fairly and that all patients have access to necessary care requires careful consideration and ethical deliberation. Addressing the ethical nuances of priority setting and the fair distribution of limited medical resources involves understanding the historical context of resource allocation and its ethical implication (O'Sullivan et al. 2022). Professional boundaries and conflicts of interest also pose ethical challenges. Healthcare professionals must maintain clear boundaries between their personal interests and patient care to avoid potential conflicts that could compromise patient welfare. CECs play a crucial role in providing guidance on maintaining these boundaries and ensuring that ethical standards are upheld.

11.4 Discussion of Implication

11.4.1 A Call for Culturally Sensitive Guidance and Responsible Oversight

Profound cultural and religious diversity across Asia necessitates the adaptation of ethical framework that are sensitive to varied norms and beliefs while upholding 'universal' ethical principles. This section delves into the broader implication of integrating CECs with a cultural sensitivity and regional needs. Challenges such as limited resources, including financial constraints and a scarcity of trained personnel, significantly impede the effectiveness of CECs in establishing comprehensive ethics training and implementing ethical decisions in patient care (Kröger et al. 2024; Louw 2016). Additionally, the variability of legal frameworks and governance across Asian countries complicates the operationalization of ethical guidelines, especially when legal and ethical standards diverge.

Despite these challenges, CECs in Asia have significant opportunities for capacity building and education, enhancing ethical awareness among healthcare professionals through region-specific training that respects local cultures and values. By addressing ethical dilemmas with transparency and effectiveness, CECs can significantly demonstrate trustworthiness in healthcare institutions, a crucial factor in navigating healthcare crises such as pandemics. Furthermore, they can influence healthcare policies, advocating for the integration of ethical considerations into national and regional healthcare legislation and policy formulation.

11.4.2 Implications for Healthcare Professionals, Institutions, and Policymakers

CECs facilitate informed and culturally sensitive decision-making processes that respect patient autonomy, promote justice, and adhere to the principles of beneficence and non-maleficence. Medical professionals require education on the cultural competencies needed to navigate these subtilities effectively. Administrators can support this by developing region-specific ethical frameworks and continuous education programmes for CEC members and healthcare staff. CECs foster trust and confidence in healthcare systems by ensuring transparency and accountability in clinical decision-making. Healthcare institutions must actively communicate the role and decisions of CECs to patients and the public, demonstrating a commitment to ethical care. For healthcare institutions, integrating CECs' guidance into clinical practices and policy formulation is a moral obligation to the society. Managers must also recognise the importance of ethical oversight in healthcare legislation and local policy development. Policymakers should tailor their guidelines and operations to accommodate local cultural and religious beliefs, ensuring decisions are globally informed and locally relevant.

11.4.3 Perspective of the Global South and Global North

The CECs in the Global South and Global North often reflect the contrasting healthcare environments, resources, and socio-cultural contexts of these regions. In the Global North, CECs tend to be well-established with robust frameworks and significant resources, including access to trained ethicists, comprehensive ethical guidelines, and advanced technologies. In contrast, CECs in the Global South frequently operate under more constrained conditions, grappling with fundamental issues such shortages of trained personnel, and significant disparities in healthcare access and quality (Elkhammas et al. 2018). Despite these challenges, CECs in the Global South bring invaluable insights into local cultural issues, often incorporating community values and traditional practices into their decision-making processes. Collaboration between CECs in these regions can foster mutual learning, where the Global North provides technical support and advanced ethical training, while the Global South offers practical approaches to resource-efficient and culturally sensitive ethical deliberation. This synergy can enhance the global effectiveness of CECs, ensuring that ethical healthcare practices are equitable and contextually relevant across diverse settings (Kowalski 2017).

11.5 Future Directions

11.5.1 Effective CECs (Clinical Ethics Committees) Models

Globally recognised models provide effective framework for CECs, including the Asian Society Bioethics and Humanities (ASBH), which is renowned for its comprehensive guidelines and structures approach to ethical decision making (American Society for Bioethics and Humanities 2024). The Clinical Ethics Network (UKCEN) in the United Kingdom (UK), emphasises a multidisciplinary approach, inclusiveness, and transparency, ensuring broad representation from various healthcare disciplines and lay members, which fosters diverse perspectives in ethical analysis and policy advisory (UKCEN 2024). The Ethox Centre's committee at the University of Oxford prioritises principles such as respect for persons, welfare, and fairness, melding ethical theory with practical application through detailed case analysis and training programmes (Ethox Centre 2024). Although the Hasting Center does not function as a clinical ethics committee itself, it provides extensive resources, research, and guidance on bioethics and clinical ethics issues. Its framework focuses on patient-centred care, shared decision-making, and cultural diversity, providing robust support for case consultation, policy review, and ethics education (The Hastings Center 2024).

The choice of a model for CEC depends on the specific needs of the healthcare institution, the typical nature of cases, and the cultural and legal context. Considerations include the primary functions required from the CEC (such as case consultation, policy development, and education), available resource (including trained ethicists and support staff), cultural alignment with regional norms, and adaptability to institutional needs (Ethox Centre 2024). For a comprehensive and adaptable framework, the ASBH committee provides a balanced approach to clinical ethics, covering case consultations, policy development, and education (American Society for Bioethics and Humanities 2024). Implementing this model involves several actionable steps:

- *Training*: Provide comprehensive bioethics and case consultation training to committee members.
- *Representation*: Include members from diverse disciplines including healthcare, law, ethicists, social workers, and patient advocate and community representatives
- *Education*: Keep the committee informed about emerging ethical issues and evolving best practices
- *Transparency*: Maintain transparency in the committee's processes and decisions

The integration of innovative technologies like AI and telemedicine necessitates a new and proactive adaptation of frameworks to address ethical challenges introduced by technological innovations. The flexibility of CECs to respond to emerging ethical issues fosters an environment of continuous learning and ethical vigilance. The proposed novel model below suggests extending ethical oversight to incorporate familial and communal considerations intrinsic to Asian societies. Emphasising 'soft autonomy', where the well-being of the family unit is considered alongside individual patient wishes, recognises the importance of collective decision-making and respects the intuitive understanding of patient wishes that may not always be explicitly stated (Gordon 2016).

11.5.2 A Novel Model for CECs (Clinical Ethics Committees) in Asia

While the ASBH model is versatile, CECs have a duty to be agile enough to navigate between traditional Asian communitarian approaches and emerging individualistic preferences in healthcare decisions, ensuring ethical deliberations are culturally sensitive and adaptable to regional societal norms and beliefs. This model should extend ethical oversight to include familial and communal considerations, embracing a 'soft autonomy' approach that values collective decision-making. Key roles of CECs in Asia (see Fig. 11.2) include enhancing cultural competencies, engaging communities, promote open dialogue, and ensuring sustainable healthcare practices

Fig. 11.2 Six roles of a novel model for Clinical Ethics Committees (CECs) in Asia

in resource-limited settings. Moreover, CECs should lead international collaborations, ensuring that ethical decision-making processes are inclusive, considering the best outcomes for humanity while minimising negative societal impacts.

Education is pivotal in efficacy of CECs. Tailored training in bioethics, cultural competencies, legal and regulatory shifts, technological literacy, and interdisciplinary collaboration is essential. A structured curriculum, expert-led sessions, active learning methods, and continuous evaluation forms backbone of this educational framework. This comprehensive approach ensures that all committee members are well-equipped to handle the ethical complexities of modern healthcare. As healthcare increasingly incorporates advanced technologies, the need to protect patients' confidentiality becomes more intricate and critical. CECs must, therefore, prioritise the development of robust protocols to address challenges related to data security and patient privacy. This entails educating healthcare providers on best practices for maintaining confidentiality in digital communications and data management.

By focusing on these areas, CECs can ensure that they not only uphold ethical standards but also lead the way in fostering trust and integrity within healthcare systems, making them indispensable in the modern medical practice.

11.5.3 Monitoring Clinical Ethics Committee (CEC) Standards and Ensuring Accountability

Regular assessments of CECs performance are crucial for maintaining high standards and accountability. These evaluations, informed by local empirical research, provides insight into the effectiveness of CEC interventions and perceptions among healthcare professionals and patients. Feedback from these assessments drives strategic improvements in CEC operation and evolving expectation, ensuring that ethical deliberations remain relevant and responsive to current healthcare challenges. Establishing rigorous standards for CECs membership through credentialing and privileging, and regular assessments of CECs' performance, informed by local empirical studies, are vital for continuous quality improvement. These evaluations also help in understanding the effectiveness of CECs' interventions and how healthcare professionals and patients perceive them, thereby driving strategic enhancements in CECs' operations and highlighting the evolving expectations from CECs.

11.6 Conclusion

Integrating CECs into the Asian healthcare system is crucial for addressing the multifaceted ethical challenges brought about by rapid technological advancements and cultural diversity. CECs play a fundamental role in upholding high ethical standards, fostering public trust, and ensuring that healthcare delivery is conducted with integrity. As healthcare evolves, the future of CECs in Asia lies in their ability to recognise potential conflicts between relativism and universalism in bioethics. With strong leadership and a commitment among to ethical standards, these committees can guide healthcare institutions in making ethically sound decisions that respect the diverse cultural and social contexts of the region. These considerations enable CECs to adapt exemplary healthcare practices to meet the needs of diverse patient populations in Asia.

Highlights

- Standardise and Integrate CECs into healthcare systems to ensure consistent ethical oversight amid rapid medical advancements.
- Develop Ethical Frameworks for AI, genomics, and telemedicine to address emerging dilemmas in patient care.
- Enhance Cultural Competency through structured training, ensuring ethical decisions align with diverse societal values.

References

Akabayashi A. Bioethics in Japan, 1980–2009: importation, development, and the future. Asian Bioeth Rev. 2009;1(3):267–78.

American Society for Bioethics and Humanities. Core competencies for healthcare ethics consultation: the report of the American Society for Bioethics and Humanities. Glenview (IL): ASBH; 2011.

American Society for Bioethics and Humanities. ASBH: homepage [Internet]; 2024 [cited 2024 Jul 5]. Available from: https://asbh.org/.
Bandini JI, Courtwright A, Zollfrank AA, Robinson EM, Cadge W. The role of religious beliefs in ethics committee consultations for conflict over life-sustaining treatment. J Med Ethics. 2017;43(6):353–8. https://doi.org/10.1136/medethics-2016-103930.
Bawa A. This Indian startup uses artificial intelligence to facilitate radiology imaging [Internet]. Marktechpost; 2021 [cited 2024 Jul 4]. Available from: https://www.marktechpost.com/2021/03/13/this-indian-startup-uses-artificial-intelligence-to-facilitate-radiology-imaging/
Carrasco-Ramiro F, Peiró-Pastor R, Aguado B. Human genomics projects and precision medicine. Gene Ther. 2017;24(9):551–61. https://doi.org/10.1038/gt.2017.77.
Chattopadhyay S, Simon A. East meets West: cross-cultural perspective in end-of-life decision making from Indian and German viewpoints. Med Health Care Philos. 2008;11(2):165–74. https://doi.org/10.1007/s11019-007-9106-y.
Chen Z, Jiang W, Xu J, Jiang L, Wang G. Ethical, legal, and social implications of genomics in China: a scoping review and implications for precision public health. China CDC Wkly. 2022;4(32):706–10. https://doi.org/10.46234/ccdcw2022.147.
Chongsuvivatwong V, Phua KH, Yap MT, Pocock NS, Hashim JH, Chhem R, et al. Health and health-care systems in Southeast Asia: diversity and transitions. Lancet. 2011;377(9763):429–37. https://doi.org/10.1016/S0140-6736(10)61507-3.
Institute Ethics Committee. AIIMS Bhubaneswar [Internet]. AIIMS Bhubaneswar; 2023 [cited 2024 Jul 4]. Available from: https://aiimsbhubaneswar.nic.in/ethicsCommittee.aspx.
Dainow B. Ethics in emerging technology. ITNOW. 2014;56(3):16–8. https://doi.org/10.1093/itnow/bwu067.
Edwards A, Ramachandran S. Around the world in healthcare systems: Asia [Internet]. Mercer; 2021 [cited 2024 Jul 4]. Available from: https://www.mercer.com/en-us/insights/us-health-news/around-the-world-in-healthcare-systems-asia/.
Elkhammas EA, Al-Mously N, Parvanov I, Youssef NA, Saad SG, Abdulhameed W. Practice of clinical ethics in developing countries: about time to revisit. Int J Med Biomed Stud. 2018;4(1):1–5. https://doi.org/10.4103/ijmbs.ijmbs_2_18.
Elliott JM, Ho CW, Lim SSN, editors. Bioethics in Singapore. Singapore: World Scientific Publishing; 2010. https://doi.org/10.1142/9789814327121_fmatter.
Ethox Centre. Support for ethics committees [Internet]; 2024 [cited 2024 Jul 5]. Available from: https://www.ethox.ox.ac.uk/ethics-support/ethics-committees.
Gordon JS. Clinical ethics consultation. London: Routledge; 2016.
Haraway D. Situated knowledges: the science question in feminism and the privilege of partial perspective. Fem Stud. 1988;14(3):575–99.
Hosny A, Parmar C, Quackenbush J, Schwartz LH, Aerts HJWL. Artificial intelligence in radiology. Nat Rev Cancer. 2018;18(8):500–10. https://doi.org/10.1038/s41568-018-0016-5.
Strategic Initiative for Developing Capacity in Ethical Review (SIDCER). Recognised ethics committees (ECs) & institutional review boards (IRBs) [Internet]; 2024 [cited YYYY Mon DD]. Available from: https://www.sidcer-fercap.org/pages/recognised-ecs-irbs.php.
Jakovljevic M, Wu W, Merrick J, Cerda A, Varjacic M, Sugahara T. Asian innovation in pharmaceutical and medical device industry—beyond tomorrow. J Med Econ. 2021;24(Suppl 1):42–50. https://doi.org/10.1080/13696998.2021.2013675.
Jamal ARA. Precision medicine: making it happen for Malaysia. Malays J Med Sci. 2021;28(3):1–4. https://doi.org/10.21315/mjms2021.28.3.1.
Jha AK, Sawka E, Tiwari B, Dong H, Oh CC, Ghaemi S, et al. Telemedicine and community health projects in Asia. Dermatol Clin. 2021;39(1):23–32. https://doi.org/10.1016/j.det.2020.08.003.
Khoo EJ. A national effort to introducing paediatric bioethics in Malaysia: a report and declaration of a national paediatric bioethics symposium. Asian Bioeth Rev. 2017;9(1):153–5.
Khoo EJ, Chua SH, Thong MK, Zilfalil BA, Lantos J. Introducing clinical ethics consultation service in Malaysia: a SWOT analysis. Clin Ethics. 2019;14(1):26–32. https://doi.org/10.1177/1477750919839913.

Kowalski CJ, Hutchinson RJ, Mrdjenovich AJ. The ethics of clinical care and the ethics of clinical research: yin and yang. J Med Philos. 2017;42(1):7–32. https://doi.org/10.1093/jmp/jhw032

Kröger C, Molewijk B, Muntinga M, Metselaar S. The diversity compass: a clinical ethics support instrument for dialogues on diversity in healthcare organizations. BMC Med Ethics. 2024;25(1). https://doi.org/10.1186/s12910-023-00992-z.

Kyung-min L. Korea issues first AI ethics checklist [Internet]. Korea Times; 2023 [cited 2024 Jul 4]. Available from: https://www.koreatimes.co.kr/www/tech/2024/01/129_352971.html.

Lantos JD, Saleem S, Raza F, Syltern J, Khoo EJ, Iyengar A, et al. Clinical ethics in pediatrics: an international perspective. J Clin Ethics. 2019;30(1):35–45.

Liu J, Shen J, Liu PZ. A survey of five first-level hospital ethics committees in Urumqi, China. J Zhejiang Univ Sci B. 2013;14(6):541–8. https://doi.org/10.1631/jzus.B1300085.

Louw B. Cultural competence and ethical decision-making for health care professionals. Humanit Soc Sci. 2016;4(2–1):41–52. https://doi.org/10.11648/j.hss.s.2016040201.17.

Magelssen M, Miljeteig I, Pedersen R, Førde R. Roles and responsibilities of clinical ethics committees in priority setting. BMC Med Ethics. 2017;18(1):68. https://doi.org/10.1186/s12910-017-0226-5.

Mahara G, Tian C, Xu X, Wang W. Revolutionising health care: exploring the latest advances in medical sciences. J Glob Health. 2023;13:03042. https://doi.org/10.7189/jogh.13.03042.

Maroju RG, Choudhari SG, Shaikh MK, Borkar SK, Mendhe H. Role of telemedicine and digital technology in public health in India: a narrative review. Cureus. 2023;15(3): e35986. https://doi.org/10.7759/cureus.35986.

McLean SA. What and who are clinical ethics committees for? J Med Ethics. 2007;33(9):497–500. https://doi.org/10.1136/jme.2007.021394.

Ministry of Health Malaysia. Clinical Ethics Consult Service (CECS) and Clinical Ethics Committee [Internet]; 2023 [cited 2024 Jul 4]. Available from: https://apps.hqe.moh.gov.my/equality/modul/pdfJSlib/web/fail/1.0_0.0_Clinical%20Ethics%20Consult%20Service(CECS)And%20Clinical%20Ethics%20Committe_11zon_11zon.pdf.

Nagao N, Takimoto Y. Clinical ethics consultation in Japan: what does it mean to have a functioning ethics consultation? Asian Bioeth Rev. 2024;16(1):15–31. https://doi.org/10.1007/s41649-023-00257-2.

O'Sullivan L, Aldasoro E, O'Brien Á, Nolan M, McGovern C, Carroll Á. Ethical values and principles to guide the fair allocation of resources in response to a pandemic: a rapid systematic review. BMC Med Ethics. 2022;23(1). https://doi.org/10.1186/s12910-022-00806-8.

Pham KTH, Nguyen LH, Vuong QH, Ho MT, Vuong TT, Nguyen HKT, et al. Health inequality between migrant and non-migrant workers in an industrial zone of Vietnam. Int J Environ Res Public Health. 2019;16(9):1502. https://www.mdpi.com/1660-4601/16/9/1502.

Ramadan Q, Zourob M. 3D bioprinting at the frontier of regenerative medicine, pharmaceutical, and food industries. Front Med Technol. 2021;2: 607648. https://doi.org/10.3389/fmedt.2020.607648.

Rudd PT. The role of clinical ethics committees. Curr Paediatr. 2001;11(5):381–5. https://doi.org/10.1054/cupe.2001.0213.

Seoul National University Hospital. Seoul National University Hospital [Internet]; 2024 [cited 2024 Jul 4]. Available from: http://www.snuh.org/global/en/find/findDeptView.do?hsp_cd=1&dp_code=HP.

SIDCER-FERCAP. Recognised ethics committees (ECs) & institutional review boards (IRBs) [Internet]; 2024 [cited YYYY Mon DD]. Available from: https://www.sidcer-fercap.org/pages/recognised-ecs-irbs.php.

The Hastings Center. Bioethics Research Institute [Internet]; 2024 [cited 2024 Jul 5]. Available from: https://www.thehastingscenter.org/.

UKCEN. UK Clinical Ethics Network [Internet]. 2024 [cited 2024 Jul 5]. Available from: https://ukcen.org/.

Yip WC. Healthcare system challenges in Asia [Internet]. Oxf Res Encycl Econ Finance; 2019 [cited 2024 Jul 4]. Available from: https://doi.org/10.1093/acrefore/9780190625979.013.508.

Yoo SH, Kim Y, Choi W, Shin J, Kim MS, Park HY, et al. Ethical issues referred to clinical ethics support at a university hospital in Korea: three-year experience after enforcement of life-sustaining treatment decisions act. J Korean Med Sci. 2023;38(24): e182. https://doi.org/10.3346/jkms.2023.38.e182.

Zuellig Pharma. Building a healthier Asia. In: 2nd Healthcare access summit, Singapore [Internet]; 2023 [cited 2024 Jul 4]. Available from: https://www.eu-asean.eu/wp-content/uploads/2023/03/Building-a-Healthier-Asia_March-2023.pdf.

Nishakanthi Gopalan is a senior lecturer in Medical Humanities and Ethics Unit (MedHEU), Faculty of Medicine, Universiti Malaya (UM), Kuala Lumpur, Malaysia. She is a member of the UM Research Ethics Committee (UMREC) and Universiti Malaya Medical Center (UMMC) Medical Research Ethics Committee (UMMC-MREC). She also coordinates the Master of Health Research Ethics (MOHRE) programme and serves as an editorial member for the Asian Bioethics Review.

Julia Patrick Engkasan is a rehabilitation medicine physician at the Department of Rehabilitation Medicine, Universiti Malaya (UM), Kuala Lumpur, Malaysia. She is a member of the Clinical Ethics Committee at Universiti Malaya Medical Centre (UMMC) as well as the Medical Research Ethics Committee (UMMC-MREC). She is actively involved in teaching and mentoring medical students and postgraduate trainees, emphasising the integration of ethics in clinical practice.

Erwin Jiayuan Khoo is a paediatrician and bioethicist at International Medical University (IMU), and a Research Fellow in Global Health and Social Medicine at Harvard Medical School. He was elected to Fellowship of the Royal College of Paediatrics and Child Health (UK) in 2019, and he completed the Harvard's Bioethics Fellowship in 2023. He also chairs the Malaysian Paediatric Association's Bioethics Task Force and serve on the Council of both the Academy of Medicine of Malaysia College of Paediatrics and the Malaysian Paediatric Association

Part III
HealthTech: Advances in Technology for Healthcare

Chapter 12
Telemedicine: The Game Changer in Future Prison Healthcare Setting

Clusters of Future Studies: Technological Trends, Past and Futures

Rumana Saifi and Yilu Qin

Abstract Globally, healthcare provision for people living in incarceration is associated with numerous challenges. This chapter explores telemedicine as an opportunity to overcome the barriers to quality health services in Malaysian prisons. We conducted a narrative review of existing studies to understand if telemedicine can increase health care access and meet the need of the prisoners. As the concept of telemedicine is still novel but has high policy relevance in Malaysia, we went through Ministry of Health and Prison Department reports; as well as WHO and UNODC resources on prison healthcare. Additionally, we shared our real-life experience from working in prison settings to discuss the pros and cons of implementing telemedicine. Telemedicine can enable healthcare providers to diagnose, monitor, and treat patients remotely. By making an alternate provision for prisoners' transportation to external medical facilities, telemedicine minimises security risks, reduces costs, and enhances safety. Telemedicine can facilitate access to a wide range of specialties, including psychiatry, cardiology, and dermatology, which are often difficult to provide on-site due to limited resources. Despite these advantages, the implementation of telemedicine in prison healthcare is not without challenges, including concerns about patient privacy, infrastructure requirements, regulatory limitations, and the need for specialised staff training.

Keywords Telemedicine · Telehealth · Prison · Prisoner · Future healthcare · Project ECHO

R. Saifi (✉)
Department of Social and Preventive Medicine, Faculty of Medicine, Universiti Malaya, Kuala Lumpur, Malaysia
e-mail: rumana@um.edu.my

Y. Qin
Yale School of Medicine, Yale University, New Haven, USA
e-mail: yilu.qin@yale.edu

R. Xiang Ng et al. (eds.), *Future of Healthcare in Asia*,
https://doi.org/10.1007/978-3-032-03183-9_12

12.1 Introduction

In recent years, technology has been playing a transformative role in healthcare settings around the globe (Beaudry et al. 2020) and Malaysia is no exception. Technological advancements in health have altered the conventional way we approach healthcare, diagnostics, treatment, and prevention. Some key categories of technological progress in healthcare include (1) telemedicine and telehealth (2) artificial intelligence and machine learning (3) robotic surgery (4) wearable and Health Monitoring Devices (5) genomics and CRISPR and (6) electronic medical records. The following section gives a brief overview of the most commonly adapted technological innovations in healthcare. Among these six classifications of technological enhancement, this chapter focuses on telemedicine as a vehicle to deliver effective health services to prison populations.

12.1.1 Telemedicine and Telehealth

The essence of telemedicine and telehealth, is the use of virtual platforms to bring patients and physicians in real time. Telehealth is a broader term that encompasses not only telemedicine (clinical healthcare) but also a wide range of healthcare services delivered remotely. In brief, while telemedicine focuses on clinical care, telehealth includes all aspects of remote healthcare, including education, prevention, and health monitoring.

Project ECHO (Extension for Community Healthcare Outcomes), a popular online training strategy, utilises telemedicine principles to link specialists with primary care providers or general practitioners in rural or underserved areas. However, it is not direct patient-to-provider telemedicine. Universiti Malaya (UM) is known as an established ECHO site which accommodates specialists and primary care providers in a common platform to exchange clinical knowledge.

12.1.2 Artificial Intelligence and Machine Learning

Artificial Intelligence (AI) in healthcare encompasses the use of advanced algorithms and machine learning techniques to the diagnosis, treatment, and management of medical conditions that normally requires human intelligence. Machine learning (ML) is a sub-set of AI that uses large datasets in healthcare to improve the accuracy of diagnosis, predict patient outcomes, and customise treatment.

12.1.3 Robotic Surgery

Robotic surgery is defined as a type of minimally invasive surgery where robotic systems provides assistance to surgeons in performing procedures with enhanced precision, flexibility, and control.

12.1.4 Wearable and Health Monitoring Devices

Health monitoring devices refer to a broad category of medical equipment used to continuously or periodically measure and record vital signs or other important health indicators.

12.1.5 Genomics and CRISPR

Genomics plays a critical role in personalised medicine, where treatments are tailored based on an individual's genetic makeup, as well as in the study of genetic disorders, cancer, and drug development. CRISPR (Clustered Regularly Interspaced Short Palindromic Repeats) is a revolutionary gene-editing technology that allows scientists to make precise, targeted changes to an organism's Deoxyribonucleic acid (DNA).

12.1.6 Electronic Medical Records

Electronic Medical Records (EMR) can provide secure, immutable storage for patient records, ensuring data privacy and confidentiality.

By definition, traditional healthcare services refer to models where patients and healthcare providers meet physically at the same location. This traditional care delivery model involves face-to-face, in-person communication and examination of patients; allowing healthcare providers to diagnose, and treat patients in real time. In order to overcome the barriers of traditional healthcare services, telemedicine emerged as a patient-centred approach intersecting mainstream healthcare approach with creative technology (Beaudry et al. 2020; Chandra et al. 2019; Fair et al. 2021). Telemedicine has the ability to increase coverage as patients can be treated without being physically present, especially patients in rural areas or those with mobility issues.

This chapter aims to assess the potentials of telemedicine in Malaysian criminal justice settings. The focus of this chapter is the prison landscape of Malaysia to implement telemedicine for healthcare services targeted to the prison population.

Recent technological improvements have made telemedicine an affordable alternative to traditional face to face health care models (https://news.va.gov/press-room/va-telehealth).

As a result, various telemedicine programmes are being tested in Malaysia following Malaysian Medical Council (MMC) guideline. The scope of existing telemedicine programmes have not been expanded from community to prisons in Malaysia though many developed countries adapted the programme to meet the need of those incarcerated.

12.1.7 Structure of the Chapter

The chapter is grouped into several sections, each identifying a theme. The chapter explains the healthcare structure in Malaysian prison systems and attempts to justify how telemedicine can be a major game changer to overcome the barriers in providing quality healthcare for the prisoners. This chapter highlights how telemedicine can emerge as a potential game changer in the healthcare settings in prisons as well as the potential challenges of implementing telemedicine in prisons.

This chapter gives a historical background of telemedicine projects around the globe. The section on the evolving history of telemedicine are followed by sections on describing overall prison healthcare systems; disease prevalence in prisons; Malaysian prison systems; Malaysian digital health interventions; benefits, barriers and facilitators of implementing telemedicine in prisons.

12.2 Review Methodology

This chapter reviews the prison healthcare in Malaysia and analyse if telemedicine can overcome the barriers in providing quality healthcare to prison population. In particular, this chapter is a narrative review of the prison situation in Malaysia and attempts to understand if telemedicine can fill the gaps to increase healthcare coverage. The narrative review in this chapter is coupled with our real-life experience gathered from Malaysian prison contexts.

This narrative review was a mix of both academic and grey literature. We used PubMed, Scopus, and Google Scholar databases for academic studies As expanding telemedicine in Malaysian prison structures has policy relevance, we reviewed Ministry of Health Malaysia and Prison Department Malaysia reports; as well as World Health Organization (WHO) and UNODC (United Nations Office on Drugs and Crime) resources on prison healthcare. The search strings for narrative review were created by using Boolean operators, e.g.: ("telemedicine" OR "telehealth") AND ("prisons" OR "correctional facilities") AND Malaysia. Research on telemedicine implementation, effectiveness, or challenges in prisons were taken into account. The key findings were synthesised into major themes.

12.3 Review Findings

12.3.1 History of Telemedicine

This section depicts the history of telemedicine as understanding this history is crucial for appreciating its significance, improving its implementation, and maximising its potential to transform global healthcare. Some of the earliest examples of telehealth are cited as the transmission of heart rhythms by telephone in the early 1900s, from sites in the Netherlands to consultation centres in Europe (Istepanian et al. 2006). Later the National Aeronautics and Space Administration (NASA) pioneered the use of remote monitoring of medical status for astronauts coupled with telecommunication to doctors for medical advice when needed (Istepanian et al. 2006; Davis et al. 2001). Around the same time, the US Veterans Health Administration (VA) in 1961 reported the testing of closed-circuit television (CCTV) technology as a video-based telehealth modality to address critical shortages and difficulty recruiting psychiatrists and neurologists (Istepanian et al. 2006; Davis et al. 2001). Fast-forward to 2018, the VA provided over 1 million telehealth encounters to around 400,000 veteran patients via real-time clinical video telehealth technology, with year-by-year increases in the use of this modality (https://news.va.gov/press/telehealth). There are a variety of terms and nomenclature within the field of digital health that are not always clearly delineated at present. According to the WHO, telemedicine involves the use of digital information and communications (ICT) technologies to overcome distance barriers in the delivery of health services (Beaudry et al. 2020; Istepanian 2006). While telehealth is sometimes used synonymously with telemedicine, other times is considered a broader use of ICT for supporting health-care services (and not necessarily only in delivery). Given the burgeoning growth and interest in digital health, WHO has suggested a classification scheme for digital health interventions based on the targeted primary user (clients/patients, healthcare providers, health systems, data services) (Fair et al. 2021). Under this taxonomy, telemedicine falls under healthcare worker interventions and includes variations such as provider-to-provider telemedicine and client-to-provider telemedicine. Figure 12.1 (see the next page) shows the various classification schemes for digital health.

The history of Malaysia's healthcare system reflects a gradual evolution of using technological advancements. It was way back in 1990's when Malaysia's government initiated healthcare reforms, incorporating technology to improve efficiency (Istepanian et al. 2006; Davis et al. 2001). Introducing electronic health records (EHR) began in 1990's in several public hospitals as pilot programmes (Istepanian et al. 2006; Davis et al. 2001). Over the last two decades, significant efforts have focused on digitising healthcare through tools like Electronic Medical Records (EMRs), teletraining platforms (e.g. project ECHO), and telemedicine platforms, though system-wide integration remains a challenge (Istepanian et al. 2006; Davis et al. 2001). EMR is a digital version of a patient's documents that contains medical and treatment history from a single healthcare provider or organisation. Globally, EMR is considered as a key component of modern healthcare systems, with a huge

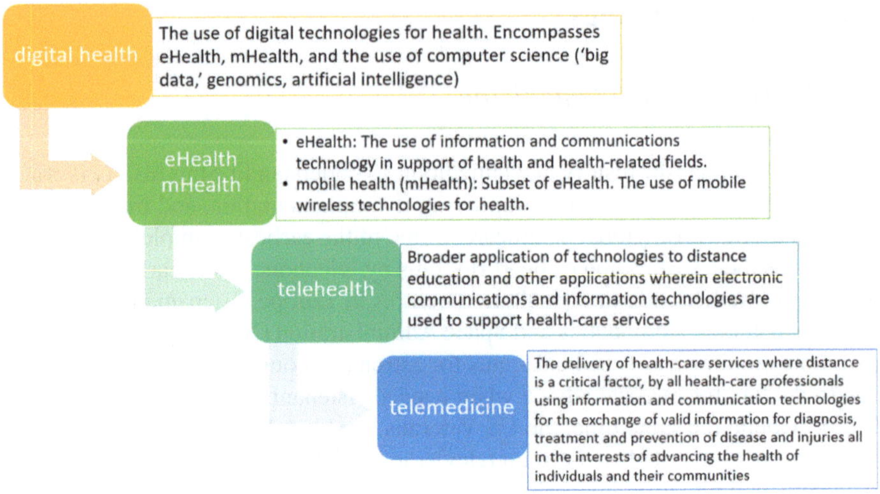

Fig. 12.1 Classification schemes for digital health

potential of healthcare coordination; accuracy of diagnoses and treatments; tracking of patient data over time, and reduction in duplication of tests or treatments. Project ECHO (Extension for Community Healthcare Outcomes), a collaborative health-care education and support model was adapted in Malaysia to improve access to specialised medical knowledge and care for healthcare providers. Developed in 2003 by Dr. Sanjeev Arora at the University of New Mexico, Project ECHO leverages on technology to create a "hub-and-spoke" model that connects specialists (the hub) with primary care providers (the spokes) in remote or resource-limited areas. While ECHO provides a teletraining platform, telemedicine expands the scope to cater to the needs of patients and link them to care through virtual platforms. Implementation of EMR, ECHO and telemedicine reflect Malaysia's commitment to optimising the use of technology for accessible, approachable, and affordable healthcare in near future.

12.3.2 Telemedicine in Prisons

Globally, an estimated 10.7 million people are incarcerated on any given day (Fair et al. 2021; Istepanian et al. 2006; Davis et al. 2001). People who experience incarceration often face significant physical and mental health challenges, leading to extreme poor health outcomes. Although International bodies such as the United Nations (UN) Human Rights Committee and the European Court of Human Rights have emphasised on the rights of people who are incarcerated, including the right to adequate health services, in traditional prison healthcare, there is often a shortage of medical professionals, particularly specialists, which limits the quality of care (Davis et al. 2001; UNODC et al. 2013; UNODC 2018). Therefore, quality of healthcare services provided in prisons often falls short when compared with community standards. Telemedicine bridges this gap by enabling remote consultations with specialists, ensuring that incarcerated individuals have access to the care they need without the need for additional in-house personnel. Several countries, e.g., US, Australia, United Kingdom, France and Canada have implemented telemedicine programmes in prisons, offering insights into how such systems can improve healthcare for prison populations (Chandra et al. 2019;. Dodds et al. 2021). Each of these implementations documents the benefits of telemedicine in terms of cost saving, security, and healthcare accessibility. A careful review of the existing literature suggests that there is a lack of evidence on telemedicine programmes in low and middle income countries as most of which have been carried out in developed countries. There are limited evidences from resource scarce countries like India and South Africa, where telemedicine has been pilot tested in prisons, primarily for mental health and chronic disease management.

Existing literature documents higher burden of diseases among prison population compared to general population. On top, prison healthcare structure, particularly in resource-limited settings, are not equipped to offer care for complex medical issues. In such scenario, telemedicine offers a promising platform for extending advanced healthcare services into prison structure, addressing several key challenges (Altice et al. 2010; Culbert et al. 2016; Chandra et al. 2019; Dodds et al. 2021). It reduces the security risks associated with transporting prisoners to external medical facilities for advance care. Transportation of prisoners for healthcare purposes often involves significant logistical challenges, including the potential for escape or violence, and requires a substantial security presence. By facilitating virtual appointments, telemedicine eliminates the need for these high-risk transfers. Finally, the cost of providing quality healthcare in prisons is a persistent issue. Telemedicine can reduce costs by minimising the need for in-person visits, transportation, and additional security measures, while also enhancing the efficiency and timeliness of care delivery (Chandra et al. 2019; Dodds et al. 2021). Table 12.1 lists the benefits and barriers of implementing telemedicine in prison.

Table 12.1 Benefits and barriers of implementing telemedicine in prison

Most common benefits cited for telemedicine in prisons
Improved Security
Personnel Safety
Costs Savings
Access to Specialists
Costs of Technology
Resistance from Medical Personnel
Lack of staff technical expertise
Difficulties coordinating

Reproduced from: Larsen et al. (2004)

Incorporating digital health technologies into prison systems ensures that incarcerated individuals receive consistent, high-quality healthcare, while simultaneously addressing the limitations of traditional prison healthcare infrastructures (Kral et al. 2018; Okeke et al. 2014).

12.3.3 Disease Prevalence in Prisons

This section illustrates the prevalence of diseases in prisons. It is crucial to understand the disease profile of incarcerated individuals to design effective telemedicine programme in closed settings. As prison health is public health, healthcare in prisons and detention centres play a central role in public health disease management. Evidence indicates that communicable diseases and mental health illnesses are more prevalent among the prison population compared to the general public. This heightened prevalence is a recognised public health concern, as it affects not only incarcerated individuals but also prison staff and the communities to which former prisoners return (UNODC et al. 2013; UNODC 2018; Altice et al. 2010; Perrin et al. 2021). Active and latent Tuberculosis (TB), HIV and hepatitis C (HCV) are the most prevalent communicable diseases in prisons. Many individuals enter prison with untreated or under-treated pre-existing health conditions, including chronic diseases (like diabetes, hypertension), infectious diseases (such as HIV, hepatitis, or TB), and substance use disorders. Additionally, people in criminal justice settings are less likely to have had access to regular health care provider prior to incarceration and compared to general population are disproportionately affected by the social determinants of health. Social determinants of health of incarcerated individuals are related with poor access to healthcare, lower socioeconomic status, and unstable living conditions: all of which contribute to worse health outcomes (Altice et al. 2022; Perrin et al. 2021). Figure 12.2 displays the prison landscape and disease prevalence. The figure illustrates the cycle of disease transmission related to prison environments. The 'Prison Environment' is depicted with figure moving in and out.

Although the Malaysian government has globally committed towards curtailing risk and reducing rates of HIV, active and latent TB, and HCV— health care in prisons and places of detention are still not held to a certain standard and priority due to punitive policies and societal stigmas violating fundamental human rights (Henny et al. 2018; Larsen et al. 2004; Poushter et al. 2016; Purnomo et al. 2018). Coupled with severe overcrowding and limited resources, this has led to high incidence and prevalence of communicable diseases in prison population and staff, exhausting the prison health system (Doran et al. 2021), and Malaysian health system at large—impacting Malaysia's ability to achieve global health targets (i.e. United Nations Sustainable Development Goal (SDGs): 3.3, 3.5 and 3.8)., particularly for TB, HIV, and HCV. The breakdowns for SDGs 3.3, 3.5 and 3.8 depicts (Bazazi et al. 2017).

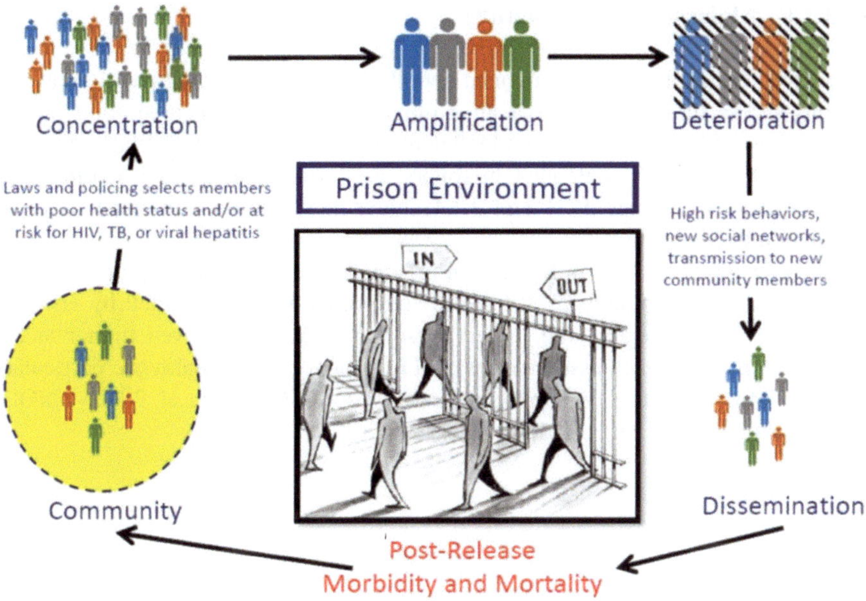

Fig. 12.2 Prison landscape and disease prevalence. Reproduced from Kamarulzaman et al. (2016)

SDG Target 3.3: End epidemics of communicable diseases

By 2030, target 3.3 aims to end the epidemics of AIDS, TB, malaria, and neglected tropical diseases, and to combat hepatitis, water-borne diseases, and other communicable diseases. This goal highlights efforts to reduce the spread and impact of these diseases, particularly in vulnerable populations, e.g. prison population.

SDG Target 3.5: Prevent and treat substance abuse

SDG 3,5 refers to strengthening the prevention and treatment of substance abuse, including narcotic drug abuse and harmful use of alcohol. This target focuses on reducing the health risks associated with drug and alcohol dependency, especially through effective interventions and rehabilitation.

SDG Target 3.8: Achieve universal health coverage (UHC)

SDG 3.8 refers to ensuring access to quality healthcare services and access to safe, effective, quality, and affordable essential medicines and vaccines for all. Universal health coverage (UHC) is key to ensuring that no one is left behind in receiving adequate health services, regardless of their financial or social situation.

These targets aim to tackle critical global health issues, from communicable diseases to non-communicable diseases, and ensure health systems are inclusive and accessible to all, including prisoners.

12.3.3.1 Malaysian Prison Settings

Annually, 10.2 million are imprisoned and 40–70 million people transition through prisons across the world. As per an estimate of 2023, the incarceration rate for Malaysia is 217 per 100,000 people. There are 35 prisons across Malaysia where approximately 72,437 sentenced prisoners are placed (Malaysia | World Prison Brief.org, n.d.). Figures 12.3 and 12.4 shows the number and rates of prison population in Malaysia.

The prevalence of HIV in Malaysian prisons is 5-fold higher than in the general population. Approximately 2% of the prison population are affected by HIV, and prevalence in the Kajang, the largest prison in Malaysia is even higher at 5%. Morbidity and mortality associated with TB in prisoners in Malaysia, particularly who are infected with HIV, is extraordinarily high (UNODC et al. 2013; UNODC 2018; Springer et al. 2011; Altice et al. 2022).

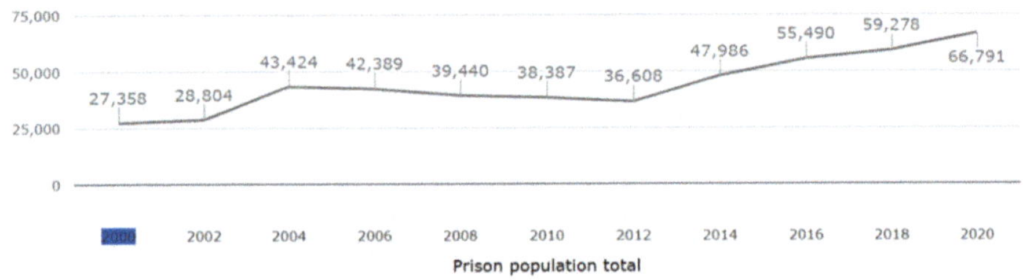

Fig. 12.3 Number of prison population in Malaysia. Reproduced from World Prison Brief (Malaysia) (n.d.)

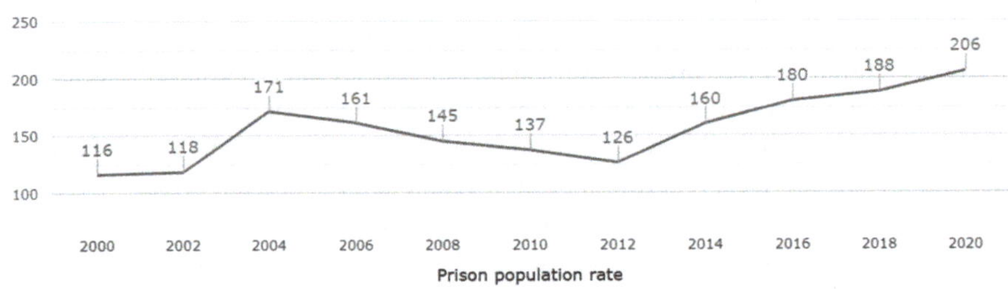

Fig. 12.4 Rates of prison population in Malaysia 2000–2020. Reproduced from World Prison Brief (Malaysia) (n.d.)

12.3.4 *Our Previous Experience in Working in Malaysian Prison Settings*

Figure 12.5 depicts the percent of latent TB infection in prison setting.

In a collaborative study conducted by UM and Yale University, it was found that Latent TB infection (LTBI) in both HIV+ and HIV− prisoners and also among prison officers are higher in comparison with the general population. In the above figure, LTBI was extraordinarily high in prisoners in both "open" (87.6%) and "closed" air prison units and markedly higher than in the general population (36.1%) (UNODC et al. 2013; UNODC 2018; Springer et al. 2011; Altice et al. 2022).

In Malaysia, criminalisation of drug use results in large numbers of people incarcerated and cycling through the criminal justice system. Although illicit opioid use in Malaysia is widespread, and the prevalence of injection drug use may be one of the highest globally, programmes for treating substance dependence and HIV within the prison care system remains limited. After prison release, the sub-optimal linkage to care, and poor adherence to treatment result high transmission of HIV and TB in the community, especially with ineffective transitional care.

TB and HCV co-infection is especially common among people who inject drugs (PWID). Standard TB treatments have been shown to increase hepatotoxicity, and require close monitoring and a longer duration of treatment (UNODC et al. 2013; UNODC 2018).

Evidence suggests that prisoners have greater health needs when compared to the general population. Although prisoners in Malaysia have a constitutional right to health, existing studies indicate that prisoners' access to health care and the quality of that care are during incarceration and post-release is lacking, when compared to

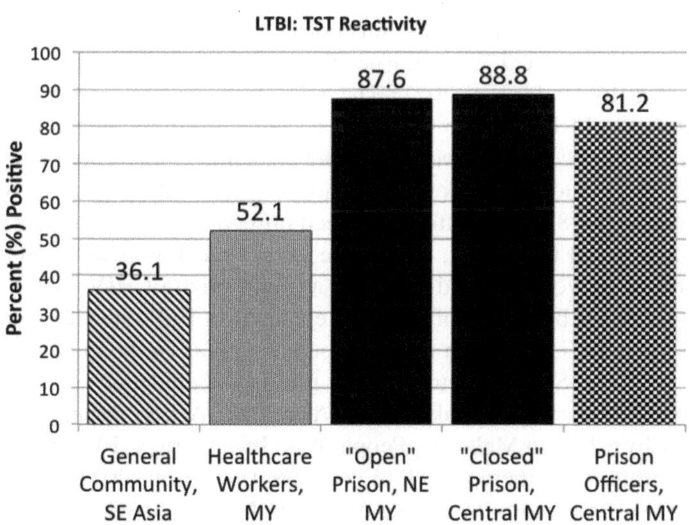

Fig. 12.5 Latent TB infection in prison setting. Reproduced from Kamarulzaman et al. (2016)

the quality of health care provided for the general population. Although prisons in Malaysia have been enhancing the delivery of healthcare by appointing in-prison 'panel doctors', the number of doctors placed at the prisons was severely disproportionate to the number of prisoners. As mentioned, in a Malaysian prison, only two prison doctors are available for 4,612 prisoners. Additionally, there is a lack of female doctors to treat female prisoners. Under such circumstances, telemedicine can serve as a platform to offer healthcare to prisoners (UNODC et al. 2013; UNODC 2018).

12.3.5 HIV Trends Globally and in Malaysia and Their Implications for Telemedicine:

The HIV care cascade, sometimes referred to as the HIV care continuum, was first described in the early 2000s as a way to understand and better capture key indicators for measuring HIV care from initial diagnosis to engagement in medical care to achieving viral suppression (Altice et al. 2022; Culbert et al. 2016). It has become an internationally recognised framework for monitoring progress and gaps in the quest to end the AIDS epidemic. In 2014, the Joint United Nations Programme on HIV/AIDS (UNAIDS) set ambitious goals coined as the 90-90-90 targets, which stipulated that by year 2020, 90% of people living with HIV (PLHIV) would know their HIV status, 90% of diagnosed individuals would be on antiretroviral therapy (ART), and 90% of those on ART will be virally suppressed (Culbert et al. 2016; Chandra et al. 2019). These targets were widely adopted and mobilised global momentum to achieve 95-95-95 targets by 2030.

In the past 10 years, remarkable progress has been made towards the global targets, going from 67-57-51 in 2015 to 86-89-93 in 2022 (Culbert et al. 2016; Dodds et al. 2021). Globally, the rate of new HIV infections has been decreasing since the 1990s and as of 2022, an estimated 39 million people around the world were living with HIV (Culbert et al. 2016; Dodds et al. 2021). The Asia–Pacific region comprises the second highest number of PLHIV in the world and accounts for more than half the world's population. Prevalence is low overall, but the epidemic is unevenly distributed and tends to be concentrated among key populations of marginalised groups (Altice et al. 2022; Kral et al. 2018). This pattern is indeed observed in Malaysia where, similar to other countries in the region, vulnerable groups such as people with a history of incarceration and/or drug addiction are at higher risk of HIV than the general population. From the beginning of the AIDS epidemic in 1981, the prevalence of HIV, HCV, HBV, and TB is higher in prison populations than in the general population (Altice et al. 2022). Several factors have a role in the epidemics of HIV, TB, and related infections in prisons. Unlike most of Asia where AIDS mortality decreased by 37%, it grew by >20% in Malaysia. People Who Inject Drugs (PWIDs) in Malaysia have low antiretroviral therapy (ART) coverage (Fair et al. 2021) and high HIV, HCV and TB-related morbidity and mortality (UNODC 2018). Opioids, the most frequently injected drugs, fuel explosive HIV pandemics in SE Asia (Istepanian et al.

2006; Davis et al. 2001). Despite opioid agonist therapy (OAT) with methadone (MMT) or buprenorphine (BMT) being effective for primary and secondary HIV prevention among PWIDs (Istepanian et al. 2006; Davis et al. 2001), yet in the 6 countries (including Malaysia) with the highest HIV-related mortality among PWIDs, OAT is inadequately scaled with <2% receiving it (UNODC 2018).

Criminalisation of drug use further concentrates PWIDs with HIV, HCV, and TB within prisons. Globally, 10.2 million are imprisoned and 40–70 million people transition through prisons annually. HIV and TB prevalence in prisoners are 5- and 100-fold higher than in the general population, respectively, resulting in explosive TB outbreaks. Prison crowding also contributes to increased TB risk, particularly among HIV+ s and PWIDs who already have elevated risk. Prison mortality is high, with 50%-80% of deaths attributed to TB, especially from HIV/TB co-infection. Nearly all prisoners return to communities where HIV and TB transmission are perpetuated, especially with ineffective transitional care. In low/middle income countries (LMIC), PWIDs, HIV, HCV, and TB are syndemic and concentrated in prisons and require innovative and integrated programmes to effectively transition them to the community (Altice et al. 2022; Kral et al. 2018).

The entry of COVID-19 in criminal justice settings has posed new challenges to the existing burden of infectious diseases among the prison population. Translating standard Covid-19 prevention, control, and management interventions in a closed community may prove insufficient as prison systems are plagued by overcrowding, poor sanitation, poor ventilation, and high admission rates in both developed and developing countries.10 In order to counteract the intersection of corona virus with existing burden of HIV, TB, HCV and substance use disorder, it is crucial to increase the capacity of prison healthcare service providers. The population to doctor ratio in Malaysian prisons is 1:454. The small number of doctors cannot provide adequate healthcare for the large number of prisoners, and that could lead to poor quality healthcare.

In addition to communicable diseases, psychiatric and substance use disorders (PSUDs), in particular, are prevalent among prison populations, especially in low and middle-income countries (LMICs) (Fair et al. 2021; Istepanian et al. 2006). A study on detainees in Malaysia found that 93.3% of study participants had at least one psychiatric disorder, whilst 53.3% had substance use disorders (SUDs) (Springer et al. 2011). However, treatment capacity of PSUDs in prisons is inadequate in the country due to various factors, including low screening, diagnosis and treatment, sub-optimal funding, prisons being largely inaccessible to expert treatment providers and data to influence policy, among other reasons (Springer et al. 2011).

12.3.6 Telemedicine: A Game Changer in Malaysian Prisons

12.3.6.1 Benefits of Telemedicine in Prisons: Increased Access to Specialist

Incarcerated individuals often have limited access to specialists, especially in resource-limited settings. Prison overcrowding is a significant health system challenge which can strain healthcare resources. According to the Human Rights Commission of Malaysia (Suhakam), the number of physicians placed in prisons is not aligned with the number of prisoners (Doran et al. 2021). Although the prison department in Malaysia is committed to ensuring the delivery of quality healthcare to the prison population, the disproportionate ratio of physicians to prisoners raises serious concerns. For instance, in the largest prison of Malaysia, only two doctors are available for 4,612 prisoners. In addition, there is a lack of female doctors to cater to the need of female prisoners (UNODC et al. 2013; UNODC 2018; Doran et al. 2021).

Telemedicine can improve quality health care coverage for prison population via consultations between patients and specialists like psychiatrists, dermatologists, or cardiologists without the need for physical transfers of prisoners from prison to specialist facilities (Bazazi et al. 2017; Muthulingam et al. 2019). As mentioned, Malaysian prisons have in-house medical clinics staffed by healthcare personnel, including general practitioners and pharmacists. Evidence suggests that most prison doctors in Malaysia lack adequate training to handle the complex health needs of prisoners. Their medical training focused primarily on the delivery of basic primary healthcare, leaving them with insufficient experience in diagnosing and treating communicable diseases, managing drug abuse, and, more critically, addressing mental health issues (Bazazi et al. 2017; Muthulingam et al. 2019). These gaps in training highlight a significant challenge in ensuring comprehensive care within the prison system. Limited healthcare personnel, coupled with a lack of training affect the overall quality and timeliness of care in prison settings.

For specialised care or more serious medical issues, prisoners are usually referred to the nearby public hospitals. Access to specialised care, especially for chronic conditions or mental health, remains an area that would require further development (Bazazi et al. 2017; Muthulingam et al. 2019; Purnomo et al. 2018; Doran et al. 2021). The advances in telecommunications (through satellites, the internet, and telephone infrastructure), combined with the capability to capture, store, transmit, and display electronic medical information, enable specialised doctors to perform many tasks remotely that were once only possible in person. Telemedicine opened the opportunities for prisoners with chronic conditions such as diabetes or hypertension to be monitored remotely, reducing the need for regular hospital visits (Low et al. 2020).

12.3.6.2 Cost Savings

In prison settings, prisoners are usually taken to specialised health care facility for complex healthcare needs, and specialist expertise. Governments and prison systems are ethically and legally obligated to ensure adequate healthcare for prisoners. If the necessary care are limited within the prison structure, it is the responsibility of the Government and prison system to arrange external healthcare for equity and justice for prisoners. Arranging healthcare beyond the prison walls often comes with challenges. Transporting prisoners to external medical facilities for advanced medical care involves significant costs, including special transportation cost, security measures and staff time (Young and Badowski 2017). Furthermore, correctional facilities are often situated in distant geographic areas where access to speciality healthcare entails huge cost involvement. Telemedicine has the potential to reduce these costs by allowing medical consultations to occur within the prison facility. Although there is no available data associated with cost savings in Malaysian prisons via telemedicine, a study conducted in a US prison shows that a Telemedicine programme averted about 35 trips for prisoners to see external specialists which facilitated a cost savings of approximately $27,500 (Muthulingam et al. 2019; www.projectsyndicate.org./reforming Malaysian prisons).

12.3.6.3 Improved Safety

Safety issues are crucial for people from a criminal justice settings due to the huge burden of mental illness and substance use disorder, communicable disease (including HIV, TB, and hepatitis), noncommunicable disease, and cognitive disability compared to the general population (Low et al. 2020). Telemedicine reduces the risk of escape or harm to guards and the public during transportation of prisoners to external healthcare facilities. It also limits potential exposure to infectious diseases by minimising contact between prisoners and external environments (Henny et al. 2018; Perrin et al. 2021; Muthulingam et al. 2019; www.projectsyndicate.org./reforming Malaysian prisons).

12.3.6.4 Timely Healthcare

Telemedicine can speed up the process of prevention, diagnosis and treatment. It allows healthcare providers to connect with patients quickly, often without the delays associated with transporting patients to external facilities or waiting for an in-person specialist visit. In situations where quick medical attention is needed, it can prevent delays caused by the logistics of transporting inmates to hospitals (Muthulingam et al. 2019; www.Suhakam.malaysianprison.com 2021).

12.3.6.5 Mental Health Services

Telemedicine is particularly useful for providing mental health care, which is often in high demand in prisons. Remote therapy, counselling, and psychiatric consultations can be offered more frequently (Muthulingam et al. 2019; www.Suhakam.malaysian prison.com, 2021).

12.3.6.6 Emergency Consultations

Telemedicine can be used for initial consultations in emergencies, allowing healthcare providers to assess whether an in-person visit is necessary (www.projectsyndicate.org./reforming Malaysian prisons; www.Suhakam.malays ianprison.com; United Nations General Assembly, 2015).

12.3.6.7 Post-Operative Care

Follow-up consultations for inmates who have undergone surgery can be managed via telemedicine, reducing the need for external appointments (www.Suhakam.mal aysianprison.com; United Nations General Assembly 2015).

12.3.6.8 Challenges of Implementing Telemedicine in Prisons

Telemedicine has the potential to significantly improve the efficiency and quality of healthcare in prisons, addressing both logistical and financial challenges while offering better care for inmates. However, its implementation requires careful planning to navigate the technical, legal, and ethical issues involved. The following section highlights a few challenges that prison systems can experience while implementing telemedicine programme (Poushter et al. 2016; Doran et al. 2021).

12.3.6.9 Infrastructure and Technology

Many prisons may lack the necessary infrastructure for telemedicine, including reliable internet connections, video conferencing equipment, and private spaces for confidential consultations (Poushter et al. 2016; Purnomo et al. 2018; Doran et al. 2021).

12.3.6.10 Privacy Concerns

Ensuring patient confidentiality in a prison setting is a challenge. Prisoners may feel uncomfortable discussing sensitive health issues over video calls, especially if the consultations are conducted in non-private settings (Poushter et al. 2016; Purnomo et al. 2018; Doran et al. 2021).

12.3.6.11 Regulatory and Legal Issues

Regulations surrounding telemedicine vary by jurisdiction, and prisons may face legal hurdles when implementing remote healthcare systems. Licensing, consent, and security protocols need to be thoroughly addressed (Poushter et al. 2016; Purnomo et al. 2018; Doran et al. 2021).

12.3.6.12 Staff Training

Prison staff, including both medical and non-medical personnel, need to be trained in the use of telemedicine technology to ensure smooth operations. Resistance to new technology may also be an issue (Poushter et al. 2016; Purnomo et al. 2018; Doran et al. 2021).

12.3.6.13 Continuity of Care

Telemedicine solutions need to integrate with the prison's existing healthcare records and systems to maintain continuity of care. Coordination between remote healthcare providers and on-site prison healthcare staff is crucial (Poushter et al. 2016; Purnomo et al. 2018; Doran et al. 2021).

12.3.7 Ethical Challenges

Telemedicine comes with several ethical challenges. Existing literature shows that ethical issues in telemedicine can be assessed from several aspects i.e., equity, doctor-patient relationship, data confidentiality and security, informed consent, patient's and family's satisfaction with telemedicine services. Compliance to ethics is a primary aspect of high quality telemedicine services. In other words, if healthcare providers abide by ethical rules, they can provide better services for patients (ECDC 2017). It is important that telemedicine platform is created with robust privacy protocols and secure platforms. Adequate training on telemedicine best practices can ensure data confidentiality and privacy; better doctor-patient relationship and sound informed consent procedures (ECDC 2017).

Healthcare practitioners are required to uphold the same ethical standards and codes of conduct, regardless of mainstream or remote health service provision (ECDC 2017). Medical advice through telemedicine should be provided only within the scope of the provider's specialty as specified in their practising certificate. Physicians must also ensure they have adequate liability protection to cover potential malpractice claims (ECDC 2017).

12.3.8 Malaysia's Digital Health Infrastructure

This section discusses if Malaysia is able to handle telemedicine in prison contexts. Malaysia has a proven track record of efficient digital health infrastructure. In response to the COVID-19 pandemic, Malaysia's digital health infrastructure strengthened using artificial intelligence/machine learning (AI/ML) for contact tracing solutions such as MySejahtera, MyTrace, and others. Additionally, virtual health clinics and e-appointment activities increased access to healthcare in the community (Young and Badowski 2017). As the rates of infectious diseases continue to rise in incarcerated settings in Malaysia, with high prevalence of HIV and TB treatment disruptions, there is a precedent to introduce telehealth and telemedicine approaches to supplement in-facility medical care to empower limited medical personnel and strengthen healthcare service and delivery for people in prison. Additionally, such services can reduce transport and related security management costs, while increasing health and clinical services for better health outcomes (Young and Badowski 2017). Malaysia's previous experience with digital health assures it's competencies to manage telemedicine in closed environments, e.g., prisons.

12.3.9 Implementation of Project ECHO® in Malaysia

Malaysia has been able to successfully implement Project ECHO which gives the country a solid basis to justify the implementation of telemedicine. UM is known as an established site for ECHO. The National Academy of Medicine and Centres for Disease Control and Prevention endorse Project ECHO® as a highly innovative strategy likely to produce significant improvements in the quality and efficiency of patient-centred integrated care (Maruschak and Berzofsky 2015). It has demonstrated efficacy for multiple chronic health conditions, including HIV prevention and PrEP, HIV treatment, hypertension, diabetes, and chronic pain (Okeke et al. 2014). Project ECHO® is popular, connecting 70,000+ clinicians in 37+ countries. The CDC is currently implementing Project ECHO® for HIV Prevention and/or Treatment in 23 countries across 6 continents, including in many low- and middle-income countries. UM is currently an academic hub for Project ECHO for HCV, addiction, and palliative care. Although ECHO is not telemedicine, it provides a platform for teletraining which has successfully been implemented in Malaysia on several occasions (https://echo.unm.edu/echos-impact).

Project ECHO® for HIV Prevention is a widely-used, evidence-based intervention for improving HIV testing, prevention, and linkage to care. Project ECHO®, which was tested originally to facilitate HCV care into non-specialty settings (https://echo.unm.edu/echos-impact), is based on established educational theories of social learning & behaviour change (Woodall and Tattersfield 2017; ECDC 2017) wherein non-specialists in community settings collaborate with specialists at academic "hubs," resulting in an innovative healthcare delivery model that yields equivalent quality care for patients in non-specialty settings (Woodall and Tattersfield 2017). ECHO's powerful coaching strengthens non-specialist care delivery and improves both capacity and access to specialty care for underserved populations. This low-cost, high-impact coaching intervention links experts with non-specialist clinicians through tele-education, in which the newly emerging experts co-manage cases and share expertise via mentoring, guidance, feedback and didactic education (Wood et al. 2018). This effectually de-monopolises specialty knowledge by enabling clinicians to develop the skills, knowledge and self-efficacy outside of their main specialisation to treat patients with common, complex diseases and thereby reducing travel time and costs, wait times, and avoidable complications. The ECHO model develops knowledge and capacity in clinicians through (a) case-based learning, (b) knowledge networks, and (c) learning loops. Knowledge networks (see Fig. 12.1 on Page 250) consist of regularly scheduled teleECHO sessions that bring together interdisciplinary specialty experts. Integrated care partners learn best practices through learning loops in which they co-manage complex patients in 'real world' situations and provide specialty care that is typically beyond the scope of their practise (Maruschak and Berzofsky 2015). Over time, these learning loops create deep knowledge, skills and self-efficacy. Unlike telemedicine where the specialist assumes clinical responsibility, ECHO is a guided practise model where clinicians retain care responsibility, operating with increasing independence as their skills and self-efficacy grow (https://echo.unm.edu/echos-impact). Figure 12.6 (see the next page) shows how Project ECHO works as a teletraining platform.

Project ECHO® is a virtual platform that uses teletechnology to train, support, and empower clinicians to provide specialty care services. It uses a "hub and spoke" model, wherein non-specialist clinicians in a community setting "spokes" learn from and collaborate with specialists at academic "hubs." UM in collaboration with the University of Delaware and Yale University used Zoom videoconferencing, which was accessed on PCs, tablets and smartphones, to connect individuals and participation linked to Continuing Medical Education, which is required annually in Malaysia and can be facilitated by the University of Malaya Centre for Continuing Education. Project ECHO® meets all of the recommended criteria for Continuing Medical Education effectiveness. Each bi-weekly multimedia Project ECHO® session included a brief didactic training by a member of the Specialist Hub, followed by a series of cases presented by clinicians from their clinics. The Project ECHO® coordinator sought cases before each session for the specialist to review. "Interactive techniques" emerged from intensive patient-case discussion, where clinicians discussed treatment options and challenges with experts. "Multiple and repeated exposures" to important clinical information is central to Project ECHO®, which

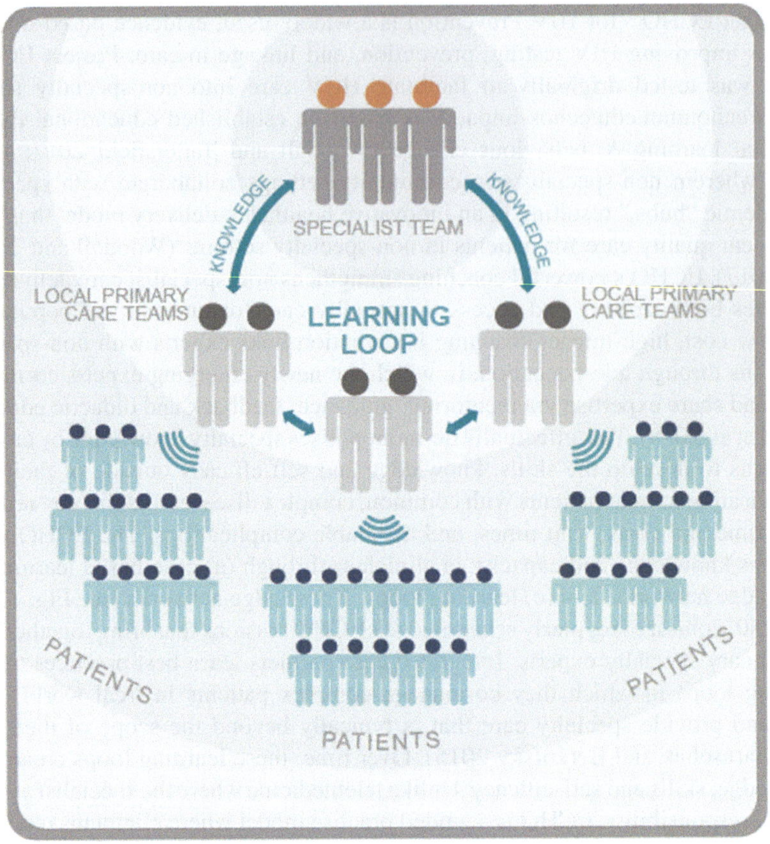

Fig. 12.6 Project ECHO (Extension for Community Healthcare Outcomes) landscape. *Source* AHRQ (2017)

repeatedly reinforces evidence-based practices via several methods (https://echo.unm.edu/echos-impact). All video sessions were recorded and archived for review by clinicians and participation was monitored centrally through a log-in (or review) process from the Project ECHO® hub. If semi-urgent questions arose in between sessions, teleconsults could be sent to the relevant expert and addressed within 24 h. Teleconsults could then be presented as cases at the next session. Feedback from prior cases were also discussed to learn about patient responses. Project ECHO is not telemedicine, however it shows the evidence that Malaysia can successfully utilise online platforms to facilitate healthcare and increase coverage (see Table 12.2 on the next page).

Table 12.2 Number of ECHO (Extension for Community Healthcare Outcomes) networks and active networks operated by hubs (Moss et al. 2024)

	Total networks launched	Active networks (sustained)
Hub 1	2	1
Hub 2	1	1
Hub 3	1	1
Hub 4	14	14
Hub 5	12	5
Hub 6	5	2
Hub 7	21	12
Hub 8	1	1
Hub 9	1	1
Hub 10	8	3
Hub 11	2	2
Hub 12	8	7
Hub 13	1	1
Total	77 Total ECHO® networks	51 active ECHO® networks at time of data collection
	Mean: 6, Min: 1, Max: 21.	Mean: 4, Min: 1, Max: 14.

12.4 Future Directions

12.4.1 Malaysian Medical Council Guideline on Telemedicine

The issuance of the new Guideline on telemedicine by the Malaysian Medical Council (MMC) represents a significant step in formalising telemedicine practices in Malaysia. Following the temporary virtual consultation advisory during COVID-19, the guideline provides a structured approach to telemedicine, ensuring that healthcare providers adhere to professional standards as set out in the Medical Act 1971 (Low and Knepper 2020). Key aspects include:

12.4.2 Professional Standards and Compliance

Medical practitioners are now required to follow specific provisions of the Medical Act, ensuring that telemedicine practices are subject to the same ethical and professional standards as in-person care. Compliance with the Code of Professional Conduct and Good Medical Practise guidelines is crucial. Practitioners are expected to deliver telemedicine services with the same diligence and ethical responsibility as traditional consultations.

12.4.3 Patient Consent and Confidentiality

Consent remains a cornerstone in telemedicine, and practitioners must ensure patients are well-informed before proceeding with virtual consultations. This includes clearly explaining the limitations and potential risks of telemedicine. Confidentiality, a critical component of telemedicine, mandates that all communications and data exchanged are secure and adhere to privacy regulations, safeguarding patients' personal information.

12.4.4 Regulation of Audio-Visual Recording

Practitioners must follow MMC (Malaysian Medical Council) protocols on audio-visual recording. Consent is required for any recording during telemedicine sessions, and the recordings must be stored and handled per confidentiality standards (Malaysian Medical Council 2024).

12.4.5 Telemedicine's Complementary Role

The guideline underscores that telemedicine should complement, not replace, in-person care where necessary. In cases where physical examinations are required, practitioners should advise patients to visit a healthcare facility. This guideline marks an essential regulatory advancement, reinforcing the importance of maintaining medical standards in telemedicine. It paves the way for safe and ethical remote healthcare, ensuring that telemedicine remains a viable, trustworthy option for patients across Malaysia.

Although the MMC (Malaysian Medical Council) guideline does not depict detailed guidance on implementing telemedicine in Malaysian prison settings, it sets the stage to implement telemedicine programmes in correctional facilities. Innovative tools like ECHO and telemedicine can support evidence-based practices such as screening, evaluation, and treatment of HIV, TB, HCV, substance use, and mental health disorders (UNODC 2018; Springer et al., 2011). Telemedicine can contribute positively to managing these complex intertwined conditions in prisons and its use has been broadly integrated into health service delivery globally, especially in post COVID-19 era. In new contexts like Malaysia where telemedicine technology is not widely employed, it is crucial to understand the landscape to guide its implementation. The absence of telemedicine is a missed opportunity for prisons in Malaysia, especially since Malaysia has been a pioneer in prison health.

One of the key features of technological advancements is it has bring the world closer. Telemedicine can connect healthcare professionals in developing regions with experts from advanced medical centres to seek second opinions or training. Medical

Institutions can collaborate on telemedicine projects to study its efficacy, share data, and train healthcare workers.

12.5 Conclusion

Wide implementation of telemedicine across Malaysian prisons can be considered as a game changer for the future of public health in Malaysia. The country has already made investments to create digital health platforms to align in-person healthcare delivery with virtual healthcare delivery including telemedicine, mobile health apps, and electronic medical records. Extending the initiative to the prison settings in Malaysia can bring huge benefits to the prisoners as well as the community at large. Effective implementation of telemedicine can also be considered as a cost-saving strategy for healthcare expenditure in Malaysia.

In summary, the overall success of telemedicine in prisons depends on addressing barriers such as technological infrastructure, training for staff, and ethical concerns surrounding privacy and consent. Moving forward, a collaborative effort among healthcare providers, policymakers, and correctional institutions is essential to fully realise the potential of telemedicine in safeguarding the health and dignity of incarcerated populations.

Highlights

- Telemedicine can reshape the landscape of prison healthcare, addressing long-standing challenges with innovative solutions.
- Prisons are often isolated from community medical centres, making timely access to specialised healthcare difficult. Telemedicine bridges this gap by enabling virtual consultations with healthcare professionals, reducing the need for costly and logistically complex prisoner transport.
- This approach is particularly impactful in managing chronic conditions, mental health, and infectious diseases—issues disproportionately affecting incarcerated populations.
- Telemedicine not only improves health outcomes but also enhances security, as it minimises movement outside the prison walls, reducing escape risks and costs.
- Despite its promise, telemedicine in prisons faces hurdles, including the need for robust technology infrastructure, staff training, and policies to ensure patient confidentiality. Addressing these barriers is crucial to fully realise the potential of telemedicine in delivering equitable healthcare.

References

AHRQ Works: Building Bridges Between Research and Practice. Association for Healthcare Research and Quality. https://www.ahrq.gov/cpi/about/impact/index.html. Published April 2017. Accessed December 3, 2019

Altice FL, Kamarulzaman A, Soriano VV, Schechter M, Friedland GH. Treatment of medical, psychiatric, and substance-use comorbidities in people infected with HIV who use drugs. BMC Infect Dis. 2022;22(1):837. https://doi.org/10.1186/s12879-022-07804-6.

Bazazi AR, Wickersham JA, Wegman MP, et al. Design and implementation of a factorial randomized controlled trial of methadone maintenance therapy and an evidence-based behavioral intervention for incarcerated people living with HIV and opioid dependence in Malaysia. Contemp Clin Trials. 2017;59:1–12. https://doi.org/10.1016/j.cct.2017.05.006.

Beaudry G, Zhong S, Whiting D, Javid B, Frater J, Fazel S. Managing outbreaks of highly contagious diseases in prisons: a systematic review. BMJ Glob Health. 2020;5(11): e003201. https://doi.org/10.1136/bmjgh-2020-003201.

Chandra DK, Bazazi AR, Solim MAN, Kamarulzaman A, Altice FL, Culbert GJ. Retention in clinical trials after prison release: results from a clinical trial with incarcerated men with HIV and opioid dependence in Malaysia. HIV Res Clin Pract. 2019;20(1):12–23.

Culbert GJ, Pillai V, Bick J, et al. Confronting the HIV, tuberculosis, addiction, and incarceration syndemic in Southeast Asia: lessons learned from Malaysia. J Neuroimmune Pharmacol. 2016;11(3):446–55. https://doi.org/10.1007/s11481-016-9676-7.

Davis P, Johnson R. NASA's contributions to telemedicine: an assessment of the feasibility of using space-based technology for the benefit of human health on Earth. NASA;2001.

Dodds S, Hess AC. Adapting research methodology during COVID-19: lessons for transformative service research. J Serv Manag. 2021;32(2):203–17. https://doi.org/10.1108/JOSM-05-2020-0153.

Doran JM, Lawson JL. The impact of COVID-19 on provider perceptions of telemental health. Psychiatr Q. 2021;92(3):1241–58. https://doi.org/10.1007/s11126-021-09899-7.

European Centre for Disease Prevention and Control (ECDC). Systematic review on active case finding of communicable diseases in prison settings: prevention and control of communicable diseases in prison settings. Stockholm: ECDC Scientific Advice;2017.

Fair H, Walmsley R. World prison population list, 13th ed. London: World Prison Brief; 2021.

Henny KD, Wilkes AL, McDonald CM, Denson DJ, Neumann MS. A rapid review of eHealth interventions addressing the continuum of HIV care (2007–2017). AIDS Behav. 2018;22(1):43–63. https://doi.org/10.1007/s10461-017-1923-2.

https://news.va.gov/press-room/va-exceeds-1-million-video-telehealth-visits-in-fy2018-one-year-achievement-represe

https://www.unodc.org/documents/hiv

https://www.unodc.org/dohadeclaration/en/prisons/publications.html

Istepanian RSH, Laxminarayan S, Pattichis CS, editors. M-health: emerging mobile health systems. Springer;2006.

Kamarulzaman A, Reid SE, Schwitters A, Wiessing L, El-Bassel N, Dolan K, Moazen B, Wirtz AL, Verster A, Altice FL. Prevention of transmission of HIV, hepatitis B virus, hepatitis C virus, and tuberculosis in prisoners. Lancet. 2016; 10;388(10049):1115–26. https://doi.org/10.1016/S0140-6736(16)30769-3.

Kral AH, Lambdin BH, Comfort M, et al. A strengths-based case management intervention to reduce HIV viral load among people who use drugs. AIDS Behav. 2018;22(1):146–57.

Larsen D, Stamm BH, Davis K, Magaletta PR. Prison telemedicine and telehealth utilization in the United States: state and federal perceptions of benefits and barriers. Telemed J E Health. 2004;10(Suppl 2):S-81–S-89. PMID: 23570218.

Low M, Knepper C. Implementing telemedicine in correctional facilities: a review of benefits, challenges, and outcomes. Correct Health. 2020;15(3):95–108.

Malaysian Medical Council. Telemedicine guidelines;2024.

Maruschak LM, Berzofsky M. Medical problems of state and federal prisoners and jail inmates, 2011–12. Bureau of Justice Statistics Special Report;2015.

Moeloek NF, Ekatjahjana W. Permenkes No. 20 Tahun 2019 tentang Penyelenggaraan telemedicine antar Fasilitas Pelayanan Kesehatan. Ministry of Health;2019. Available from: https://www.researchgate.net/publication/338428993.

Moss P, Hartley N, Russell T. Project ECHO: a global cross-sectional examination of implementation success. BMC Health Serv Res. 2024;24(1):583. https://doi.org/10.1186/s12913-024-10920-5.

Muthulingam D, Bia J, Madden LM, Farnum SO, Barry DT, Altice FL. Using nominal group technique to identify barriers, facilitators, and preferences among patients seeking treatment for opioid use disorder. J Subst Abuse Treat. 2019;100:18–28. https://doi.org/10.1016/j.jsat.2019.01.019.

Okeke NL, Ostermann J, Thielman NM. Enhancing linkage and retention in HIV care: a review of interventions. Curr HIV/AIDS Rep. 2014;11(4):376–92. https://doi.org/10.1007/s11904-014-0233-9.

Perrin A. Mobile technology and home broadband 2021. Pew Research Centre;2021 [cited 2022 Oct 26]. Available from: https://www.pewresearch.org/internet/2021/06/03/mobile-technology-and-home-broadband-2021/.

Poushter J. Smartphone ownership and internet usage continues to climb in emerging economies. Pew Research Centre's Global Attitudes Project;2016. Available from: https://www.pewresearch.org/global/2016/02/22/smartphone-ownership-and-internet-usage-continues-to-climb-in-emerging-economies/.

Project ECHO. ECHO's Impact [Internet];2019 [cited 2019 Dec 17]. Available from: https://echo.unm.edu/echos-impact.

Project Syndicate. Reforming Malaysian Prisons [Internet];(n.d.). Available from: https://www.projectsyndicate.org/reforming-Malaysian-prisons.

Purnomo J, Coote K, Mao L, et al. Using eHealth to engage and retain priority populations in the HIV treatment and care cascade in the Asia-Pacific region. BMC Infect Dis. 2018;18(1):82. https://doi.org/10.1186/s12879-018-2972-5.

Springer SA, Spaulding AC, Meyer JP, Altice FL. Public health implications for adequate transitional care for HIV-infected prisoners. Clin Infect Dis. 2011;53(5):469–79. https://doi.org/10.1093/cid/cir446.

Suhakam. Suhakam report Malaysian prison;2021. Available from: https://www.suhakam.malaysianprison.com.

United Nations. Sustainable development goals;2015. Available from: https://sdgs.un.org/goals.

United Nations General Assembly. Transforming our world: the 2030 agenda for sustainable development. A/RES/70/1. United Nations;2015.

United Nations Office on Drugs and Crime. Introductory handbook on the prevention of recidivism and the social reintegration of offenders. United Nations;2018.

UNODC, WHO, UNAIDS. HIV prevention, treatment and care in prisons and other closed settings: a comprehensive package of interventions;2013. Available from: https://www.unodc.org/documents/hiv-aids/HIV_comprehensive_package_prison_2013_eBook.pdf.

Wood B, Mann M, Martinez-Paz N, et al. Project ECHO: telementoring to educate and support prescribing of HIV pre-exposure prophylaxis. Sex Health. 2018;15(6):601–5.

Wood BR, Unruh KT, Martinez-Paz N, et al. Impact of a telehealth programme that delivers remote consultation and mentorship to HIV providers. Open Forum Infect Dis. 2016;3(4):ofw123. https://doi.org/10.1093/ofid/ofw123.

Woodall J, Tattersfield A. Perspectives on the implementation of telemedicine within a UK prison setting: a qualitative evaluation. Int J Environ Res Public Health. 2017;14(6):693.

World Prison Brief (Malaysia) (n.d.). London: Institute for Crime & Justice Policy Research, Birkbeck, University of London. Retrieved Dec 16, 2025 Available from: https://www.prisonstudies.org/country/malaysia

Young JD, Badowski ME. Telehealth: increasing access to high quality care in correctional medicine. J Clin Med. 2017;6(2):20.

Rumana Akhter Saifi (MSS, MA, MPH, PhD) scholarly work involves research across diversified cultural contexts from cross-border areas of irregular migrants to conservative Muslim rural areas to urban metropolitan to urban slum to hard- to- reach areas; and across various disadvantaged population groups from young adolescents in early marriage to MSM to sex workers to transgender to injection drug users to prisoners to fishermen. Dr Saifi is working as a Faculty at Department of Social and Preventive Medicine, Universiti Malaya, Kuala Lumpur. Dr Saifi holds a Doctorate in Demography and Masters in Public Health, Population and Reproductive Health Research and Economics.

Yilu Qin (MD) is a Yale Fogarty fellow at the Universiti Malaya, studying telemedicine models for individuals struggling with addiction and transitioning back to community after prison. Dr. Qin was a Fulbright scholar studying HIV self-testing in China. She completed her residency training at the Yale Internal Medicine HIV training track in 2020, where she took care of patients living with HIV and also researched PrEP for women with substance use disorders. She served as a telemedicine doctor at the VA hospital in West Haven, Connecticut, from 2020 to 2023, to provide care to rural, remote veterans.

Chapter 13
Hospital Care Delivered at Home: The Future of Healthcare in Asia with Hospital-at-Home

Cluster of Future Studies: Technological Trends

Farah Azemee, Stephanie Ko, and Terence Ong

Abstract The rise in ageing demographics, non-communicable diseases and care needs seen in Asia will pose significant challenges on healthcare delivery and hospital-level resources. Hospital-at-Home (HaH) is a substitute to hospital inpatient care provided at the person's home. It is a potential solution to this demand–supply gap. This narrative review synthesised evidence drawn from existing published scientific literature, relevant policy documents, and specialist societies' statements related to HaH. A predefined framework guided this review covering (1) an overview of HaH, its implementation and delivery (2) the evidence supporting it, (3) drivers for its growth outside Asia, and (4) its place in Asia's future. This review also detailed specific considerations for HaH's adoption in Asia. The evidence to date has concluded that HaHs are generally safe, deliver comparable outcomes to usual inpatient care, and is acceptable by the public. Cost evaluation studies remain limited. In Asia, there is only one country that has an established nationally funded HaH service. For HaH to be widely implemented in this region, there are unique considerations to be considered (rural–urban setting, language, culture, technological and healthcare aspects). HaH can be a viable strategy to manage acute care demands in Asia.

Keywords Hospital-at-Home · Home healthcare · Virtual ward · Telehealth · Telemedicine · Community

F. Azemee
Department of Social and Preventive Medicine, Faculty of Medicine, Universiti Malaya, Kuala Lumpur, Malaysia

S. Ko
Department of Medicine, National University Hospital, Kent Ridge, Singapore
e-mail: stephanie_qianwen_ko@nuhs.edu.sg

T. Ong (✉)
Department of Medicine, Faculty of Medicine, Universiti Malaya, Kuala Lumpur, Malaysia
e-mail: terence.ong@ummc.edu.my

R. Xiang Ng et al. (eds.), *Future of Healthcare in Asia*,
https://doi.org/10.1007/978-3-032-03183-9_13

13.1 Introduction

Asia is experiencing significant demographic shifts, particularly in Japan, China, South Korea, and Singapore, where populations are ageing rapidly. The proportion of the population 60 years or greater is expected to increase from 12% in 2016 to over a quarter by 2050—amounting to an estimated 1.3 billion people (United Nations ESCAP 2017). While this transition to an aged society took 85 years for Sweden and 115 years for France, it is expected to take countries like Singapore, Thailand and China only 22–25 years (United Nations 2022). This also means that older people living with multimorbidity will increase. These trends will pose significant burden on healthcare delivery and increase demand for hospitalisation. Additionally, increasing healthcare needs and expenditures, coupled with shortages of hospital beds and workforce have led to hospital capacity constraints. Such constraints decrease the efficiency and quality of healthcare delivery, leads to delays in elective and emergency treatment, increase mortality risk, lower patient and staff satisfaction, and contribute to higher cost of healthcare (Leong et al. 2021).

To fill this demand–supply gap and extend healthcare capacity, some countries have explored substitutes to inpatient hospitalization—specifically the Hospital-at-Home (HaH) care model. HaH aims to provide hospital-level acute care in people's homes as a substitute to conventional hospitalisation (Leff et al. 1997). This allows treatment to be delivered in someone's home. It would rationalise the use of inpatient beds, improve hospital patient flow, and increase the effectiveness and efficiency of resources within hospitals (Chappell et al. 2024).

Early versions of the programme have been around in Europe for the last half century. The first reference to HaH was during the late 1970s in United Kingdom (Patel and West Jr 2021). It has slowly gained traction internationally since the 1990s, and is currently at various stages of implementation globally. The recent COVID-19 pandemic has accelerated the implementation of HaH in many parts of the world. However, its implementation is highly variable despite similar pressures on hospital resources worldwide. Most literature on HaH has originated from high-income countries in Europe, America, Australia, and New Zealand with a paucity of services in Asia (Edgar et al. 2024; Wallis et al. 2024). This chapter aims to provide an overview of HaH from its definition, operational processes, evidence-to-date, its driver for growth outside Asia and then focus in on its potential role in Asia.

13.2 Review Methodology

This chapter aims to make the case for HaH as an important element for healthcare in Asia's future. A narrative review methodology allows for breadth and a wider search scope of the topic on HaH. It also means that a wide variety of source materials can contribute to this chapter. Additionally, narrative reviews enable authors to provide their own insights and perspectives on relatively new and novel topics (Edgar et al.

2024; Wallis et al. 2024). In this narrative review, we synthesised evidence drawn from existing published scientific literature, relevant policy documents, and specialist societies' statements related to HaH model of care. Literature or documents were identified in electronic bibliographic databases and search engines using variations of the terms 'virtual ward' and 'hospital-at-home'. Searches were also performed based on the authors' knowledge of the topic. The search was iterative, involving multiple cycles of searching (including citation searching), analysis and interpretation. A predefined framework guided this review. It covered (1) details of HaH and it is operated, (2) the evidence supporting it, (3) drivers for its growth outside Asia, and (4) its adoption in Asia. Findings were presented narratively to fit within the framework.

13.3 Review Findings

13.3.1 The Hospital-At-Home Healthcare Model

This section provides an overview of HaH. It introduces HaH from a broad global perspective. It covers what defines HaH, how it is delivered and the outcomes associated with such services.

13.3.1.1 Defining What is Hospital-At-Home (HaH)

The evolution of HaH has been inconsistent, with varied approaches leading to different definitions for the service. The World Hospital at Home Congress in 2023 provided a consensus definition that HaH is a specialist-led, acute hospital-substitutive care service (including diagnostics, therapeutics and observation) governed by a strict framework, delivered 24 h a day, 7 days a week. This definition distinguishes true HaH services from programmes that focused on admission prevention (rather than substitution), community-based chronic disease management or solely virtual care and telemonitoring services (Hospital At Home 2023). Figure 13.1 (see the next page) depicts where HaH falls in the healthcare delivery spectrum.

There are different names for HaH. It has been called Hospital in the Home (HITH), Acute Hospital Care at Home (AHCaH), Mobile Acute Care Teams (MACT), and Mobile Inpatient Care@Home (MIC@Home) in different health systems. A term that is sometimes used alongside HaH is 'virtual ward'. In the United Kingdom (UK), the term 'virtual ward' is used interchangeably with HaH. However, outside the UK, virtual wards are a distinct model of care that provides virtual-only services, with remote monitoring and virtual interactions. In contrast, HaH may (but not necessarily exclusively) use technology and remote monitoring. The emphasis remains on providing hospital-level care to individuals in their homes.

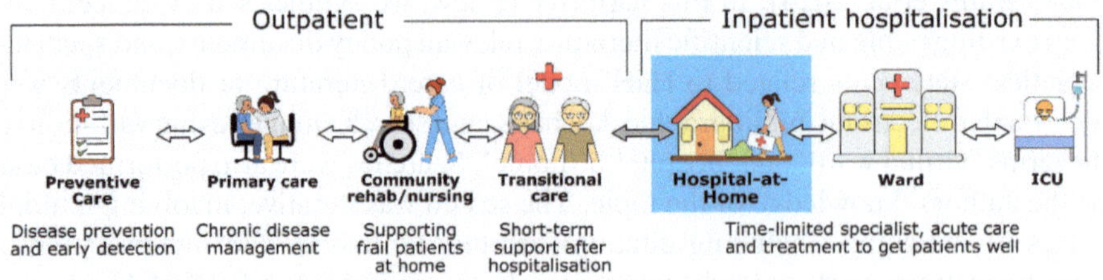

Fig. 13.1 Hospital-at-Home in the healthcare delivery spectrum

For example, some care models in US currently requires twice-daily home visits as a condition for reimbursement (Kanagala et al. 2023).

13.3.1.2 Patient Eligibility

HaH targets patients who require short-term acute care, are not critically unwell, and have support at home. It has proven to benefit older people living with frailty, patients with chronic obstructive pulmonary disease (COPD), heart failure and stroke (Kanagala et al. 2023). There are no overarching eligibility criteria of who would be suitable for HaH. The way the service operates and resources available would be the determining factors.

Broad criteria of the type of patients that would be suitable have been suggested and are summarised as follows:

- Clinically stable, or if not, in line with goals of care
- Requires ongoing hospitalisation
- A suitable home environment (some programmes may accept patients living in nursing homes or skilled nursing facilities)
- Lives within the catchment area set by the HaH programme
- Willing to engage with the HaH programme, and its requirements.

13.3.1.3 Operational Process

HaH typically operates through one of these three possible pathways based on the referral entry point for the individual: (1) from the ward, shortening the hospital length of stay, (2) from the emergency department (ED), an alternative substitute to ward admission, and (3) directly from the community setting. Entry points from the ward are sometimes referred to as "early supported discharge" (ESD), or "step-down". Entry points from the ED or community are sometimes referred to as "admission avoidance" (AA) or "step-up" (Leong et al. 2021, Norman et al. 2023). These operational pathways are depicted in Fig. 13.2.

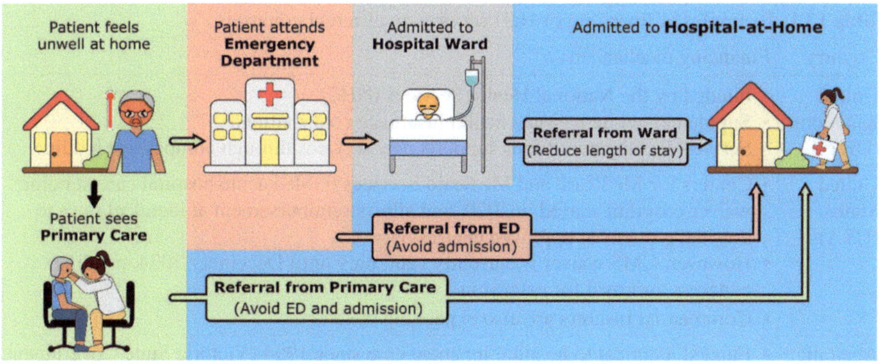

Fig. 13.2 Referral sources for Hospital-at-Home

Once admitted, the HaH episode is typically considered a hospital admission, with a defined admission and discharge date. During this episode of care, healthcare teams provide selected services to patients that are similar to what they would require in the hospital setting. A typical scope includes:

- Daily reviews by HaH specialists (often by a hospitalist or internal medicine specialist, conducted remotely or at the individual's own home)
- Vital sign monitoring
- Parenteral and enteral therapeutics
- Diagnostics (e.g. laboratory tests, radiology)
- Allied health services (e.g. physiotherapy, wound dressing, medication supply and delivery).

For the service to be safe and effective, there must be an adequate workforce. This includes the clinical staff (e.g. physicians, nurses, pharmacists, and therapists), management and support staff (e.g. administrative staff and managers). Workforce requirement will be dependent on the scope of the respective HaH service and the volume of patients being treated. Additionally, the multidisciplinary team working in HaH services must have appropriate training and competency (NHS England 2022; Shaver 2022). Training requirements range from performing clinical assessment virtually, using remote monitoring equipment, risk assessment and ensuring familiarity with relevant pathways and processes. However, there is no consensus on what these competencies should look like.

13.3.1.4 Financing Models

Several financing mechanisms exists for HaH. This varies according to each country's health financing infrastructure (Table 13.1).

Table 13.1 Examples of financing of HaH services in different countries

Country	Financing mechanism
United Kingdom (UK)	• Funded by the National Health Services (NHS) • Supplemented by national health insurance contributions • Service development funds supports new services through competitive bidding
United States (USA)	• Centers for Medicare and Medicaid Services (CMS) acute hospital care at home waiver program started in 2020, and allows reimbursement at identical rates to hospital patients at approved centres only • However, CMS waiver is currently valid only until December 2024, pending congress approval for extension • Commercial insurers are also expanding coverage
Australia	• Funded identical to hospital inpatient care since 1994 (Victoria State Government 2011) • Quantum determined by case-mix of activity based-funding, based on diagnostic-related group (DRG)
Spain	• Funded identical to hospital inpatient care in Catalonia since 2006 (González-Colom et al. 2024) • Reimbursement schemes based on patient DRG
Singapore	• Funded identical to hospital inpatient care from 2024 • Combination of government subsidies, patient health savings (Medisave) and insurance payouts (Medishield Life and Integrated Shield Plans)

13.3.1.5 Leadership, Governance and Legislation

Leadership and governance in developing HaH services must ensure frameworks exist, oversight in place and accountability to stakeholders is demonstrated (World Health Organization 2010). Governance structures in existing HaH services are determined by the health provider, funding arrangement, and any existing laws and regulations. There have to be clear lines of clinical responsibility in place, with an organisational board directing the service and a separate independent body overseeing it. Additionally, there needs to be managerial support for financial governance, service sustainability, staff and service planning. A recommendation for doing this was provided by Healthcare Improvement Scotland (Healthcare Improvement Scotland 2020). These were the recommended steps to run a safe and effective HaH service:

- Service protocols or standard operating procedures ensure safe and consistent working. These should cover triage, clinical assessment and patient clerking, senior medical review, diagnostics and prescribing, care and support and review including the daily ward round or multi-disciplinary team meeting.
- Clear accountability for preventing and responding to risks, adverse events and performance reporting.
- A supportive culture for innovation that allows testing, adaptation and development.
- An embedded culture of reflection and learning, for example, the use of morbidity and mortality meetings or Schwartz rounds (staff members getting together to

discuss and focus on the emotional impact of patient care experienced by them), to learn from things that have gone well as well as those that have not.
- Clear links to local governance reporting structures.
- A realistic medicine approach that puts person-centred care and shared decision-making at the heart of care planning from the outset.

The performance of the service must be measurable, visible and accountable to all those involved. This provides assurance that the HaH service is safe and effective. It also facilitates quality improvement. Examples of metrics that could be measured include referral load, proportion of patients admitted, length of stay, mortality and readmission (Healthcare Improvement Scotland 2020; World Health Organization 2010).

HaH occupies a unique position where conventionally hospital inpatient care is delivered in a community outpatient setting. The duty of care, payment and insurance coverage are usually separated into 'inpatient' and 'outpatient' settings, i.e. whether the person being treated is within the walls of the hospital or not. Due to this distinction, changes have to be made to overcome this through policy and legislation, such as considering acute hospital treatment delivered in a home environment as part of inpatient services for payment purposes (Adams et al. 2023). Furthermore, the use of telemedicine is not covered by many countries' health legislation. Telemedicine involves using technology to communicate between a patient and healthcare practitioners located in different places. Ideally, to ensure safe telemedicine practice, the area of informed consent, data protection, confidentiality, and how to deal with malpractice and liability arising from telemedicine needs to be considered (Nittari et al. 2020).

13.3.1.6 Clinical Outcomes

There have been several systematic reviews and meta-analyses evaluating the clinical impact of different HaH models compared to routine inpatient care. One review on the impact of 'admission avoidance' (AA) HaH services included 20 randomised controlled trials recruiting 3100 participants evaluated (Edgar et al. 2024). Another review focused on 'early supportive discharge' (ESD) HaH services included 32 trials with 4746 participants. Conditions included in these meta-analyses were stroke, COPD, general medical condition and post-operative care.

Key findings compared to conventional hospitalisation included:

- HaH reported similar mortality rates compared with conventional hospitalisation.
- Both have comparable patient-reported health outcomes
- There is little or no difference in readmission rates
- HaH reduces the patient's likelihood of being instituted post-treatment.

Furthermore, a systematic review of reviews synthesised data from ten reviews of mostly moderate to high quality evaluated HaH based on its role for ESD, AA, or both (Leong et al. 2021). ESD and AA reviews generally revealed comparable mortality and readmissions to inpatient care. For studies that delivered a combination of ESD and AA service, there was a lower mortality and readmission rate for patients with chronic obstructive pulmonary disease.

Hence, although HaH shows comparable clinical outcomes to hospital care and is safe for certain cohorts, it is still too early to assume HaH would be the panacea for the pressures experienced by hospital services, especially on a larger systemic scale. The impact on bed occupancy is not easy to work out as it could remain high even if hospital beds were freed up, given the huge pressures on the system. It would mean any available bed would be quickly occupied. HaH themselves could also provide 'beds' for patients who would have previously refused to go to hospital for care (Chappell et al. 2024).

13.3.1.7 Cost Outcomes

Hospitals and hospitalisation are expensive and constitute a significant portion of healthcare expenditure (Cygańska et al. 2023). The evidence regarding HaH is a cost-saving intervention remains uncertain, with the published findings mixed. A review of 12 admission avoidance HaH studies involving 2148 participants suggested that these services may be less costly than hospital admissions (Edgar et al. 2024). The reduction in direct costs per admission varied across different studies. This possible cost-saving service was attributed to several factors:

- Episodic rather than round-the-clock nursing staffing
- Substitution of some in-person care with the integration of telemedicine
- Reduced utilisation of healthcare services (e.g. laboratory tests and imaging requests)
- Opportunity cost, by freeing up hospital beds in congested health systems
- Lower capital costs of building hospitals or wards to increase inpatient capacity.

A separate UK-based retrospective study found that HaH beds costs nearly twice as much as hospital beds (Jalilian et al. 2024). However, this cost comparison was made when the HaH programme operated at a very low 25% occupancy, which would have influenced the findings. Furthermore, a systematic review of reviews of different HaH services (ESD, AA, or both) observed that the cost–benefit for HaH remains uncertain, regardless of the service type (Leong et al. 2021). Some findings from the review also showed conflicting results regarding its cost-saving benefits.

Economic evaluation of HaH service is still lacking. The variation in costing methodologies and follow-up durations complicate direct cost comparisons. Additionally, there is a lack of evidence on its impact on indirect costs, such as care and societal costs. Hence, high-quality studies are needed to truly evaluate the cost–benefit potential of HaH, taking into account various occupancy levels, operational scales and financing mechanisms.

13.3.1.8 Patient and Caregiver Satisfaction

Individuals have reported high levels of satisfaction with HaH due to the following factors (Levine et al. 2021).

- Conducive recovery environment
- Better access to the clinical team and easier communication
- Simpler admission process
- The comfort of familiar surroundings.

However, the caregiver's experience varies. HaH does require some aspect of care to be delivered by the patient's caregivers, potentially leading to uncertainty around medication management and caregiver fatigue (Edgar et al. 2024). More research is necessary to fully understand the impact of HaH on both patients and caregivers, particularly in managing the increased care demands at home.

13.3.2 Drivers for Growth

In this section, the chapter looks at what drove the adoption for HaH services globally. There were many drivers for this and not just because of a global pandemic. This section will look at three main factors—stakeholders pushing for it, technology supporting such innovation and public acceptance for treatment to be delivered at home.

13.3.2.1 Alignment of Multiple Stakeholders' Priorities

A HaH service aims to provide hospital-substitute care. However, it is unlikely that many (if any) new services will be able to deliver complete substitution by providing a full range of hospital care at home. Furthermore, the coverage of HaH services and who may be eligible for them could also impact access to this service. Hence, the first step in implementing HaH requires careful identification and engagement of potential stakeholders, ranging from commissioners, policymakers, funding agencies, and healthcare services in both primary and secondary care. This facilitates key discussion on how such a service should started, coordinated, delivered, expanded and performance-managed.

This multistakeholder engagement plays a key role in HaH's growth. An example of this was seen in England's National Health Service virtual ward (HaH) programme which was launched in April 2022. It set an ambitious target of at least 40 virtual ward beds per 100,000 people (Chappell et al. 2024). Every local health service was tasked with delivering this within their catchment area. Almost half a billion pounds was made available for the first two years of its rollout to deliver on this objective. To achieve this efficiently and safely, the National Institute for Health and Care Excellence (NICE) which provides evidence-based recommendations in England

and Wales, actively contributed to this work through guidelines, recommendations and economic evaluations of this endeavour (Salmon 2023). Clear guidance of what this service is, its scope of care, and the resources needed to be successful was explicitly defined. Throughout the delivery and expansion of this service, the evaluation of its impact on the public, patients, and healthcare providers was continuously conducted by health providers and stakeholders such as the Nuffield Trust and the Health Foundation. Such real-world data is key to understanding HaH's impact.

Besides that, stakeholders need to consider how to ensure that HaH services remain accessible and equitable. HaH should not be available only to those who can afford the service, come from more fortunate socioeconomic circumstances, or have a more secure home environment. As part of setting up HaH, frameworks such as the health equity implementation framework or the healthcare disparities framework must be considered (Edgar et al. 2024). One study reported that those from more disadvantaged socioeconomic background were more likely to benefit from HaH because healthcare teams were able to closely observe their health status and care received in the patient's own environment (Siu et al. 2022).

13.3.2.2 Role of Technology

These recent years have also seen the growth and acceptance of telehealth. Telehealth refers to any electronic and telecommunications technologies that allows healthcare to be delivered remotely. The need for healthcare during the pandemic where movement restrictions were in place and concern over infection risk has led healthcare providers and users to embrace telehealth. Many parts of the world now have wider high-speed internet coverage. Devices now have better-performing video and audio tools for communication. More providers have also entered the healthcare space offering online virtual consultations, health apps, smartphone messaging and remote patient monitoring (Pandit et al. 2024; Shaver 2022).

Relevant to HaH, healthcare providers have become more familiar with adopting telehealth, especially in managing acute and sub-acute illnesses. As HaH runs as a multidisciplinary service, telehealth can support clear communication channels and cooperation among healthcare professionals. Electronic health records also have the potential to allow sharing of patient information, which would support care. Although not specific to HaH, robust health information systems can support large amounts of data to be generated, compiled, analysed and shared much quicker among stakeholders involved in HaH. This includes data on the service and facilities, patient flow, their characteristics and outcomes.

This technology space will continue to grow. For instance, point-of-care testing of laboratory tests, portable ultrasound scanning, and home x-rays are advances that will support patient care in their own homes.

13.3.2.3 Public Opinion of Care at Home

Another facilitator to HaH's success is the acceptance and perceived benefit by the staff involved, patients and caregivers. The UK's Health Foundation surveyed over 7000 members of the public and 1200 healthcare staff on their views of this. Overall, the UK public is supportive of such services (45% in support against 36% who do not, with the rest undecided), and 71% of the public surveyed were open to being treated by HaH in the right circumstances. Healthcare staff were even more supportive of this (63% in support) (Thornton 2023).

All this has been drawn from what is delivered and reported in countries outside the Asia Pacific. The drive for HaH was initiated to address local healthcare pressures around hospital capacity, based on growing evidence that in the right context, it is a safe substitute for inpatient care. Findings on what facilitated its implementation and delivery successfully were driven by evidence generated mostly from Europe, North America and Australia. Questions remain unanswered as to whether HaH in Asia will mirror the trajectory seen elsewhere.

13.4 Future Directions

13.4.1 Hospital-At-Home Implementation in Asia

In many Asian communities, some form of health treatment at a person's home exists mainly as bespoke services to the wealthy. Equitable and accessibility was addressed in the earlier section of this chapter. Singapore stands out in Asia for integrating HaH into mainstream healthcare delivery (see Table 13.2 on the next page). The country's public hospitals began HaH programmes between 2019 and 2020, with the COVID-19 pandemic accelerating its adoption. Between 2022 and 2024, the Ministry of Health funded a 2-year sandbox, which funded subsidies equivalent to inpatient care. By 2024, HaH was officially mainstreamed, demonstrating the programme's feasibility, safety and patient satisfaction.

Studies done in Singapore examined the willingness of multiracial Asian populations to consider HaH. Prior to the initiation of a HaH programme, patients interviewed were willing to consider HaH if clinically suitable (Lai et al. 2021), and providers were open to the idea but concerned about medicolegal implications (Chua et al. 2022). An early pilot study showed that HaH was feasible, safe and associated with high levels of patient satisfaction (Ko et al. 2022). Qualitative interviews again revealed high levels of patient satisfaction, but concerns about carer burden and stress (Ko et al. 2023). At the time of writing this article, there were no studies yet about cost-effectiveness.

Elsewhere in Asia, the authors understand from professional networks that HaH services are developing in Malaysia, Thailand, Japan and Taiwan.

Table 13.2 Future directions for Hospital-at-Home in Asia

Key features	Examples
Institutions	
Government structures	• Government support to enable HaH care to be financed similarly to in-hospital care is critical for HaH expansion • For example, in Singapore, the Ministry of Health funded a 2-year sandbox between 2022 and 2024, providing subsidies equivalent to inpatient care. By 2024, HaH was officially mainstreamed
Policy legacies	• Each health system will need to adapt HaH financing frameworks based on existing policy legacies • Singapore and Malaysia's healthcare systems originate from the NHS, with government financing and self-payment components. South Korea and Taiwan use National Health Insurance systems with income-based premiums, encouraging a mature private healthcare sector
Policy networks	• Policy networks including government, implementor and other stakeholders can be helpful in advocating for the above policy shifts
Interests	
Interest groups	• As HaH does not clearly fall within an existing professional group, professional networks are forming around HaH in Singapore, Malaysia, Thailand, Japan, and Taiwan
Civil society	• HaH movements have largely been professional led, however there may be role for civil groups e.g. caregivers of patients with dementia, to advocate for service expansion
Ideas	
Values	• High prevalence of multi-generational cohabitation in Asia influences the dynamic of HaH • Filial piety increases family involvement in caregiving, facilitating HaH uptake but also causing caregiver burden • Domestic helpers in Hong Kong, Singapore, and Malaysia may be expected to support HaH care and may require training and support
Research evidence	• Studies in Singapore showed willingness among multiracial Asian populations to consider HaH if clinically suitable • Early pilot studies in Singapore indicated that HaH is feasible, safe, and associated with high patient satisfaction • Qualitative interviews revealed positive feedback from patients but highlighted caregiver burden
External factors	
Political changes	• Bed capacity pressures such as due to pandemics or ageing populations can often be a politically charged topic which may influence politicians to support alternatives to expand bed capacity such as HaH
Economic change	• HaH programs tend to be potentially less costly to payors and patients, benefiting hospitals, insurers, governments, and patients if implemented successfully
Release of major reports	• Reports in favour of HaH expansion by major international organisations or influential health systems may encourage further investment into HaH program development

(continued)

Table 13.2 (continued)

Key features	Examples
Technological changes	• High mobile phone penetration and cultural openness to home-based services present opportunities for HaH expansion • Telemedicine platforms on chat apps like WeChat may facilitate implementation of HaH care delivery
New diseases	• The COVID-19 pandemic accelerated the adoption and mainstreaming of HaH programs in Singapore
Media coverage	• Media can play a role in highlighting the benefits and challenges of HaH programs, influencing public and policy perception

* HaH, Hospital-at-Home

13.4.2 *Challenges and Opportunities for Hospital-At-Home Implementation in Asia*

When considering adopting HaH in Asian health systems, the heterogeneity of the region due to the wide variation in demographics, geography and cultures must be recognised. HaH programmes, like any healthcare service, must be tailored to local needs and the healthcare system. These considerations have been summarised using the 3I + E framework in Table 13.2, with specific points discussed in detail as follows.

13.4.2.1 Health System Considerations

Geographically, many Asian countries have both rural and urban areas, with reduced access to healthcare services in remote rural areas. Mayo Clinic's Advanced Care at Home (ACH) programme has described a hybrid HaH model with a physician-staffed command centre in a rural area and a vendor-based in-person medical supply chain in a rural setting, showing feasibility and safety (Paulson et al. 2023). With increasing technology penetration across Asia, these hybrid models for HaH may be a viable solution.

Healthcare financing in Asia ranges from entirely out-of-pocket expenses to universal healthcare models, which will influence how HaH programmes are financed. Singapore and Malaysia as former British colonies had healthcare systems originating from the National Health Service (NHS), with a large component of government financing and a component of self-payment. South Korea and Taiwan instead adopt National Health Insurance systems with income-based premiums (Smullen and Hong 2015), encouraging the development of a more mature private healthcare sector. As discussed earlier, since HaH programmes tend to be potentially less costly to payors and patients, HaH adoption may be beneficial to hospitals, insurers, governments and patients if implemented successfully.

Next, the multitude of languages and dialects across Asian communities can complicate communication between healthcare providers and patients. These challenges are also faced in a physical hospital environment, and digital tools may present opportunities where HaH systems can excel over hospital environments with the integration of interpreters into remote consultations, such as the use of virtual interpretation during the COVID-19 pandemic (Regenstein et al. 2023). Diversity in languages may present opportunities where HaH systems can excel, using translators who can dial into remote consultations.

13.4.2.2 Cultural Considerations

Filial piety, a Confucian value, emphasises children's respect and care for their parents, including living with the parent and providing care whether ill or healthy (Cheng and Chan 2006). Unlike Western cultures which put patient autonomy at the forefront of medical decision-making, Asian families are often highly involved in the medical decision-making process. They are also more likely to live with patients, and therefore must be engaged before admission to HaH. In a study from a HaH programme in Singapore, patients' families often feel responsible for providing care for their parents, and would both honour their wishes (e.g. to be at home) but equally feel responsible and burdened by their care (Ko et al. 2023). On the other hand, high levels of family involvement in caregiving could facilitate the uptake of HaH, as home-based care aligns with cultural preferences and dependent patients are more likely to have available and willing caregivers.

In many urban Asian communities, care homes (or nursing homes) for the dependent older person is on the rise. Such patients often suffer from the complications of immobility, contributing to significant acute hospital utilisation. In Asia, cultural beliefs about food as being more than just biological sustenance but a reflection of social and cultural values have led to a high prevalence of artificial nutrition by way of long-term nasogastric tube feeding (Wong and Krishna 2014). This phenomenon has been reported in Hong Kong (Luk et al. 2017), Malaysia and Singapore. In Singapore, one study suggested more than half of residents in long-term care facilities receive nasogastric tube feeding which is an additional risk factor for pneumonia (Wong et al. 2019; Yuen et al. 2022). This patient cohort in care homes may benefit greatly from the availability of HaH services which can offer a trial of treatment in-situ and transition to end-of-life care when the need arises.

Live-in domestic workers are commonly employed in families in many Asian countries, originally to assist in childcare and household tasks, and are increasingly involved in the care of seniors. In regions like Hong Kong, Singapore and Malaysia, these domestic helpers typically originate from Southeast Asian countries like Indonesia, Philippines and Myanmar (Liem et al. 2023). In China and India, the domestic helpers tend to be local but originate from rural areas and poorer households. In the context of HaH, domestic helpers already involved in senior care would naturally be expected by their employers to care for patients while being hospitalised at home, and are an additional stakeholder in the HaH care delivery. In Singapore, in a recent survey over half had to support healthcare needs of older persons but had inadequate knowledge and lacked confidence in these tasks (Ang et al. 2022). The role of domestic helpers in HaH care and acceptability requires further evaluation.

13.4.2.3 Technology Considerations

Asia's high mobile phone penetration and cultural openness to home-based services present unique opportunities for HaH expansion. Asian communities have one of the highest levels of mobile phone penetration in the world, even in rural areas, which will lend itself well to technology-enabled HaH. A survey of nearly 10,000 smartphone users across Asian countries found that nearly 90% used smartphones to seek health-related information and did not correspond to income inequality (Wang et al. 2022). Many Asian economies and businesses run on chat platforms like WhatsApp, Line, WeChat, KakaoTalk, IMO and Viber. WeChat for example, already hosts multiple telemedicine platforms that connect doctors with patients and is the most common platform used for telemedicine services in China (Ye et al. 2023). These chat platforms may be adapted to be a key enabler for HaH care, reducing the barrier to entry for patients.

Apart from one-to-one communication, penetration of food, grocery and wellness services delivered directly to one's home through Superapps like Gojek Tokopedia, Grab, WeChat mean that culturally, many Asian communities are gravitating to the concept of having services delivered to their doorstep. The advent of these Superapps has also been successful in creating a common platform that combines many disparate small players in the e-commerce and transport space. There are parallels in these spaces to the healthcare space, with multiple small local players that when combined can be used for successful local delivery of HaH.

While established HaH programmes were initiated in the era prior to mobile health, the penetration of these enablers is unique to Asia and may even be a facilitator for new Asian HaH programmes to be more efficient and more widely adopted than in more established programmes. Table 13.2 summarises the future directions for HaH in Asia.

13.5 Conclusion

The demand for hospital-level care is increasing. HaH, an innovative hospital-substitute service is an option to address this demand–supply mismatch, as seen in countries like the United Kingdom, America, Australia and New Zealand. The evidence does support HaH's as safe, and in some instances, it delivers clinical outcomes comparable to hospital inpatient care. Whether it truly relieves pressures on healthcare demands remain uncertain.

HaH has not been widely adopted in Asia, with Singapore the only country with an established service (see case study below). To establish a HaH service in this part of the world, nuances specific to this region must be considered. The conditions treated and services provided in each HaH programme must be contextualised to what is typically treated in the hospital in each community. System, cultural and technological considerations also suggest that the implementation of HaH in Asian communities will be different from the existing published literature. However, as long as they fulfil the same overall objectives of acute hospital care substitution, there will be more similarities than differences, and more lessons that can be learnt from each other. The establishment of training programmes and professional societies for HaH in Asia could support the development and standardisation of HaH services alongside policy advocacy and sharing of best practices. This can also be done collaboratively with international HaH societies, and established services outside Asia. Furthermore, patient readiness and awareness of HaH must be built, and clinician mindset and trust in HaH must be established.

How this will pan out in the future remains to be seen. As Asian countries navigate the challenges brought on by the increasing demand for acute hospitalisation and the potential enablers for HaH across Asian communities suggest great potential for HaH programmes to contribute to health system transformation across Asia.

Example of a Hospital-at-Home service in Asia

Case study: NUHS@Home

NUHS@Home is the largest HaH programme in Singapore, having treated nearly 4000 patients since its initiation in 2020. It is the HaH programme of the National University Health System, an academic health system in Western Singapore.

Patient selection: All patients must fulfil all of the following

1. Clinically stable or in line with goals of care
2. Able to self-care or have a caregiver
3. Require ongoing hospitalisation
4. Treatment plan can be provided at home
5. Lives in Western Singapore.

Patient journey: Patients go through three phases

1. Admission: assessment of suitability, patient explanation and consent
2. In-flight: patient's start their day with a video call with a doctor, followed by visits by health professionals as their situation requires e.g. nurses for blood tests and intravenous treatment, doctors for physical examination, physiotherapy for mobility. Medications are delivered to patients as required. Patients needing imaging are transferred back to hospital and back home again.
3. Discharge: when patient's no longer need hospital-level treatment, they are discharged and care transitioned back to their community and primary care providers.

Involvement of technology

1. Remote vital signs monitoring: patients are provided with thermometers, oximeters and blood pressure meters and taught how to use them. Those able to operate a smartphone enter the readings into an online form. Those unable to operate

a mobile phone are given Bluetooth-connected devices that integrates readings directly into a dashboard.
2. Video conferencing: two-way live video conferencing between patients and the care team.
3. Text messaging: chat groups are used between patients, caregivers and the care team in order to facilitate communication of logistics.
4. Smart glasses: two-way audio communication and first-person view, worn by a junior staff and supervised by a senior staff.

Who's in a team

1. Medical team: the medical team is led by a hospital-based consultant in internal medicine, with additional training in home-based hospital care delivery.
2. Nurses: the nursing team is a dedicated nursing team that comprises senior nurses with backgrounds spanning emergency, intensive care, ward and community care. They are supported by 3rd party vendor nurses to help with capacity management.
3. Pharmacist: the team is supported by a hospital-based pharmacist who checks medication orders and helps to optimise medical management
4. Allied health: physiotherapists, occupational therapists, speech therapists, dieticians and medical social workers are available to support the team.
5. Operations team: an operations team is critical to coordinate the various services and items delivered to the patient's home and the patient's transport to and from the hospital as necessary.

Highlights

- Currently, Hospital-at-home (HaH) is not widely available in Asia despite its potential as an effective service to complement inpatient hospital care.
- Interest is growing from stakeholders to invest and adopt it to futureproof healthcare sustainability.
- Systems, governance, cultural and technological alignment is needed to implement HaH services.

References

Adams D, Wolfe AJ, Warren J, Laberge A, Richards AC, Herzer K, Fleisher LA. Initial findings from an acute hospital care at home waiver initiative. JAMA Health Forum. 2023;4:e233667. https://doi.org/10.1001/jamahealthforum.2023.3667.

Ang KW, Koh YLE, Wang XF, Yusoff AM, Tan AM, Poh XYC, et al. Pilot cross-sectional study of foreign domestic workers supporting the basic healthcare needs of older persons in the community. BMJ Open. 2022;12: e051877. https://doi.org/10.1136/bmjopen-2021-051877.

Chappell PCM, Hardie T, Lloyd T, Tallack C, Gerhold M, Mayers C. What do virtual wards look like in England? The Health Foundation;2024. https://www.health.org.uk/reports-and-analysis/working-papers/what-do-virtual-wards-look-like-in-england. Accessed 4 May 2024.

Cheng ST, Chan ACM. Filial piety and psychological well-being in well older Chinese. J Gerontol B Psychol Sci Soc Sci. 2006;61:262–9. https://doi.org/10.1093/geronb/61.5.P262.

Chua CMS, Ko SQ, Lai YF, Lim YW, Shorey S. Perceptions of stakeholders toward "Hospital at Home" program in Singapore: a descriptive qualitative study. J Patient Saf. 2022;18:e606–12. https://doi.org/10.1097/pts.0000000000000890.

Cygańska M, Kludacz-Alessandri M, Pyke C. Healthcare costs and health status: insights from the SHARE survey. Int J Environ Res Public Health. 2023;20:1418. https://doi.org/10.3390/ijerph20021418.

Edgar K, Iliffe S, Doll HA, Clarke MJ, Gonçalves-Bradley DC, Wong E, Shepperd, S. Admission avoidance hospital at home. Cochrane Database Syst Rev. 2024;3:CD007491. https://doi.org/10.1002/14651858.CD007491.pub3.

NHS England. Supporting information: virtual ward including hospital at home;2022. https://www.england.nhs.uk/publication/supporting-information-virtual-ward-including-hospital-at-home/. Accessed 4 May 2024.

Gonçalves-Bradley DC, Iliffe S, Doll HA, Broad J, Gladman J, Langhorne P, et al. Early discharge hospital at home. Cochrane Database Syst Rev. 2017;(6):CD000356. https://doi.org/10.1002/14651858.CD000356.pub4.

González-Colom R, Carot-Sans G, Vela E, Espallargues M, Hernández C, Jiménez FX, et al. Five years of hospital at home adoption in Catalonia: impact, challenges, and proposals for quality assurance. BMC Health Serv Res. 2024;24:154. https://doi.org/10.1186/s12913-024-10603-1.

Healthcare Improvement Scotland. Hospital at Home. Guiding principles for service development;2020. https://ihub.scot/media/6928/2020205-hospital-at-home-guiding-principles.pdf. Accessed 12 May 2024.

Hospital At Home. What is Hospital at Home?;2023. https://www.hospitalathome.org.uk/whatis. Accessed 4 May 2024.

Jalilian A, Sedda L, Unsworth A, Farrier M. Length of stay and economic sustainability of virtual ward care in a medium-sized hospital of the UK: a retrospective longitudinal study. BMJ Open. 2024;14: e081378. https://doi.org/10.1136/bmjopen-081378.

Kanagala SG, Gupta V, Kumawat S, Anamika F, McGillen B, Jain R. Hospital at home: emergence of a high-value model of care delivery. Egypt J Intern Med. 2023;35:21. https://doi.org/10.1186/s43162-023-00206-3.

Ko SQ, Chua CMS, Koh SH, Lim YW, Shorey S. Experiences of patients and their caregivers admitted to a hospital-at-home program in Singapore: a descriptive qualitative study. J Gen Intern Med. 2023;38:691–8. https://doi.org/10.1007/s11606-022-07765-1.

Ko SQ, Goh J, Tay YK, Nashi N, Hooi BMY, Luo N, et al. Treating acutely ill patients at home: data from Singapore. Ann Acad Med Singap. 2022;51:392–99. https://doi.org/10.47102/annals-acadmedsg.2021465.

Lai YF, Lim YW, Kuan WS, Goh J, Soong JTY, Shorey S, Ko SQ. Asian attitudes and perceptions toward hospital-at-home: a cross-sectional study. Front Public Health. 2021;9: 704465. https://doi.org/10.3389/fpubh.2021.704465.

Leff B, Burton L, Bynum JW, Harper M, Greenough WB, Steinwachs D, Burton JR. Prospective evaluation of clinical criteria to select older persons with acute medical illness for care in a hypothetical home hospital. J Am Geriatr Soc. 1997;45:1066–73. https://doi.org/10.1111/j.1532-5415.1997.tb05968.x.

Leong MQ, Lim CW, Lai YF. Comparison of hospital-at-home models: a systematic review of reviews. BMJ Open. 2021;11: e043285. https://doi.org/10.1136/bmjopen-2020-043285.

Levine DM, Pian J, Mahendrakumar K, Patel A, Saenz A, Schnipper JL. Hospital-level care at home for acutely ill adults: a qualitative evaluation of a randomized controlled trial. J Gen Intern Med. 2021;36:1965–73. https://doi.org/10.1007/s11606-020-06416-7.

Liem A, Anggraini L, Bariyah, Nasrikah, Lestari E. A long overdue recognition: domestic workers as caregivers for older people in Asia. Lancet Healthy Longev. 2023;4:e129–30. https://doi.org/10.1016/s2666-7568(23)00033-8.

Luk JK, Chan FH, Hui E, Tse CY. The feeding paradox in advanced dementia: a local perspective. Hong Kong Med J. 2017;23:306–10. https://doi.org/10.12809/hkmj166110.

Mahase E. Hospitals are urged to increase use of virtual wards to cut admissions. BMJ. 2024;385: q1123. https://doi.org/10.1136/bmj.q1123.

Nittari G, Khuman R, Baldoni S, Pallotta G, Battineni G, Sirignano A, et al. Telemedicine practice: review of the current ethical and legal challenges. Telemed J E Health. 2020;26:1427–37. https://doi.org/10.1089/tmj.2019.0158.

Norman G, Bennett P, Vardy E. Virtual wards: a rapid evidence synthesis and implications for the care of older people. Age Ageing. 2023;52:afac319. https://doi.org/10.1093/ageing/afac319.

Pandit JA, Pawelek JB, Leff B, Topol EJ. The hospital at home in the USA: current status and future prospects. NPJ Digit Med. 2024;7:48. https://doi.org/10.1038/s41746-024-01040-9.

Patel HY, West JDJ. Hospital at home: an evolving model for comprehensive healthcare. Glob J Qual Saf Healthc. 2021;4:141–6. https://doi.org/10.36401/JQSH-21-4.

Paulson MR, Shulman EP, Dunn AN, Fazio JR, Habermann EB, Matcha GV, et al. Implementation of a virtual and in-person hybrid hospital-at-home model in two geographically separate regions utilizing a single command center: a descriptive cohort study. BMC Health Serv Res. 2023;23:139. https://doi.org/10.1186/s12913-023-09144-w.

Regenstein M, Park YH, Krips M. The use of interpreters in health centers: a mixed-methods analysis. J Health Care Poor Underserved. 2023;34:224–45. https://doi.org/10.1353/hpu.2023.0015.

Salmon M. Supporting the NHS's ambition for virtual wards. National Institute for Health and Care Excellence;2023. https://www.nice.org.uk/news/blog/supporting-nhs-england-s-ambition-for-virtual-wards. Accessed 13 Apr 2024.

Shaver J. The state of telehealth before and after the COVID-19 pandemic. Prim Care. 2022;49:517–30. https://doi.org/10.1016/j.pop.2022.04.002.

Siu AL, Zhao D, Bollens-Lund E, Lubetsky S, Schiller G, Saenger P, et al. Health equity in hospital at home: outcomes for economically disadvantaged and non-disadvantaged patients. J Am Geriatr Soc. 2022;70:2153–6. https://doi.org/10.1111/jgs.17759.

Smullen A, Hong PK. Comparing the health care systems of high-performing Asian countries. Asia Pac Policy Stud. 2015;2:347–55. https://doi.org/10.1002/app5.76.

Thornton N, Horton T, Hardie T. The Health Foundation;2023. https://www.health.org.uk/news-and-comment/charts-and-infographics/how-do-the-public-and-nhs-staff-feel-about-virtual-wards. Accessed 4 May 2024.

United Nations. World population prospects;2022. https://population.un.org/wpp/. Accessed 11 May 2024.

United Nations ESCAP. Ageing in Asia and the Pacific: overview;2017. https://www.unescap.org/resources/ageing-asia-and-pacific-overview. Accessed 11 May 2024.

Victoria State Government. Hospital in the home guidelines. Victoria;2011. https://www.health.vic.gov.au/publications/hospital-in-the-home-guidelines. Accessed 11 May 2024.

Wallis JA, Shepperd S, Makela P, Han JX, Tripp EM, Gearon E, et al. Factors influencing the implementation of early discharge hospital at home and admission avoidance hospital at home: a qualitative evidence synthesis. Cochrane Database Syst Rev. 2024;3:CD014765. https://doi.org/10.1002/14651858.CD014765.pub2.

Wang X, Shi J, Lee KM. The digital divide and seeking health information on smartphones in Asia: survey study of ten countries. J Med Internet Res. 2022;24: e24086. https://doi.org/10.2196/24086.

Wong A, Sowa PM, Banks MD, Bauer JD. Home enteral nutrition in Singapore's long-term care homes—incidence, prevalence, cost, and staffing. Nutrients. 2019;11:2492. https://www.mdpi.com/2072-6643/11/10/2492.

Wong V, Krishna L The meaning of food amongst terminally ill Chinese patients and families in Singapore. JMED Res. 2014:1–8. https://doi.org/10.5171/2014.670628.

World Health Organization. Monitoring the building blocks of health systems: a handbook of indicators and their measurement strategies;2010. https://iris.who.int/bitstream/handle/10665/258734/9789241564052-eng.pdf. Accessed 22 Nov 2024.

Ye J, He L, Beestrum M. Implications for implementation and adoption of telehealth in developing countries: a systematic review of China's practices and experiences. NPJ Digit Med. 2023;6:174. https://doi.org/10.1038/s41746-023-00908-6.

Yuen JK, Luk JKH, Chan TC, Shea YF, Chu ST, Bernacki R, et al. Reduced pneumonia risk in advanced dementia patients on careful hand feeding compared with nasogastric tube feeding. J Am Med Dir Assoc. 2022;23:1541–7. https://doi.org/10.1016/j.jamda.2022.03.011.

Farah Azemee is a Medical Officer under the Ministry of Health, Malaysia. She earned her medical degree from the National University of Ireland, Galway in 2013. She completed a Master of Public Health (MPH) at Universiti Malaya in 2023 and is currently pursuing a Doctorate in Public Health (DrPH) at Universiti Malaya. Her ongoing research focuses on the implementation and evaluation of the Virtual Ward program at the University Malaya Medical Centre (UMMC).

Stephanie Ko is a Consultant in Advanced Internal Medicine at National University Hospital, Singapore. She completed her medical education at the University of Cambridge and University College London in 2013, after which she returned to clinical practice in Singapore where she was later Chief Resident of Internal Medicine Residency. She completed a Masters of Public Health in Clinical Effectiveness at the Harvard TH Chan School of Public Health in 2017. In 2020, she led the development of NUHS@Home, a hospital-at-home programme. NUHS@Home is now planning for the next phase of development to expand the service to more patient profiles.

Terence Ong is an associate professor at Universiti Malaya and consultant geriatrician at the Universiti Malaya Medical Centre, Malaysia. He completed his training at Nottingham University Hospitals NHS Trust and the University of Nottingham, United Kingdom. He currently leads the virtual ward programme at his hospital which provides teleconsultation and remote patient monitoring for patients treated at home. His clinical and academic interests are in musculoskeletal health, geriatric medicine liaison and healthcare service research.

Chapter 14
Breaking Taboos on Sexual Health in Asia Through Telemedicine: A Conceptual Framework

Clusters of Future Studies: Past and Futures; Technological Trends

Hoe Leong Sii, Eng Keat Goh, and Rong Xiang Ng

Abstract In Asian societies, sexual health remains essential yet often stigmatised, with issues ranging from HIV among men who have sex with men (MSM) to erectile dysfunction (ED), premature ejaculation (PE), and teenage pregnancy. Despite telemedicine's long-standing presence, the COVID-19 pandemic has emphasised its capacity to bridge healthcare gaps, especially in sexual medicine. The pandemic has shifted how patients approach sexual health, with many preferring remote consultations over traditional in-person visits due to privacy concerns and convenience. Drawing upon a comprehensive review of existing literature, this chapter presents a conceptual framework for leveraging telemedicine to address sexual health barriers in Southeast Asia, a region where cultural stigma around sexual health is often rooted in conservative beliefs, religious influences, and traditional values, especially when compared to Western Pacific or East Asian regions. The proposed framework emphasises five key elements: cultural sensitivity, anonymity, accessibility, trust-building, and quality assurance of healthcare providers (CAAT-Q). By embedding these elements, telemedicine platforms can encourage open discussions on sexual health, enhance care access, and foster healthier communities. This approach acknowledges cultural nuances and promotes a safe, supportive environment, aiming to improve overall sexual health across Asian societies.

H. L. Sii
Department of Family Medicine, Raffles Medical Group, Singapore, Singapore
e-mail: sii_hoeleong@rafflesmedical.com

E. K. Goh
SingHealth Polyclinic, Singapore, Singapore
e-mail: goh.eng.keat@singhealth.com.sg

R. X. Ng (✉)
Department of Medicine, Faculty of Medicine, Universiti Malaya, Kuala Lumpur, Malaysia
e-mail: rxng@um.edu.my

University of Glasgow, Glasgow, UK

R. Xiang Ng et al. (eds.), *Future of Healthcare in Asia*,
https://doi.org/10.1007/978-3-032-03183-9_14

Keywords Telemedicine · Sexual health · Asian · HIV · Men who have Sex with Men (MSM)

14.1 Introduction

Sexual health is integral to overall well-being, yet discussions and treatments are often hindered by deeply rooted cultural norms, religious beliefs, and social stigma (Guttmacher Institute 2010; Luo et al. 2023a, b; Nicolosi et al. 2005; Wong 2012). This can prevent individuals from seeking the care and education they need, increasing their risk for untreated sexual health issues, unwanted pregnancies, and the spread of sexually transmitted infections (STIs), including HIV infection (Rade et al. 2018b; Li et al. 2022; Yamani et al. 2025; Zablotska et al. 2016). Limited access to sexual health information, coupled with pervasive stigma, disproportionately affects marginalised groups and vulnerable populations across this diverse region. Hence, addressing social taboos around sexual health in the Asia–Pacific region requires a conceptual framework that acknowledges cultural sensitivities while promoting accessible healthcare solutions.

A multifaceted framework to tackle these taboos should integrate community-based education, digital health innovations, and culturally adapted telemedicine solutions. Education initiatives should be tailored to engage communities through trusted channels, encouraging open conversations around sexual health in a manner that respects local values. Telemedicine presents a particularly promising approach, offering discreet and accessible care that reduces the fear of judgment often associated with in-person consultations (Saigí-Rubió et al. 2022). This is particularly important in the Asia–Pacific region, where cultural taboos often prevent individuals from seeking help for sensitive sexual health concerns. By leveraging mobile health platforms, governments and healthcare providers can bridge gaps in knowledge and resources, reaching individuals who might otherwise remain isolated due to geographical or social barriers (Salsabilla et al. 2021).

To ensure the success of this framework, policies and regulations must prioritise privacy, data security, and inclusivity in healthcare delivery. Developing culturally relevant guidelines and training healthcare professionals to approach sexual health issues with sensitivity can further help dismantle the barriers preventing individuals from accessing care. Addressing social taboos on sexual health in the Asia–Pacific region, therefore, demands a comprehensive approach that combines education, technology, and policy reform. By fostering a supportive environment for open dialogue and accessible healthcare, this framework can contribute to improved sexual health outcomes and a gradual shift in societal perceptions.

14.2 Methodology

This article employs a conceptual approach to demonstrate the landscape of sexual health telemedicine in Asian countries and explore the potential of telemedicine to navigate sexual taboos and improve sexual health across Asian countries (MacInnis 2011). It relies on comprehensive review of existing literature, focusing on sexual health challenges, cultural barriers, and the application of telemedicine in the Asia–Pacific region rather than empirical data. Following the "3Es" of exposure, expertise, and experience, it identifies key insights, gaps, and inconsistencies in research on sexual health and social taboos in the Asia–Pacific region, focusing on the role of telemedicine in overcoming these challenges (Kraus et al. 2022). Design for conceptual article is based on the approaches entailed by Jaakkola (Jaakkola 2020). This article combines (1) theory synthesis, integrating current knowledge on sexual health barriers; (2) theory adaptation, applying cultural adaptation theory to suggest telemedicine solutions; (3) typology, categorising social taboos across Asian societies; and (4) model development, proposing a framework linking telemedicine access, cultural adaptation, and improved health outcomes. This framework sets the stage for future empirical research on digital health interventions that address culturally rooted barriers to sexual health in diverse Asian communities.

14.3 Key Concepts and Issues

This section presents the conceptual framework proposed to address the challenges of sexual health access in Asia. To establish a strong foundation for this framework, we will first explore the existing landscape of sexual health issues and the current state of telemedicine adoption in the region. This will be followed by a detailed explanation of the proposed framework, including its key components and their rationale.

14.3.1 Understanding the Landscape: Sexual Health Challenges and Telemedicine Adoption

14.3.1.1 Sexual Health Challenges in Asian Society

Sex is a vital aspect of life that spans across various age groups, from adolescence to the elderly. While sexual activity may increase with age, it remains prevalent among both adolescents and the elderly. The Global Study of Sexual Attitudes and Behaviours (GSSAB) revealed that 75–89% of men and 51–67% of women aged 40–80 had engaged in sexual intercourse within the past 12 months (Nicolosi et al. 2006). Additionally, a technical report published by the Malaysian Institute for Behavioural Health Research in 2021 indicated that 30.4% of adolescents had their sexual debut before the age of 15 (Panting 2021). Furthermore, a study conducted in China surveyed 688 older Chinese individuals (over 60 years old) and found that 32–51% of men and 26–41% of women reported engaging in some form of sexual activity (Yang and Yan 2016).

The World Health Organization defines sexual health as a state of physical, emotional, mental, and social well-being in relation to sexuality; it is not merely the absence of disease, dysfunction, or infirmity (World Health Organization 2002). While this topic is vast, we will focus on highlighting a few key issues in this article. According to the UNAIDS Global AIDS Update 2023, there are 6.5 million people living with HIV in the Asia–Pacific region alone (Joint United Nations Programme on HIV/AIDS 2023). Young people accounted for around a quarter of new HIV infections in the region in 2022. Since 2010, the estimated numbers of new HIV infections men who have sex with men (MSM) have increased, indicating that this key populations continue to have disproportionately high HIV incidence (Korenromp et al. 2024). In the Philippines, a study reported an HIV incidence of 2.7 per 100 person-years among MSM from 2014 to 2018 (Ditangco and Mationg 2022). In Vietnam, HIV prevalence among MSM increased from 6.6% in 2015 to 13.8% in 2020, with HIV incidence rising from 1.36 to 2.61 per 100 person-years over the same period (Thanh Nguyen et al. 2024). In China, HIV incidence among MSM in Jiangsu province remained high, ranging from 5.0% to 9.0% between 2016 and 2021 (Zhou et al. 2023).

On the contrary, traditional conservative Muslim practices in the Middle East can have both beneficial and negative impacts on HIV prevalence, influencing behaviours related to HIV transmission (Abu-Raddad et al. 2010; Alageel et al. 2024; Kteily-Hawa et al. 2022; Obermeyer 2006). Muslim practices include male circumcision, Nikah (Islamic marriage), and prohibition of alcohol use may reduce risky sexual behaviours (DeJong et al. 2005). However, barriers to quality sexual healthcare treatment include lack of awareness, attitudes and practices among physicians in the region. Some research reveal that Iranian physicians lack sufficient knowledge about HIV/AIDS and have negative attitudes towards these patients (Hedayati-Moghaddam et al. 2012; Jaafari et al. 2022; Karimi et al. 2025). In addition, women in Arab countries are at risk of being stigmatised by physicians when they attempt to seek medical advice regarding sexual health such as contraceptive or HIV treatment due to high religious and social beliefs (Alomair et al. 2020, 2023a, b; Obermeyer 2006).

Another major health issue in the Asia is erectile dysfunction (ED). A multicentre study published in 2007, involving 10,934 adult men aged 20–75 years from five regions of Asia (China, Japan, Korea, Malaysia, and Taiwan) (Tan et al. 2007), confirmed that ED is a prevalent condition. The prevalence of ED varied significantly among the regions, ranging from 3% in Malaysia to 14% in Japan (Tan et al. 2007). More recent studies, however, suggest that the prevalence of ED may be increasing. For instance, a 2023 study in Malaysia reported that moderate to severe ED affects 31.6% of men aged 18 and above, indicating a significant rise compared to earlier data (Rezali et al. 2023). This study utilised the International Index of Erectile Function (IIEF) to assess ED prevalence and associated factors, highlighting the growing burden of this condition in the region.

Premature ejaculation (PE) is another common sexual health issue, often more prevalent than ED and overlooked. Data from 2011 examining the Asia–Pacific region showed that the overall prevalence of PE, diagnosed using the Premature Ejaculation Screening Tool (PEST), was 16%, with rates as low as 3% in Indonesia and as high as 33% in South Korea (McMahon et al. 2012). More recent research supports these findings and suggests potential increases in PE prevalence over time. For example, a 2019 study in South Korea found that self-identified PE increased from 19.0% in 2006 to 21.6% in 2016, defined by an estimated intravaginal ejaculation latency time of less than 3 min (Song et al. 2019). These trends underscore the ongoing challenges posed by PE in the region.

Together, these studies highlight the persistent and evolving nature of sexual health issues such as ED and PE in the Asia–Pacific region. While earlier studies established baseline prevalence rates, newer research provides updated insights into the changing landscape of these conditions, emphasising the need for continued attention and intervention.

Another issue is adolescent pregnancy. In low- and middle-income countries (LMICs) in Asia, this remains a significant health issue, associated with increased risks of adverse health outcomes for both young mothers and their offspring. A systematic review highlighted that pregnant adolescents face increased risks of anaemia, preterm birth, low birth weight, and preeclampsia, compounded by limited access to appropriate antenatal care and support services (Sabet et al. 2023). These risks are particularly concerning given the high prevalence of unintended adolescent pregnancies in the region. Adolescent pregnancy rates in South Asia vary significantly, with the highest rate in Afghanistan (71.1%, or 711 per 1000 ever-married women) and the lowest in the Maldives (39.5%, or 395 per 1000) (Poudel et al. 2023). Factors contributing to adolescent pregnancy include low socioeconomic status, rural residency, limited access to family planning, and patriarchal family structures. (Poudel et al. 2023).

Next, it is about contraceptive use. Contraceptive use among adolescent women in LMICs in Asia remains suboptimal, further exacerbating the issue. A 2021 study using Bayesian hierarchical modelling estimated that among 300 million women aged 15–19 years in 2019, only 29.8 million used any contraception, and 15.0 million had unmet needs for family planning (Kantorová et al. 2021). The proportion of contraceptive needs satisfied by modern methods was significantly lower among adolescents compared to older women, particularly in regions like Western Asia and Central and Southern Asia. Challenges faced by adolescent women in accessing contraception include stigma, lack of confidentiality, and judgmental attitudes from healthcare providers. Similarly, a 2021 study in Myanmar found that modern contraceptive prevalence rates were low among youth, with significant geographical variation and lower utilisation among rural youth (Lun et al. 2021).

These findings underscore the urgent need for targeted interventions to improve access to contraceptive services and comprehensive sexual education for adolescents in LMICs in Asia. Addressing these gaps could help reduce unintended pregnancies and mitigate the associated health risks for young mothers and their children.

Although the prevalence of sexual problems in Asia is relatively high, only a small proportion of people experiencing these issues seek medical care. The issue of not seeking help for sexual health problems has been present for the past 20 years (Nicolosi et al. 2006) and continues to persist today. Recent studies confirm this trend: a 2023 study by Luo et al. found that up to 66% of Chinese women felt ashamed of sexual health-related disorders and another systematic review by Rade et al. highlighted that migrant from Southeast Asia living in high-income countries experienced numerous barriers to accessing sexual health services, including stigma and lack of cultural competency within health services (Luo et al. 2023a, b; Rade et al. 2018a).

The cultural and environmental diversity across Asia means that barriers to seeking care may vary slightly from one region to another. However, common themes persist, including perceptions of problem severity, lack of awareness regarding available treatments, pervasive stigma, discrimination, limited access to affordable care, and bureaucratic obstacles.

14.3.1.2 Role of Telemedicine in the Asia Pacific Region

Accessibility of internet and mobile service experiences a notable surge on a global scale, particularly within emerging economies across the Asia Pacific region. The influence of mobile connectivity is demonstrated by its economic contribution in the region. In 2022, the collective contribution of mobile technologies and services amounted to just under 5% of the Gross Domestic Product (GDP) in the Asia Pacific region, yielding a remarkable economic value added of $810 billion (GSMA 2023). By the end of 2030, there will be approximately 1.4 billion 5G connections, equivalent of 41% of total mobile connections (GSMA 2023).

In 2020, the COVID-19 pandemic underscored deficiencies within healthcare delivery systems and emphasised the critical role of resilient telemedicine platforms in fortifying public health responses during emergencies (Smith et al. 2020). Telemedicine is defined as the delivery of health care services through information and communication technologies (ICT). Telemedicine serves as a pivotal tool in meeting the evolving demands of healthcare systems by bridging the gap between patients and physicians, offering accessible and cost-effective medical care, and substantially enhancing health outcomes (Nnamoko et al. 2016). At the primary care level, telemedicine facilitates a seamless continuum of care, ensuring comprehensive and timely interventions to address individual health needs effectively (Gudi et al. 2021; Salsabilla et al. 2021).

Recent studies highlight the substantial cost-saving benefits of telemedicine across various medical fields, for example, high income Asian countries like Japan has adopted telemedicine to connect healthcare providers to specialists in fields, such as dermatology, radiology, and pathology to help mitigate the rising costs of healthcare (Akiyama and Yoo 2016). Tele-dermatology reported a cost-saving of $330 per person per week (Dekio et al. 2010), and another study reported a cost-saving of $144 per reading via tele-radiology by avoiding unnecessary enhanced CT scans

(Takada et al. 2003). A recent scoping review by Liu et al. reported significant cost reductions in six out of twelve studies examining telemedicine interventions, underscoring its potential to enhance resource utilisation, confirming the previous studies (Liu et al. 2025). This benefit extends to palliative field. Home palliative care utilising telemedicine was estimated to save $5000 annually per person compared to traditional home care methods by reducing the number physician visits, underscoring the importance of telemedicine in ageing Japanese population (Aoki et al. 2006). Tele-dermatology is also being adopted at the Institute of Mental Health Singapore among a group of psychiatric patients with chronic skin diseases, and it is as effective as face-to-face consultation in diagnostic accuracy (Seghers et al. 2015). Cheng et al. in Taiwan revealed that an interactive tele-dermatology program for psoriasis care in remote regions was cost-effective, with similar yearly costs to in-person clinic visits, while maintaining high patient satisfaction (Cheng et al. 2022). Collectively, these studies underscore the substantial and consistent cost-saving and cost-effective benefits of telemedicine across diverse healthcare settings.

Furthermore, the National Telemedicine Centre of China, a cross-regional system established over 20 years ago at the First Affiliated Hospital of Zhengzhou University, has shown rapid growth in teleconsultation from 2,322 cases in 2014 to 22,242 cases in 2019, representing a ten-fold increase within the span of 5 years, significantly bolsters healthcare accessibility among underserved populations in an upper middle income country (Jha et al. 2021; Mars 2018).

COVID-19 pandemic also revealed an exponential increase in telehealth services in Malaysia when the country incorporates electronic prescription after the amendment was made to the Poisons Act 1952 (Loo et al. 2023). There was a total of 835,826 teleconsultation conducted over a 3-year period between 2019 and 2021, involving 88.8% (n = 741,236) assisted consultations and 11.2% (n = 94,590) direct consultations (Loo et al. 2023). According to the Malaysian study, assisted consultations involve pharmacists guiding users to consult virtual physicians to obtain their medication by providing supporting information such as medical histories, blood pressure and blood glucose readings. Common diagnoses for telehealth services include essential hypertension, diabetes mellitus, dyslipidemia and ischemic heart disease (Loo et al. 2023). This study indicates the significance of telemedicine in the management of chronic conditions.

14.3.1.3 Role of Telemedicine in Sexual Health in Asian Society

Several studies highlight the promise of telemedicine for HIV prevention in Asia. A systematic review found that eHealth interventions conducted in Asia Pacific countries (Australia, China, India, New Zealand, Pakistan, Taiwan and Vietnam) can be effective in engaging priority populations across the HIV treatment and care cascade (Purnomo et al. 2018). A Malaysian study has also examined the potential of mHealth for reaching specific populations at high risk for HIV, including MSM and transgender women (Krishnan et al. 2021). In China, a pilot study protocol demonstrates the feasibility of using a community-engaged mHealth intervention to increase uptake of PrEP, a highly effective HIV prevention medication among MSM (Li et al. 2022). These studies provide valuable evidence for the potential of telemedicine platforms to play a crucial role in HIV prevention strategies across Asia. It is important to promote awareness and education about HIV transmission, prevention methods, and available treatment options to patients, and facilitating access to Pre-exposure Prophylaxis (PrEP) for individuals at high risk. Telemedicine platforms can streamline the process of connecting users with qualified healthcare providers who can prescribe PrEP and offer ongoing monitoring.

While there has been widespread use of telemedicine in Asia Pacific region, its application in sexual health still encounters significant challenges and the literature on this remains scarce (Purnomo et al. 2018; Salsabilla et al. 2021). Issues such as cultural taboos, privacy concerns, and regulatory restrictions pose significant barriers to the effective delivery of sexual health services through telemedicine platforms (Purnomo et al. 2018). Many individuals in Asia–Pacific countries may feel uncomfortable discussing sensitive sexual health topics remotely, leading to underutilisation of telemedicine services in this area. Additionally, disparities in internet access and digital literacy among certain demographic groups may exacerbate inequalities in accessing sexual health information and services through telemedicine. Despite these limitations, efforts to address cultural sensitivities, enhance privacy measures, and establish clear regulatory frameworks are crucial to expanding the scope and effectiveness of telemedicine in addressing sexual health needs across the Asia–Pacific region.

14.3.2 Proposed Telemedicine Framework for Sexual Health

This chapter proposes a novel framework for utilising telemedicine platforms to overcome cultural barriers and improve access to sexual healthcare in Asia". It focuses on addressing a range of sexual health issues in the region, including:

- Sexually transmitted infections (STIs)
- Unintended pregnancy
- Erectile dysfunction (ED)
- Premature ejaculation (PE)
- Access to sexual and reproductive health information.

Besides, this framework also highlighting the role of telemedicine in HIV Prevention, acknowledging the significant public health concern of HIV infection in Asia. The framework recognises the potential of telemedicine to play a crucial role in HIV prevention strategies, particularly through heightening awareness and constant education about HIV transmission, prevention methods, and available treatment options to encourage people living with HIV enter the care cascade. Not forgetting, this framework also facilitates access to Pre-exposure Prophylaxis (PrEP) for individuals at high risk. Telemedicine platforms can streamline the process of connecting users with qualified healthcare providers who can prescribe PrEP and offer ongoing monitoring.

Key Components of the Proposed Framework (CAAT-Q):

- Cultural Sensitivity: The platform design and communication strategies must be tailored to respect diverse cultural norms and values in Asia by offering content in multiple languages and considering local sensitivities regarding sexual health terminology and images.
- Anonymity: The platform should guarantee user anonymity about sensitive sexual health concerns. This can be achieved through secure communication channels and options for anonymous consultations.
- Accessibility: The platform should be designed to be accessible to diverse populations with varying levels of digital literacy and technology access, considering low-bandwidth functionality for users with limited internet connectivity.
- Trust-Building: Strategies to build trust with users include utilising qualified healthcare providers, ensuring secure communication channels, and incorporating positive user testimonials. Additionally, the platform should be transparent about data privacy practices and user confidentiality.
- Quality Assurance: This component ensures the qualifications and expertise of healthcare providers offering consultations on the platform. This could involve credential verification processes, ongoing training requirements, and user feedback mechanisms to monitor provider performance.

By incorporating these key components, the proposed framework aims to empower individuals in Asia to prioritise their sexual health, overcome cultural barriers, and access essential healthcare services through telemedicine platforms. Figure 14.1 (see the next page) illustrates the components of the proposed telemedicine framework.

14.4 Discussion of Implications

This study reveals the benefits, challenges, and barriers of telemedicine adoption in Asian countries considering the substantial gap regarding the absence of a telemedicine framework in sexual health in the region. Telemedicine serves as a valuable resource for patients seeking consultation on sensitive and private sexual health matters, given the high prevalence of sexual dysfunction and low engagement in sexual health-seeking behaviours. Luo et al. (2023a, b) found that up to 66% of Chinese women felt ashamed of sexual health-related disorders, which significantly hindered their willingness to seek help. Online medical consultations could also provide men with a platform to discuss sexual dysfunction openly, reducing the stigma associated with face-to-face interactions (Li et al. 2022). These findings resonate with earlier research by Nicolosi et al. (2005), who reported that only 21% of individuals experiencing sexual dysfunction actively sought medical assistance.

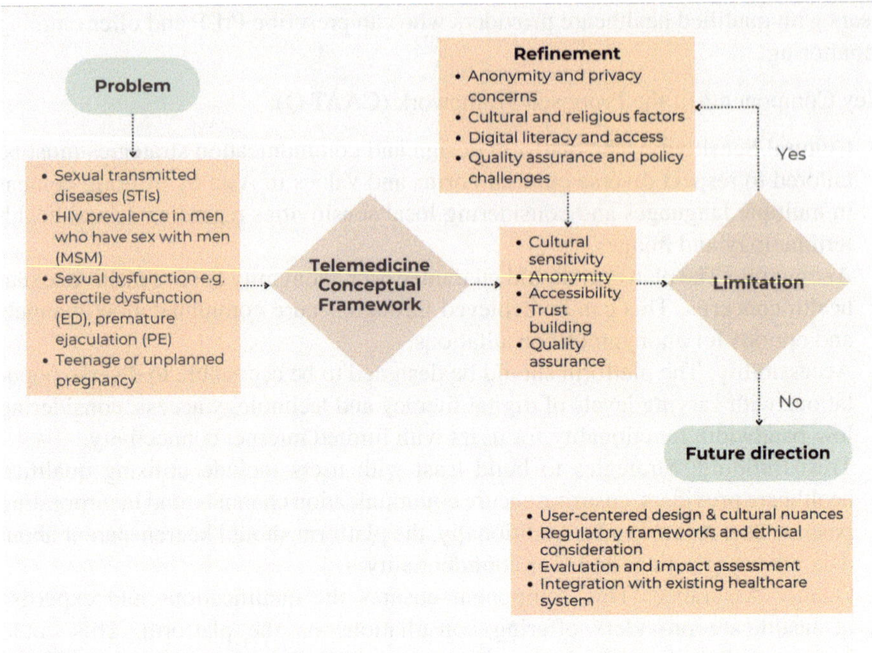

Fig. 14.1 Telemedicine framework for addressing sexual health challenges in Asia

Given these circumstances, it is imperative for telemedicine platforms to prioritise user anonymity concerning sensitive sexual health issues. Verma and Kerrison's review underscores patient concerns regarding the privacy and confidentiality of their discussions with healthcare providers (Verma and Kerrison 2022). Ensuring secure communication channels and offering options for anonymous consultations can address these apprehensions effectively.

A recent systematic review encompassing nearly 2,239 studies involving over 20,000 patients in the WHO European Region, indicates that resistance to telemedicine primarily stemmed from individual factors such as language barriers, lack of motivation or support, and sociocultural aspects (Saigí-Rubió et al. 2022). Cultural and religious beliefs significantly shape attitudes towards sexual health in Asian societies. In East Asian nations like Korea, China, Taiwan, and Japan, Confucian traditions emphasise propriety, sexual restraint, and the primacy of family, which can lead to conservative attitudes towards sexual health and reluctance to seek medical advice for sexual issues (Lieber et al. 2009). In Malaysia and Indonesia, Islamic beliefs influence sexual and reproductive health behaviours, with barriers such as lack of basic reproductive knowledge, misconceptions about contraception, and fear of stigmatisation affecting access to sexual health services (Alomair et al. 2020; Arousell et al. 2016). Buddhism in Thailand and Singapore also impacts sexual health attitudes, promoting modesty and restraint, which can hinder open discussions about sexual health (Nicolosi et al. 2005). These cultural sensitivities affect sexual reproductive and contraceptive knowledge among multiethnic populations. For example, in Malaysia, with multiracialism, Malay Muslims had lower knowledge scores compared to other ethnicities (Wong 2012). This disparity is influenced by religious and cultural beliefs that limit access to sexual health information and services. Thus, telemedicine providers should acknowledge difference in cultural and religious beliefs so that patients feel safe and empowered to discuss personal issues related to sexual health.

Another frequently cited concern regarding teleconsultation, affecting both physicians and patients, is the lack of confidence in and access to technology and technical skills (Isautier et al. 2020, Verma and Kerrison 2022, Vosburg and Robinson 2022). This challenge is particularly pronounced for patients residing in rural areas with inadequate connections and substandard audio quality (Verma and Kerrison 2022). In Asia, culturally and linguistically diverse (CaLD) communities face significant digital health inequities, with barriers such as limited access to technology and low digital literacy as well (Brady et al. 2024; Gallegos-Rejas et al. 2025). An Australian national study also identified that slightly over a quarter of 1,369 participants reported that telehealth services were too complicated, despite nearly 80% of participants attended universities (Isautier et al. 2020). Hence, the platform must be designed to cater to diverse populations with varying degrees of digital literacy and access to technology, including the incorporation of low-bandwidth features to accommodate users with limited internet connectivity. Healthcare providers should use simple descriptions to avoid confusion around diagnoses of condition and treatment and appropriate interpreter should be readily available to assist in teleconsultation.

In many Asian countries with significant HIV prevalence among MSM, quality assurance systems for sexually transmitted infection (STI) and HIV testing services may have gaps or be underdeveloped but it is actually the entry point for HIV prevention and treatment (World Health Organization 2015). Punitive laws are still being enforced against marginalised groups in certain Southeast Asians countries such as Philippines, Myanmar and Cambodia, resulting in poor PrEP uptakes (Draper et al. 2017; Espera et al. 2025; Oliveri et al. 2024; Restar et al. 2022; Zablotska et al. 2016). Furthermore, antiretroviral therapy (ART) coverage for people living with HIV reported to be very poor in the region such as India (57%), Myanmar (45%) and significantly lower in Indonesia (18%) (Zablotska et al. 2016). Conversely, countries like Thailand and Malaysia have been working on the PrEP policy development such as the Thailand National Guidelines on HIV/AIDS Treatment and Prevention and the Malaysian National Strategic Plan for Ending AIDS 2016–2030 (Zablotska et al. 2016). Telemedicine can further support PrEP service delivery. In Thailand, the inclusion of PrEP in the Universal Coverage Scheme (UCS) has been supported by telemedicine initiatives that facilitate service integration and improve access for high-risk populations (Rayanakorn et al. 2022; Viriyathorn et al. 2023). While in Malaysia, telemedicine can address barriers identified in PrEP uptake among MSM, such as costs, limited knowledge, and social stigma, by offering remote consultations and educational resources (Rosen et al. 2023). In the current development of PrEP usage, Asian-Pacific countries must work with closely with stakeholders and policymakers to introduce telemedicine services with quality assurance in place. Policy for HIV testing, prevention and treatment should be standardised, and healthcare providers should be trained on timely basis.

By creating a safe and accessible platform for open communication and by promoting effective HIV prevention strategies, telemedicine has the potential to transform sexual health outcomes and well-being across Asia (Hung et al. 2022; Phan et al. 2022; Purnomo et al. 2018).

14.5 Future Directions

While this framework offers a promising approach, further research and development are crucial for its successful implementation and long-term impact.

1. User-Centred Design and Cultural Nuances

Conducting in-depth user studies remains imperative to comprehend diverse user needs, preferences, and cultural sensitivities concerning telemedicine platforms for sexual health across various Asian regions. Personalising and modifying platforms according to cultural nuances, such as those in China, Korea, Japan, Southeast Asia, India, and the Middle East, informs design and communication strategies for optimal cultural appropriateness and user satisfaction. Additionally, research and development efforts must focus on effective methods for delivering multilingual content and functionalities to accommodate the linguistic diversity prevalent in the region.

2. Regulatory Frameworks and Ethical Considerations

Advocating for clear and supportive regulatory frameworks for telemedicine platforms in Asia is essential to ensure quality, safety, and user privacy. This entails investigating and addressing ethical implications, including data privacy, informed consent, and the potential misuse of information, associated with telemedicine for sexual health in the region.

3. Evaluation and Impact Assessment

Developing and refining appropriate metrics is paramount to evaluating the framework's effectiveness in achieving its goals. Metrics may encompass user uptake, changes in sexual health knowledge and behaviour, and reductions in STI and HIV infection rates, possibly through online testing or surveys. Pilot studies conducted in real-world settings with diverse Asian populations, employing methodologies such as in-depth interviews, serve to assess effectiveness, user experience, and areas for improvement prior to widespread implementation.

4. Integration with Existing Healthcare Systems

Exploring strategies for effective collaboration between telemedicine platforms and existing healthcare systems in Asia is essential for ensuring continuity of care, referral pathways, and integration with electronic health records. Moreover, offering training programs for healthcare providers facilitates the acquisition of necessary skills and knowledge to deliver high-quality sexual health consultations through telemedicine platforms.

By addressing these future directions, researchers and policymakers can refine the proposed framework, ensure its ethical and effective implementation, and ultimately empower Asian communities to leverage telemedicine for improved sexual health and well-being.

14.6 Conclusion

Sexual health concerns are prevalent across Asia, yet cultural taboos and limited access to care often deter individuals from seeking help. Traditional healthcare models may fall short in adequately addressing these challenges. This paper presents a comprehensive literature review highlighting the landscape of sexual health issues in Asia and the potential of telemedicine to bridge existing gaps.

The proposed conceptual framework offers a novel approach to leveraging telemedicine platforms for overcoming cultural barriers and improving sexual health outcomes in Asia. The framework focuses on several key components, including cultural sensitivity, anonymity, accessibility, trust-building, and quality assurance. By incorporating these elements, the framework aims to empower individuals to prioritise their sexual health and access essential healthcare services through telemedicine platforms effectively.

The potential benefits of this framework are significant. By facilitating access to sexual health information and providing consultations for various sexual health concerns such as HIV prevention, erectile dysfunction, premature ejaculation, and teenage pregnancy, telemedicine has the capacity to break down taboos surrounding sexual health and transform sexual health outcomes and overall well-being across Asia.

However, further research and development are crucial for the successful implementation of telemedicine solutions in the Asian context. User-centred design, clear regulatory frameworks, ethical considerations, and evaluation metrics are all essential for ensuring the framework's effectiveness and long-term impact. It is imperative to address cultural nuances, regulatory challenges, and ethical considerations to tailor telemedicine solutions to the diverse needs of Asian communities effectively.

Highlights

- Asia's digital growth positions telemedicine to bridge sexual health access across the region.
- The CAAT-Q framework offers trusted remote care for diverse Asian populations.

- Telemedicine's discreet platforms can overcome cultural taboos and reduce stigma around sexual health in Asia.
- Future success in Asia requires user-focused design, integrated digital systems, and supportive regulations.
- Strategic telemedicine adoption, guided by frameworks, promises improved sexual well-being throughout Asia.

References

Abu-Raddad LJ, Hilmi N, Mumtaz G, Benkirane M, Akala FA, Riedner G, et al. Epidemiology of HIV infection in the Middle East and North Africa. AIDS. 2010;24 Suppl 2:S5–23. https://doi.org/10.1097/01.aids.0000386729.56683.33.

Akiyama M, Yoo BK. A systematic review of the economic evaluation of telemedicine in Japan. J Prev Med Public Health. 2016;49(4):183–96. https://doi.org/10.3961/jpmph.16.043.

Alageel S, Alsadhan NM, Alkhaldi G, Alkasabi R, Alomair N. Public perceptions of HIV/AIDS awareness in the gulf council cooperation countries: a qualitative study. Int J Equity Health. 2024;23(1):269. https://doi.org/10.1186/s12939-024-02346-6.

Alomair N, Alageel S, Davies N, Bailey JV. Factors influencing sexual and reproductive health of Muslim women: a systematic review. Reprod Health. 2020;17(1):33. https://doi.org/10.1186/s12978-020-0888-1.

Alomair N, Alageel S, Davies N, Bailey JV. Muslim women's knowledge, views, and attitudes towards sexually transmitted infections in Saudi Arabia: A qualitative study. PLoS ONE. 2023a;18(6): e0286822. https://doi.org/10.1371/journal.pone.0286822.

Alomair N, Alageel S, Davies N, Bailey JV. Muslim women's perspectives on the barriers to sexually transmitted infections testing and diagnosis in Saudi Arabia. Front Public Health. 2023b;11:1248695. https://doi.org/10.3389/fpubh.2023.1248695.

Aoki N, Dunn K, Johnson-Throop KA, Turley JP. Triangulation analysis of tele-palliative care implementation in a rural community area in Japan. Telemed J E Health. 2006;12(6):655–62. https://doi.org/10.1089/tmj.2006.12.65.

Arousell J, Carlbom A. Culture and religious beliefs in relation to reproductive health. Best Pract Res Clin Obstet Gynaecol. 2016;32:77–87. https://doi.org/10.1016/j.bpobgyn.2015.08.011.

Brady B, Saberi G, Santalucia Y, Gorgees P, Nguyen TT, Le H, et al. 'Without support CALD patients will be left behind': a mixed-methods exploration of culturally and linguistically diverse (CALD) client perspectives of telehealth and those of their healthcare providers. J Telemed Telecare. 2024;30(9):1493–506. https://doi.org/10.1177/1357633X231154943.

Cheng YW, Wu CY, Wang BC, Lan KC, Ou SY, Lin LM, et al. A desperate need for psoriasis health care in remote regions as revealed by a live interactive teledermatology program serving Penghu Islands in Taiwan Strait. Telemed J E Health. 2022;28(8):1109–16. https://doi.org/10.1089/tmj.2021.0349.

DeJong J, Jawad R, Mortagy I, Shepard B. The sexual and reproductive health of young people in the Arab countries and Iran. Reprod Health Matters. 2005;13(25):49–59. https://doi.org/10.1016/S0968-8080(05)25181.

Dekio I, Hanada E, Chinuki Y, Akaki T, Kitani M, Shiraishi Y, et al. Usefulness and economic evaluation of ADSL-based live interactive teledermatology in areas with shortage of dermatologists. Int J Dermatol. 2010;49(11):1272–5. https://doi.org/10.1111/j.1365-4632.2010.04572.x.

Ditangco R, Mationg ML. HIV incidence among men who have sex with men (MSM) in Metro Manila, the Philippines: A prospective cohort study 2014–2018. Medicine (Baltimore). 2022;101(35): e30057. https://doi.org/10.1097/MD.0000000000030057.

Draper BL, Fowkes FJI, Oo ZM, Thein ZW, Aung PP, Veronese V, et al. Willingness to use HIV pre-exposure prophylaxis among gay men, other men who have sex with men and transgender women in Myanmar. J Int AIDS Soc. 2017;20(1):21885. https://doi.org/10.7448/IAS.20.1.21885.

Espera JR, Reyes CA, Sarmiento EG, Sitchon CM, Zantua R, Carandang RR. Acceptability and feasibility of HIV pre-exposure prophylaxis (PrEP) in Southeast Asia: a scoping review. Int J STD AIDS. 2025;36(4):260–74. https://doi.org/10.1177/09564624241306158.

Gallegos-Rejas VM, De Guzman KR, Kelly JT, Smith AC, Thomas EE. Strategies to improve telehealth access for culturally and linguistically diverse communities: a systematic review. J Public Health (Oxf). 2025:fdaf030. https://doi.org/10.1093/pubmed/fdaf030.

GSMA. The mobile economy Asia Pacific 2023;2023. Available from: https://www.gsma.com/solutions-and-impact/connectivity-for-good/mobile-economy/wp-content/uploads/2023/07/Mobile-Economy-Report-Asia-Pacific-2023.pdf. Accessed 1 Apr 2024.

Gudi N, Konapur R, John O, et al. Telemedicine supported strengthening of primary care in WHO South East Asia region: lessons from the COVID-19 pandemic experiences. BMJ Innov. 2021;7(3):580–5. https://doi.org/10.1136/bmjinnov-2021-000699.

Guttmacher Institute. Facts on the sexual and reproductive health of adolescent women in the developing world. New York: Guttmacher Institute;2010. Available from: https://www.guttmacher.org/fact-sheet/facts-sexual-and-reproductive-health-adolescent-women-developing-world. Accessed 13 Mar 2024.

Hedayati-Moghaddam MR, Marjaneh MM, Mashhadi IE. Knowledge and attitudes of physicians in private practice towards HIV/AIDS in Mashhad, Iran. Int J STD AIDS. 2012;23(8):e11–6. https://doi.org/10.1258/ijsa.2009.009447.

Hung CC, Banerjee S, Gilada I, Green K, Inoue Y, Kamarulzaman A, et al. Impact of COVID-19 on the HIV care continuum in Asia: insights from people living with HIV, key populations, and HIV healthcare providers. PLoS ONE. 2022;17(7): e0270831. https://doi.org/10.1371/journal.pone.0270831.

Isautier JMJ, Copp T, Ayre J, et al. People's experiences and satisfaction with telehealth during the COVID-19 pandemic in Australia: cross-sectional survey study. J Med Internet Res. 2020;22(12): e24531. https://doi.org/10.2196/24531.

Jaafari Z, McFarland W, Eybpoosh S, Tabatabaei SVA, Bafti MS, Ranjbar E, et al. Barriers and facilitators of access to HIV prevention, care, and treatment services among people living with HIV in Kerman, Iran: a qualitative study. BMC Health Serv Res. 2022;22(1):1097. https://doi.org/10.1186/s12913-022-08483-4.

Jaakkola E. Designing conceptual articles: four approaches. AMS Rev. 2020;10(1):18–26. https://doi.org/10.1007/s13162-020-00161-0.

Jha AK, Lin L, Savitz LA. Telemedicine and community health projects in Asia. Dermatol Clin. 2021;39(1):23–32. https://doi.org/10.1016/j.det.2020.08.003.

Joint United Nations Programme on HIV/AIDS. The path that ends AIDS: UNAIDS global AIDS update 2023. Geneva: Joint United Nations Programme on HIV/AIDS; 2023. Available from: https://www.unaids.org/sites/default/files/media_asset/2023-unaids-global-aids-update_en.pdf. Accessed 4 Apr 2024.

Kantorová V, Wheldon MC, Dasgupta ANZ, Ueffing P, Castanheira HC. Contraceptive use and needs among adolescent women aged 15–19: regional and global estimates and projections from 1990 to 2030 from a Bayesian hierarchical modelling study. PLoS ONE. 2021;16(3): e0247479. https://doi.org/10.1371/journal.pone.0247479.

Karimi K, Azizpour Y, Shafaati M, Nejad SM, Ehsani R, Nikfarjam A, et al. Assessing HIV transmission knowledge and rapid test history among the general population in Iran. Sci Rep. 2025;15(1):4944. https://doi.org/10.1038/s41598-025-88844-1.

Korenromp EL, Sabin K, Stover J, Brown T, Johnson LF, Martin-Hughes R, et al. New HIV infections among key populations and their partners in 2010 and 2022, by world region: a multisources estimation. J Acquir Immune Defic Syndr. 2024;95(1S):e34–45. https://doi.org/10.1097/QAI.0000000000003340.

Kraus S, Breier M, Lim WM, Dabić M, Kumar S, Kanbach D, et al. Literature reviews as independent studies: guidelines for academic practice. Rev Manag Sci. 2022;16:2577–95. https://doi.org/10.1007/s11846-022-00588-8v.

Krishnan A, Ferris DG, Mitchell JW, Cheah WH, Lim SH, Chan RKW, et al. Assessing mobile technology use and mHealth acceptance among HIV-positive men who have sex with men and transgender women in Malaysia. PLoS ONE. 2021;16(3): e0248705. https://doi.org/10.1371/journal.pone.0248705.

Kteily-Hawa R, Hawa AC, Gogolishvili D, Al Akel M, Andruszkiewicz N, Vijayanathan H, et al. Understanding the epidemiological HIV risk factors and underlying risk context for youth residing in or originating from the Middle East and North Africa (MENA) region: a scoping review of the literature. PLoS ONE. 2022;17(1): e0260935. https://doi.org/10.1371/journal.pone.0260935.

Li C, Xiong Y, Muessig KE, Zhang W, Liu X, Wang H, et al. Community-engaged mHealth intervention to increase uptake of HIV pre-exposure prophylaxis (PrEP) among gay, bisexual and other men who have sex with men in China: study protocol for a pilot randomised controlled trial. BMJ Open. 2022;12: e055899. https://doi.org/10.1136/bmjopen-2021-055899.

Lieber E, Chin D, Li L, et al. Sociocultural contexts and communication about sex in China: Informing HIV/STD prevention programs. AIDS Educ Prev. 2009;21(5):415–29. https://doi.org/10.1521/aeap.2009.21.5.415.

Liu S, Kumluang S, Gaewkhiew P, et al. Assessing the impact of telemedicine interventions on health care costs and utilization: a scoping review. Telemed J E Health. 2025. https://doi.org/10.1089/tmj.2024.0565.

Loo JS, Yow HY, Ten YY, Govindaraju K, Megat Mohd Zubairi MH, Oui HC, et al. Exploring the rise of telehealth services in Malaysia: a retrospective study. Digit Health. 2023;9:20552076231216275. https://doi.org/10.1177/20552076231216275.

López-Liria R, Valverde-Martínez MÁ, López-Villegas A, et al. Teledermatology versus face-to-face dermatology: an analysis of cost-effectiveness from eight studies from Europe and the United States. Int J Environ Res Public Health. 2022;19(5):2534. https://doi.org/10.3390/ijerph19052534.

Lun CN, Aung T, Mya KS. Utilization of modern contraceptive methods and its determinants among youth in Myanmar: analysis of Myanmar demographic and health survey (2015–2016). PLoS ONE. 2021;16(10): e0258142. https://doi.org/10.1371/journal.pone.0258142.

Luo L, Huang J, Li H. Barriers to sexual health-seeking behaviours for Chinese women. Arch Womens Ment Health. 2023a;26(5):581–8. https://doi.org/10.1007/s00737-023-01348-7.

Luo L, Zhang Y, Wang X, et al. Barriers to seeking help for sexual health issues among Chinese women: a cross-sectional study. BMC Womens Health. 2023b;23(1):123. https://doi.org/10.1186/s12905-023-02456-7.

MacInnis DJ. A framework for conceptual contributions in marketing. J Mark. 2011;75(4):136–54. https://doi.org/10.1509/jmkg.75.4.136.

Mars M. 1st International Conference on eHealth and Telemedicine (ICEHAT) Kathmandu 1st-3rd November 2018. J Int Soc Telemed eHealth. 2018;6:eS2(1–17). https://doi.org/10.29086/JISfTeH.6.eS2.

McMahon CG, Lee G, Park JK, Adaikan PG. Premature ejaculation and erectile dysfunction prevalence and attitudes in the Asia-Pacific region. J Sex Med. 2012;9(2):454–65. https://doi.org/10.1111/j.1743-6109.2011.02507.x.

Nicolosi A, Laumann EO, Glasser DB, Moreira ED Jr, Paik A, Gingell C. Sexual behaviour and dysfunction and help-seeking patterns in adults aged 40–80 years in the urban population of Asian countries. BJU Int. 2005;95(4):609–14. https://doi.org/10.1111/j.1464-410X.2005.05348.x.

Nicolosi A, Laumann EO, Glasser DB, Moreira ED Jr, Paik A, Gingell C. Sexual behaviour, sexual dysfunctions and related help seeking patterns in middle-aged and elderly Europeans: the global study of sexual attitudes and behaviors. World J Urol. 2006;24:423–8. https://doi.org/10.1007/s00345-006-0088-9.

Nnamoko N, Arshad F, Hammond L, Mcpartland S, Patterson P. Telehealth in primary health care: analysis of Liverpool NHS experience. In: Al-Jumeily D, Hussain A, Mallucci C, Oliver C, editors. Applied computing in medicine and health. London: Morgan Kaufmann;2016. p. 269–86. https://doi.org/10.1016/B978-0-12-803468-2.00013-8.

Obermeyer CM. HIV in the Middle East. BMJ. 2006;333(7573):851–4. https://doi.org/10.1136/bmj.38994.400370.7C.

Oliveri C, Eang R, Ségéral O, Mora M, Puth L, Prom S, et al. High PrEP acceptability and need for tailored implementation in Cambodian key populations: results from a qualitative assessment. AIDS Care. 2024;36(Suppl 1):109-16. https://doi.org/10.1080/09540121.2024.2308029.

Panting AA. Technical report: Adolescent sexual and reproductive health in Malaysia: A scoping review (NMRR-18-840-40524). In: Amirudin N, Jaafar N, Yong TSM, editors. Shah Alam, Selangor: Institute for Health Behavioural Research, Ministry of Health Malaysia;2021. Available from: https://library.nih.gov.my/cgi-bin/koha/opac-detail.pl?biblionumber=25370&shelfbrowse_itemnumber=56703. Accessed 14 Apr 2024.

Phan JM, Kim S, Linh ĐTT, Cosimi LA, Pollack TM. Telehealth interventions for HIV in low- and middle-income countries. Curr HIV/AIDS Rep. 2022;19(6):600–9. https://doi.org/10.1007/s11904-022-00630-0.

Poudel S, Dobbins T, Razee H, Akombi-Inyang B. Adolescent pregnancy in South Asia: a pooled analysis of demographic and health surveys. Int J Environ Res Public Health. 2023;20(12):6099. https://doi.org/10.3390/ijerph20126099.

Purnomo J, Coote K, Mao L, Fan L, Gold J, Ahmad R, et al. Using eHealth to engage and retain priority populations in the HIV treatment and care cascade in the Asia-Pacific region: a systematic review of literature. BMC Infect Dis. 2018;18(1):82. https://doi.org/10.1186/s12879-018-2972-5.

Rade DA, Crawford G, Lobo R, Gray C, Brown G. Sexual health help-seeking behavior among migrants from Sub-Saharan Africa and South East Asia living in high income countries: a systematic review. Int J Environ Res Public Health. 2018a;15(7):1311. https://doi.org/10.3390/ijerph15071311.

Rade K, Hjelm K, Apeland A, et al. Barriers to accessing sexual health services among Southeast Asian migrants in high-income countries: a systematic review. Int J Environ Res Public Health. 2018b;15(10):2234. https://doi.org/10.3390/ijerph15102234.

Rayanakorn A, Chautrakarn S, Intawong K, Chariyalertsak C, Khemngern P, Olson D, et al. A comparison of attitudes and knowledge of pre-exposure prophylaxis (PrEP) between hospital and key population led health service providers: lessons for Thailand's universal health coverage implementation. PLoS ONE. 2022;17(5): e0268407. https://doi.org/10.1371/journal.pone.0268407.

Restar A, Surace A, Adia A, Goedel W, Ogunbajo A, Jin H, et al. Characterizing awareness of pre-exposure prophylaxis for HIV prevention in Manila and Cebu, Philippines: web-based survey of Filipino cisgender men who have sex with men. J Med Internet Res. 2022;24(1): e24126. https://doi.org/10.2196/24126.

Rezali MS, Anuar MFM, Razak MAA, et al. Prevalence and associated factors of moderate to severe erectile dysfunction among adult men in Malaysia. Sci Rep. 2023;13(1):21483. https://doi.org/10.1038/s41598-023-48778-y.

Rosen AO, Wickersham JA, Altice FL, Khati A, Azwa I, Tee V, et al. Barriers and facilitators to pre-exposure prophylaxis by men who have sex with men and community stakeholders in Malaysia. Int J Environ Res Public Health. 2023;20(9):5669. https://doi.org/10.3390/ijerph20095669.

Sabet F, Prost A, Rahmanian S, et al. The forgotten girls: the state of evidence for health interventions for pregnant adolescents and their newborns in low-income and middle-income countries. Lancet. 2023;402(10412):1580–96. https://doi.org/10.1016/S0140-6736(23)01682-3.

Saigí-Rubió F, Borges do Nascimento IJ, Robles N, Ivanovska K, Katz C, Azzopardi-Muscat N, et al. The current status of telemedicine technology use across the World Health Organization European region: an overview of systematic reviews. J Med Internet Res. 2022;24(10):e40877. https://doi.org/10.2196/40877.

Salsabilla A, Azzahra AB, Syafitri RIP, Supadmi W, Suwantika AA. Cost-effectiveness of telemedicine in Asia: a scoping review. J Multidiscip Healthc. 2021;14:3587–96. https://doi.org/10.2147/JMDH.S332579.

Seghers AC, Prens EP, van der Zee HH, Boer J, van Doorn MBA. A prospective study on the use of teledermatology in psychiatric patients with chronic skin diseases. Australas J Dermatol. 2015;56(3):170–4. https://doi.org/10.1111/ajd.12280.

Smith AC, Thomas E, Snoswell CL, Haydon H, Mehrotra A, Clemensen J, et al. Telehealth for global emergencies: implications for coronavirus disease 2019 (COVID-19). J Telemed Telecare. 2020;26(5):309–13. https://doi.org/10.1177/1357633X20916567.

Song WH, Yoo S, Oh S, et al. Ten-year interval changes in the prevalence of self-identified premature ejaculation and premature ejaculation based on an estimated intravaginal ejaculation latency time of <3 minutes in the general population: The Korean Internet Sexuality Survey (KISS) 2016. J Sex Med. 2019;16(4):512–21. https://doi.org/10.1016/j.jsxm.2019.02.003.

Takada A, Kasahara T, Kinosada Y, Hosoba M, Nishimura T. Economic impact of real-time teleradiology in thoracic CT examinations. Eur Radiol. 2003;13(7):1566–70. https://doi.org/10.1007/s00330-002-1684-8.

Tan HM, Marumo K, Yang DY, Hwang TI, Ong ML. Prevalence and correlates of erectile dysfunction (ED) and treatment seeking for ED in Asian men: the Asian men's attitudes to life events and sexuality (MALES) study. J Sex Med. 2007;4(6):1582–92. https://doi.org/10.1111/j.1743-6109.2007.00602.x.

Thanh Nguyen HT, Nguyen LT, Thanh Hoang HT, Bui DH, Thu Phan HT, Van Khuu N, et al. Increase in human immunodeficiency virus and syphilis prevalence and incidence among men who have sex with men, Vietnam 2015–2020. Int J STD AIDS. 2024;35(3):197–205. https://doi.org/10.1177/09564624231214583.

Verma P, Kerrison R. Patients' and physicians' experiences with remote consultations in primary care during the COVID-19 pandemic: a multi-method rapid review of the literature. BJGP Open. 2022;6(2):BJGPO.2021.0192. https://doi.org/10.3399/BJGPO.2021.0192.

Viriyathorn S, Sachdev S, Suwanwela W, Wangbanjongkun W, Patcharanarumol W, Tangcharoensathien V. Procedural fairness in benefit package design: inclusion of pre-exposure prophylaxis of HIV in Universal Coverage Scheme in Thailand. Health Policy Plan. 2023;38(Suppl 1):i36-48. https://doi.org/10.1093/heapol/czad061.

Vosburg RW, Robinson KA. Telemedicine in primary care during the COVID-19 pandemic: provider and patient satisfaction examined. Telemed J E Health. 2022;28(2):167–75. https://doi.org/10.1089/tmj.2021.0174.

Wong LP. An exploration of knowledge, attitudes and behaviours of young multiethnic Muslim-majority society in Malaysia in relation to reproductive and premarital sexual practices. BMC Public Health. 2012;12:865. https://doi.org/10.1186/1471-2458-12-865.

World Health Organization. Defining sexual health: Report of a technical consultation on sexual health, 28–31 January 2002. Geneva: World Health Organization;2006. Available from: https://www3.paho.org/hq/dmdocuments/2009/defining_sexual_health.pdf. Accessed 3 Oct 2024.

World Health Organization. Improving the quality of and access to HIV, syphilis and hepatitis B and C testing: laboratory gap analysis in selected countries of the Western Pacific Region. Manila: WHO Regional Office for the Western Pacific;2015. Available from: https://www.who.int/publications/i/item/9789290617365. Accessed 10 Oct 2024.

Yamani LN, Astutik E, Qurniyawati E, Lusida MI, Getaneh Y, Kelly M. Associations between sociodemographics, sexual knowledge and behaviour and sexually transmitted infections among reproductive-age women in Southeast Asia: demographic health survey results. BMC Public Health. 2025;25(1):738. https://doi.org/10.1186/s12889-025-21962-7.

Yang S, Yan E. Demographic and psychosocial correlates of sexual activity in older Chinese people. J Clin Nurs. 2016;25(5–6):672–81. https://doi.org/10.1111/jocn.12998.

Zablotska I, Grulich AE, Phanuphak N, Anand T, Janyam S, Poonkasetwattana M, et al. PrEP implementation in the Asia-Pacific region: opportunities, implementation and barriers. J Int AIDS Soc. 2016;19(7 Suppl 6):21119. https://doi.org/10.7448/IAS.19.7.21119.

Zhou Y, Chen Y, Lu J, Zhang Z, Sun Q, Liu X, et al. Sustained high HIV incidence among men who have sex with men in Jiangsu province, China: based on the limiting-antigen avidity EIA method and six consecutive surveys, 2016–2021. Front Public Health. 2023;11:1277570. https://doi.org/10.3389/fpubh.2023.1277570.

Hoe Leong Sii graduated with Bachelor of Medicine and Bachelor of Surgery (MBBS) from the University of Malaya in 2018 with the Best Overall Achievement Award (Dean's List). He completed Master of Preventive Cardiovascular Medicine (MSc) through distance learning in 2021 from the University of South Wales under a full sponsorship from Merck Foundation. Aspiring to be a Family Physician, he is currently pursuing Graduate Diploma in Family Medicine (GDFM) in Singapore under sponsorship. His interests include chronic disease management, public health, and primary care dermatology. To date, he has published numerous research papers on clinical medicine and public health.

Eng Keat Goh is a medical doctor who graduated with an MBBS from the University of Malaya in 2018, receiving the Best Overall Achievement Award as one of the top MBBS graduates. He is currently working as a Resident Physician at Singhealth Polyclinic in Singapore. He has held various medical officer positions in both Malaysia and Singapore. His professional interest lies in family medicine, where he aims to provide holistic and patient-centered care.

Rong Xiang Ng is an Infectious Diseases physician and medical lecturer at Universiti Malaya. His work focuses on optimising clinical pathways, antimicrobial stewardship, and exploring the role of digital tools in infectious disease care. He has contributed to hospital treatment guidelines, research on AI chatbots for HIV-related opportunistic infections, multidrug-resistant organisms, COVID-19, and the potential of telemedicine in sexual health services, such as MY-HERO project, a gamified mobile health initiative supporting HIV prevention among key populations in Malaysia. Dr. Rong Xiang is a Fellow of the Asia Pacific AIDS and Co-infections Conference (APACC), an affiliate of the Centre of Excellence for Research in Infectious Diseases and AIDS (CERIA, Universiti Malaya), and serves as an affiliate lecturer in University of Glasgow.

Chapter 15
Artificial Intelligence in Healthcare: Standard of Care Under the Law of Medical Negligence in Malaysia and Singapore

Cluster of Future Studies: Past and Futures

Chee Ying Kuek

Abstract Artificial intelligence (AI) has been increasingly making inroads to our daily lives and its application in the healthcare industry is particularly promising. It has brought substantial improvements to the clinical practice in terms of time, costs, efficiency, decision making, and patient outcomes. However, AI, its user, AI developer and manufacturer are not immune from errors. The use of AI in healthcare may result in harm or injuries to the patients. This conceptual paper aims to examine on how the standard of care may adapt to the use of AI in healthcare, the potential liabilities of medical practitioners and the possible options of remedies for patients. A doctrinal research method was adopted in this study, in which court decisions in Malaysia and Singapore as well as the Civil Law Act 1909 (Singapore) were examined. The relevant legal elements were then synthesised to establish the adaptation of the standard of care and its impact on the liability of medical practitioners. Currently, Malaysia and Singapore have issued guidelines or framework to ensure responsible use of AI. Any possible enactment of a new law regulating AI in the future should seek to ensure patients' safety without impeding the development of technology.

Keywords Artificial Intelligence (AI) · Healthcare · Malaysia · Law of medical negligence · Singapore · Standard of care

C. Y. Kuek (✉)
Faculty of Law, Universiti Malaya, Kuala Lumpur, Malaysia
e-mail: cykuek@um.edu.my

R. Xiang Ng et al. (eds.), *Future of Healthcare in Asia*,
https://doi.org/10.1007/978-3-032-03183-9_15

15.1 Introduction

Artificial intelligence (AI) has been increasingly making inroads to our daily lives and its application in the healthcare industry is particularly promising. The common applications of AI in healthcare such as automating routine medical tasks, managing medical data, conducting AI-assisted medical diagnosis, performing AI-assisted robotic surgery, making medical risk prediction and treatment recommendations have brought substantial improvements to the clinical practice in terms of time, costs, efficiency, decision making, and patient outcomes. For instance, AI assists in diagnosis by expediting image interpretation and automated illness identification (Anderson et al. 2020; Mehta 2024). Early detection of any disease allows the medical practitioners to take the necessary measures to manage the disease, leading to better patient outcomes. Based on the clinical data and algorithms programmed, AI is able to predict the possible outcomes of treatment options, recommend treatment and dosages which the doctor may validate. Hence, it plays a supportive role in clinical decision making. In addition, the surgery robot enables a more precise, delicate and minimally invasive surgical operation compared to what is performed by a human hand (Griffin 2021).

In Malaysia, Watson for Oncology cognitive computing system has been adopted by private healthcare provider to help oncologists to develop treatment plans or deliver evidence-based treatments for cancer patients (Zainul 2017). Robotic technologies such as da Vinci Xi Surgical System, SmartRobotics and ROSA Robotic System are used by the private hospitals in Malaysia to perform surgeries with higher precision, smaller incisions, better visualisation, enhanced manoeuvrability, and less blood loss that lead to faster recovery, shorter hospital stay and minimal scarring (The Star 2022). It was reported that as of 2022, there were less than 40 robotic surgery systems in Southeast Asia. However, Singapore and Thailand accounted for more than half of these systems (Hospital Management Asia 2022). In fact, a private hospital in Singapore has used da Vinci Xi system to conduct complex multi-quadrant operations across different areas of the human body in the same surgery, eliminating the need for multiple procedures (The Straits Times 2022).

However, AI, its user especially the medical practitioner and AI developer and manufacturer are not immune from errors. The application of AI in healthcare may result in harm or injuries to the patients. It poses a peculiar challenge to the issue of liability in the law of medical negligence due to the complex nature of AI and the possible human misconduct. Nevertheless, most of the medical AI serves to aid and support the medical practitioner in diagnosis, treatment or clinical decision making. It is up to the medical practitioner to use the AI or adopt the treatment recommendation. Hence, the medical practitioner may be exposed to the risk of medical negligence suit filed by the patient, even if the injury was caused by the defective AI or wrong treatment recommendation made by AI which was followed by the medical practitioner.

This conceptual paper discusses how the standard of care under the law of medical negligence may adapt to the use of AI, its impact on the potential liabilities of the medical practitioners in Malaysia and Singapore, and discusses the current and other possible options of compensating patients who may be injured or harmed by the use of AI.

This chapter is structured into several parts. Section 15.1 provides an introduction to the study, followed by the methodology adopted as explained in Sect. 15.2. It then examines the concept of the standard of care under the law of medical negligence in Malaysia and Singapore, as can be seen in Sect. 15.3. This leads to the analysis on how the standard of care may adapt to the use of AI and its impact on the potential liabilities of the medical practitioners of these two jurisdictions in Sect. 15.4. Subsequently, Sect. 15.5 discusses the future directions on the liability system, the possible options of compensating patients in a medical negligence case involving AI in Malaysia or Singapore, and the current approach taken by Malaysia and Singapore in promoting responsible use of AI through guidelines or framework, instead of resorting to command and control regulation. Finally, Sect. 15.6 of this chapter concludes that any possible enactment of new law on AI in the future should seek to balance between patients' safety and the development of technology.

15.2 Methodology

The method adopted by this study is doctrinal research, which seeks to examine critically the essential elements of legislation and case law of a particular legal category, synthesise them to establish a statement of law on a matter in question. Doctrinal research also analyses the relationship between the rules, explains areas of difficulty and predicts future developments (Hutchinson 2018). Therefore, this study examined the standard of care expected of the medical practitioners as ruled in the court judgments in Malaysia and Singapore (which adopted the tests in the English cases and an Australian case), and as prescribed in the Civil Law Act 1909 (Singapore). The relevant legal elements were then synthesised to establish how the existing standard of care of medical practitioners may adapt to the use of AI and its impact on the potential liabilities of the medical practitioners. Analysis was also done on the possible options of compensating patients who may be injured or harmed by the use of AI.

Malaysia and Singapore are selected for this study as both are Asian countries with common law system and both have embraced AI in healthcare. Searches for primary sources of law namely the cases and statute relating to medical practitioners' standard of care under the law of medical negligence, as well as secondary sources of law on the same subject matter (including potential liabilities arising from the use of AI in healthcare) such as journal articles, books, newspaper articles and websites were conducted. The search spanned databases and websites namely CLJPRIME, Lexis Advance, HeinOnline, Springer e-Books, SpringerLink, ScienceDirect, Singapore Statute Online and online newspapers (The Star, The Straits Times and The Edge). Key terms for searching included 'standard of care', 'medical negligence', 'professional negligence', 'artificial intelligence', 'medical AI' and 'diagnosis, treatment and advice'. However, this study has limitations that should be acknowledged. First, there is no reported court decision on medical negligence involving AI in Malaysia and Singapore, probably because the technology is relatively new. Therefore, there is no direct case law discussing the standard of care of the medical practitioners using AI in these two jurisdictions. The searches conducted in the study were confined to sources in the English language, and thus there is a possibility of omitting good available literature in other languages.

15.3 Key Concepts and Issues

15.3.1 Standard of Care Under the Law of Medical Negligence

To succeed in a tortious claim for medical negligence, a plaintiff must satisfy three elements: (a) a duty of care was owed by the medical practitioner to the plaintiff; (b) the medical practitioner breached the duty of care; and (c) the breach of duty of care caused the plaintiff's injury or loss. The existence of a duty of care between a medical practitioner and his or her patient is normally straightforward and not questioned. The medical practitioner owes a duty to exercise reasonable care and skill in diagnosis, treatment and provision of advice to the patient. The duty of care was breached when the medical practitioner acted below the reasonable standard of care expected of him. A distinction is made between the standard of care in diagnosis and treatment on one hand, and standard of care in advice on the other hand, which will be explained in Sects. 15.3.1.1 and 15.3.1.2. There must be a causal link between the breach of duty and the resulting injury or loss. If the plaintiff succeeds in his or her medical negligence claim, he or she will be entitled to damages or compensation. The conceptual framework of the law of medical negligence is illustrated in Fig. 15.1 (see the next page).

15.3.1.1 Malaysia

In Malaysia, the Federal Court in *Zulhasnimar bt Hasan Basri & Anor v Dr Kuppu Velumani P & Ors* (2017) distinguished between the standard of care for diagnosis and treatment on one hand, and the standard of care for provision of advice on the other hand. In regard to diagnosis and treatment, the standard of care expected of a medical practitioner is the *Bolam-Bolitho* test. A medical practitioner is not negligent if he has 'acted in accordance with a practice accepted as proper by a responsible body of medical men' (*Bolam v Friern Hospital Management Committee* 1957). Such practice must be capable of withstanding logical basis (*Bolitho v City and Hackney HA* 1997). In other words, a medical practitioner will not be negligent if his treatment or diagnosis is supported by a body of medical opinion as sound medical practice and

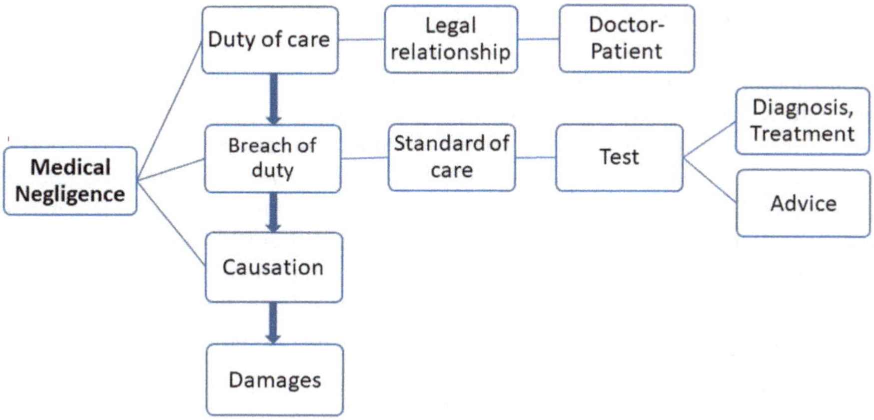

Fig. 15.1 Conceptual framework of the law of medical negligence

defensible. However, the court can reject the expert evidence if it is unreasonable, irresponsible or illogical. The medical practitioner will also be judged according to the current state of knowledge at the time of the incident (*Roe v Minister of Health* 1954; *Rathanamalah d/o Shunmugam v Chia Kok Hoong* 2018).

As for provision of advice, the test in the Australian case of *Rogers v Whitaker* (1992) will be applicable. It is a more patient-centric test in which a medical practitioner owes a duty to warn or disclose to a patient about a material risk in a recommended treatment. The risk is material if it is significant to a reasonable person in the patient's position, or the medical practitioner is or should reasonably be aware that the patient would attach significance to it. These two tests for different aspects of medical negligence were affirmed by the Federal Court in *Dr. Hari Krishnan v Megat Noor Ishak bin Megat Ibrahim* (2018). The subsequent Court of Appeal case in *Ahmad Zubir bin Zahid (suing by himself and as the administrator of the estate of Fatimah bt Samat, deceased) v Datuk Dr. Zainal Abidin Abdul Hamid* (2019) clarified that the broad test of duty to advise in *Montgomery v Lanarkshire Health Board* (2015), which is similar to the *Rogers v Whitaker* test but with an additional element of disclosure of material risk in any reasonable alternative treatments, is not applicable to Malaysia. Although there were a few cases such as *Gurisha* (2020), *Dayalan Sathiamutty* (2023) and *Gan Hong Heng* (2023) which attempted to depart from this legal position and held that *Montgomery* test applied in Malaysia, those decisions were decided by the High Court, which is lower than the Court of Appeal and the Federal Court in the Malaysian courts' hierarchy. Therefore, it is submitted that the duty to advise remains the *Rogers v Whitaker* test and not the *Montgomery* case which placed a higher threshold for the medical practitioners.

15.3.1.2 Singapore

In Singapore, the law of medical negligence is governed by a mixture of common law and statutory provision. Similar as Malaysia, Singapore adopts the *Bolam-Bolitho* test for the standard of care in diagnosis and treatment, as decided by the Court of Appeal case of *Khoo James and Another v Gunapathy d/o Muniandy and Another Appeal* (2002). As for the standard of care for provision of medical advice, Singapore has changed its law from the *Bolam-Bolitho* test applied in *Khoo James and Another v Gunapathy d/o Muniandy and Another Appeal* to the modified *Montgomery* test or three-stage common law test in *Hii Chii Kok v Ooi Peng Jin London Lucien and Another* (2017), which was subsequently replaced with the codified test in Section 37 of the Civil Law Act 1909.

The new test under Section 37(1) of the Civil Law Act 1909 requires a healthcare professional to satisfy two conditions so as to meet the statutory standard of care in providing medical advice. Firstly, the healthcare professional's act is accepted by the medical opinion of a respectable body or peer professional opinion as reasonable professional practice. Secondly, the peer professional opinion must be logical, and the logic test is provided in Section 37(5). These two conditions reflect the *Bolam-Bolitho* test. Pursuant to Section 37(2) of the same Act, the healthcare professional must give the information that a person in the patient's position would reasonably require to make an informed consent on a treatment or follow a medical advice, and information which is material for that particular patient to make such informed consent. The materiality of the information must relate to the patient's specific concern or query which is expressly communicated by him, or what ought to be apparent from his medical records that the health professional has reasonable access to and ought reasonably to review, although not expressly communicated by the patient. In short, the statutory standard of care in respect of provision of medical advice is based on the peer professional opinion but it also upholds the patient autonomy (Ministry of Health, Singapore 2020).

15.4 Discussion of Implications

15.4.1 Potential Liabilities of the Medical Practitioners

The increasing use of AI in healthcare has raised the issue on how the standard of care required of the medical practitioner under the law of medical negligence may adapt, and its impact on the potential liabilities of the medical practitioners.

Pursuant to the *Bolam-Bolitho* test, a medical practitioner in Malaysia and Singapore will not be liable for negligence in diagnosis or treatment if he meets the minimal threshold of a practice accepted by a responsible body of medical opinion, and the opinion can be logically defended. In *De Freitas v O'Brien* (1995), even a small number of 1.1% (11 out of 1000) is sufficient to form the responsible body. If a medical practitioner departs from conventional treatment and uses a particular innovative or ground-breaking AI technology for diagnosis or treatment which is yet to be adopted as practice by the medical community, he may risk violating the standard of care if the use of that AI technology results in the patient's injury (Mehta 2024; Chan 2021). However, deviation from conventional treatment and the use of novel AI technology will not necessarily make a medical practitioner negligent if the application of the AI brings therapeutic effect or leads to good patient outcome. In contrast, if a specific AI technology has been recognised in medical practice or its use has become the mainstream, failure to embrace such technology may render a medical practitioner acting below the required standard of care if it is unreasonable not to adopt it. Assuming a medical practitioner relies on the treatment recommendation made by the AI, he may be negligent if he does not evaluate such recommendation properly and critically. He may incur liability if he unreasonably followed the AI treatment recommendation which was wrong and caused the patient to experience a poor outcome. He may also be liable if he unreasonably disagreed and rejected AI treatment recommendation which was correct, resulting in a bad patient outcome.

Litigants who bring medical negligence claims bear the onus of proving their cases through expert evidence, while the medical practitioners being sued will need to establish their defences. However, the operation of the AI techniques and the construction of the algorithm developed by medical AI are difficult to be understood by the medical practitioners and patients. The complexity, opacity and lack of interpretability of AI that uses machine learning or even deep learning will pose an evidentiary challenge to the patients who file medical negligence suits. For instance, the machine-learning algorithm is plastic (able to self-learn) and opaque in nature. It constantly weighs new inputs in determining the output and tweaks its rules in its process of learning (Molnar-Gabor 2020). Therefore, it would be very difficult or almost impossible to envisage how the AI reached a certain outcome. It follows that a medical practitioner may be unable to take the necessary steps to mitigate the risks of using the system, or sometimes finds it challenging in validating the output of the AI. Furthermore, AI is highly dependent on data. The quality of the training data will affect the accuracy of the AI's output. Hence, it would be difficult for the patient to demonstrate his injury was caused by the medical practitioner's negligence in using the AI, or the medical practitioner's acceptance or rejection of the treatment recommendation by the AI was unreasonable, as it involves highly technical aspects of the AI system and complex fact-finding exercise. The complication of proving causation has been compounded by the intertwinement of human medical practitioners and AI (Griffin 2021) since both have a role to play in the diagnosis or treatment.

For a medical practitioner who uses AI as a diagnostic or treatment tool, question arises as to whether there is a need to advise the patient on the potential system risk of the AI before obtaining the patient's consent. It has to be borne in mind that the mere signing of the consent form by a patient may not be sufficient if the patient did not understand the nature of the treatment (*Gurmit Kaur a/p Jawant Singh v Tung Shin Hospital & Anor* 2012) and the risk involved. The consent must be an informed consent. Applying the test in *Rogers v Whitaker* (in the Malaysian context) or the test under Sect. 37 of the Civil Law Act 1909 (in the context of Singapore), the medical practitioner may need to disclose the potential system risk to the patient if it is a material risk, or the information is material for the patient to give an informed consent, or it relates to the patient's specific concern about the proposed treatment or medical advice. A medical practitioner's conduct will only be assessed based on the state of medical knowledge and literature at the time of the alleged negligence. Therefore, he may not be negligent if a particular risk of using an AI technology or the wrong prediction of the AI would not have been known at the time of the incident, or at the time the AI prediction was relied upon. Therefore, it is submitted that a medical practitioner will not be negligent if he fails to disclose or advise on a risk which is not known at the time of the diagnosis or treatment using the AI technology.

15.5 Future Directions

15.5.1 Liability System and Possible Options of Compensating Patients

Under the current fault-based tort system, a plaintiff who succeeds in a medical negligence suit is entitled to recover monetary compensation for the damage or injury caused to him. The damages awarded by the court is compensatory in nature and seek to restore the plaintiff to the original state had the negligence not occurred. The types of damages that may be awarded by the court are special damages, general damages and even aggravated damages. Special damages represent the plaintiff's actual pecuniary loss between the date of the negligent act and the date of the Court's judgment. They consist of the out-of-pocket expenses that are specifically pleaded and proven. General damages represent the loss to the plaintiff that cannot be quantified with precision. They comprise pain and suffering, and loss of amenities. The court will decide on the amount of damages based on precedent or estimation. Aggravated damages may be awarded as a separate head of damages for injury to feelings, mental distress, anger and indignation in a medical negligence case (*Dr. Hari Krishnan v Megat Noor Ishak bin Megat Ibrahim* 2018).

Tort as the conventional liability system may not be well equipped to address the unique challenges arising from the medical negligence involving AI due to the complex, opaque and plastic nature of AI. Furthermore, AI itself may have malfunctioned. Scholars have suggested a departure from the conventional tort liability system to distribute liability among the stakeholders such as the medical practitioner, the healthcare provider and the AI developer or software development company or manufacturer. Under this model, the risk of using medical AI will be shared among stakeholders through indemnification contracts or insurance (Maliha et al. 2021; Mehta 2024). How the stakeholders divide their liability will depend on their negotiation. Some scholars describe this model as 'risk pooling' through insurance (Smith and Fotheringham 2020), while some proposed similar model called 'enterprise liability'. The enterprise liability model allows spreading of fault and losses among different actors within the enterprise namely the medical practitioner, the hospital and the software development company. They will be covered by an insurance jointly paid by them or separate policies. The liability is divided pro rata according to the control and financial benefit (Smith 2020; Mehta 2024). Whether this risk pooling through insurance or enterprise liability model will be adopted as an alternative to the fault-based tort system remains to be seen.

Alternatively, there were suggestions to introduce the no-fault system which the patient would be compensated for the injuries caused without the need of proving the fault by the medical practitioner or the healthcare provider. The no-fault system will operate by a comprehensive state insurance scheme. Medical practitioners and healthcare providers will be shielded from liability, in which the government will take over the full or part of the risk (Maliha et al. 2021; Mehta 2024). However, the main obstacle in implementing no-fault system is the funding problem, which may lead to undesirable move of increase of taxes or imposition of fees on the stakeholders. Therefore, the feasibility of no-fault system is questionable due to practical reason.

15.5.2 Responses from Malaysia and Singapore Towards Artificial Intelligence (AI)

The application of AI in healthcare in Malaysia and Singapore is still relatively new and growing. Although both jurisdictions do not have any specific law on AI, both have issued guidelines or framework to ensure responsible use of AI. Indeed, the application of AI in healthcare has raised ethical concerns. Among the main concerns are the safety to the patients, transparency, data privacy, algorithmic biases and the informed consent from the patients. In Malaysia, the National Guidelines on AI Governance and Ethics (AIGE) was launched on 20 September 2024. One of its objectives is 'to manage risks caused by the development and deployment of AI technology' (Ministry of Science, Technology and Innovation 2024). The AIGE guides the stakeholders including AI developers, designers, technology providers and suppliers to adopt responsible AI practices by adhering to seven AI principles. These principles include prioritising the reliability, safety and control of the AI systems through rigorous testing, proactive identification of potential risks and assessment, so that harmful outcome can be avoided. The AIGE is of general application. Specific guideline such as the MMC Guideline on the Ethical Use of Artificial Intelligence (AI) in Medical Practice was issued by the Malaysian Medical Council in 2025. Under this MMC Guideline, registered medical practitioners should only use AI tools and/or technologies which are reliable and meet the standards of 'reasonable confidence'. The criteria for 'reasonable confidence' include trustworthiness, appropriateness, transparency and conformity to standards (Malaysian Medical Council 2025).

Singapore, on the other hand, has been focusing on promoting responsible use of AI through the Model Artificial Intelligence Governance Framework ('Model

Framework') (The Legal 500). This Model Framework, which was launched in January 2019, translates ethical principles into practical recommendations for organisations to apply AI responsibly. The ethical principles include accuracy, auditability, explainability, fairness, human centricity and well-being, human rights alignment, inclusivity, progressiveness, responsibility, accountability, transparency, robustness, security and sustainability (Info-communications Media Development Authority 2020). Apart from the Model Framework, the Artificial Intelligence in Healthcare Guidelines (AIHGle) jointly developed by the Ministry of Health, Singapore, Health Sciences Authority (HSA) and the Integrated Health Information Systems (IHiS), and published in October 2021 also serves as a guide to encourage the safe development and implementation of AI-medical devices and other AI used in healthcare settings in Singapore (Ministry of Health et al. 2021). The AIHGle is applicable to the developers and implementers comprising organisations or individuals who use AI-medical device in delivering healthcare services (with focus on organisational implementers). Hence, both jurisdictions seek to promote responsible use and safe implementation of AI through guidelines or framework, rather than resorting to command and control regulation which imposes standard backed by criminal sanctions (Baldwin et al. 2012), as a response to the use of this emerging technology particularly in healthcare. Being living documents, guidelines can be updated periodically to correspond with the rapid development of medical AI. This will be more flexible and less cumbersome compared to the command and control regulation where revision of regulation is more complicated and less frequent.

15.6 Conclusion

AI is transforming the medical industry in the delivery of healthcare. Significant benefits have been brought by the adoption of this technology. However, as AI becomes more widely embraced and may potentially outperform medical practitioners, it generates new liabilities issues. At present, Malaysia and Singapore do not have legislation that specifically addresses the application of AI, including the medical AI. The law on medical negligence in these two jurisdictions continues to form the foundation for liability framework. Deviation from standard of care may occur if a medical practitioner used AI when it offered unconventional treatment that resulted in the patient's injury, or he refused to use AI that provided mainstream treatment and his refusal led to poor patient outcome. However, the standard of care for medical practitioners in Malaysia and Singapore is expected to evolve with the increasing use of medical AI. While the peculiar nature of AI may pose an evidential challenge to a plaintiff in the fault-based tort system, suggestion to change to no-fault system seems unlikely for practical reason. The model of risk sharing among stakeholders through indemnification contracts or insurance may be a possible remedy but whether it is a viable option in the future is not known. Currently, in the absence of any specific law governing AI, Malaysia and Singapore introduce guidelines or framework to ensure responsible use of AI. However, any possible enactment of new law regulating AI including medical AI in the future, should seek to ensure patient's safety without impeding the development of technology.

Highlights

- The standard of care of medical practitioners in Malaysia and Singapore adapts and evolves with the use of AI in healthcare
- Tort may not address well the challenges from medical negligence involving AI
- Distributing liability among stakeholders through indemnification contracts or insurance may be a possible remedy

References

1000 Robot-assisted surgeries and counting. The Star [Internet]. 2022 Dec 22 [cited 2024 Nov 17]. Available from: https://www.thestar.com.my/metro/metro-news/2022/12/22/1000-robot-assisted-surgeries-and-counting.

Ahmad Zubir bin Zahid (suing by himself and as the administrator of the estate of Fatimah bt Samat, deceased) v Datuk Dr Zainal Abidin Abdul Hamid & Ors (2019) 5 MLJ 95 (CA).

Anderson T, Torreggiani WC, Munk PL, Mallinson PI. The impact of the introduction of artificial intelligence in radiology and its potential legal implications in the UK and Ireland. BJR Open. 2020;2(1):20200030. https://doi.org/10.1259/bjro.20200030.

Baldwin R, Cave M, Lodge M. Understanding regulation, strategy and practice. 2nd ed. Oxford: Oxford University Press; 2012.

Bolam v Friern Hospital Management Committee [1957] 1 WLR 582.

Bolitho v City and Hackney HA [1998] A.C. 232.

Chan GKY. Medical AI, standard of care in negligence and tort law. In: Chan GKY, Yip M, editors. AI, data and private law: translating theory into practice. Oxford: Hart Publishing; 2021. p. 173–98.

Dayalan Sathiamutty v. Gleneagles Hospital (Kuala Lumpur) Sdn Bhd & Ors [2023] 10 CLJ 253 (HC).

De Freitas v O'Brien [1995] BMLR 51.

Dr. Hari Krishnan v Megat Noor Ishak bin Megat Ibrahim [2018] 3 MLJ 281 (FC).

Gan Hong Heng (suing as the administrator of the estate of Gan Hong Wee, Deceased) v Dr Chandran a/l Gnanappah & Assunta Hospital [2023] 1 LNS 1487.

Griffin F. Artificial intelligence and liability in health care. Health Matrix. 2021;31:65–106.

Gurisha Taranjeet Kaur (an infant suing by her father and litigation representative, Taranjeet Singh s/o Bhagwan Singh) & Anor v Dr Premitha Damodaran & Anor [2020] 9 MLJ 409.

Gurmit Kaur a/p Jaswant Singh v Tung Shin Hospital & Anor [2012] 4 MLJ 260.

Hii Chii Kok v Ooi Peng Jin London Lucien and Another [2017] SGCA 38.

Hospital Management Asia. Why Malaysia's Sunway Medical Centre invested in the latest robotic surgical system [Internet]. Hospital Management Asia; 2022 Mar 10 [cited 2024 Nov 17]. Available from: https://www.hospitalmanagementasia.com/tech-innovation/why-malaysias-sunway-medical-centre-invested-in-the-latest-robotic-surgical-system/.

How robots are changing surgery. The Straits Times [Internet]. 2022 June 27. Available from: https://www.straitstimes.com/multimedia/graphics/2022/06/singapore-surgery-robots-medical-operations-hospital/index.html?shell.

Hutchinson T. Doctrinal research: researching the jury. In: Watkins D, Burton M, editors. Research methods in law. 2nd ed. Oxon: Routledge; 2018. p. 8–39.

Info-communications Media Development Authority. Model artificial intelligence governance framework, 2nd ed. [Internet]. Info-communications Media Development Authority; 2020 January [cited 2024 Nov 17]. Available from: https://www.pdpc.gov.sg/-/media/files/pdpc/pdf-files/resource-for-organisation/ai/sgmodelaigovframework2.pdf.

Khoo James and Another v Gunapathy d/o Muniandy and Another Appeal [2002] 2 SLR(R) 1024.

Malaysian Medical Council. MMC guideline on the ethical use of artificial intelligence (AI) in medical practice [Internet]. Malaysian Medical Council; 2025 [cited 2025 October 13]. Available from:https://mmc.gov.my/wp-content/uploads/2025/09/MMC-Guideline-on-The-Ethical-Use-of-Artificial-Intelligence-AI-.pdf.

Maliha G, Gerke S, Cohen IG, Parikh RB. Artificial intelligence and liability in medicine: balancing safety and innovation. Milbank Q. 2021;99(3):629–47.

Mehta D. The role of artificial intelligence in healthcare and medical negligence. Liverpool Law Rev. 2024;45(1):125–42. https://doi.org/10.1007/s10991-023-09340-y.

Ministry of Health, Singapore. Opening speech for second reading of the Civil Law (Amendment) Bill and Medical Registration (Amendment Bill) by Mr Edwin Tong, second Minister for Law on 6 October 2020 [Internet]. Ministry of Health, Singapore; 2020 Oct 6 [cited 2024 Nov 17]. Available from: https://www.moh.gov.sg/news-highlights/details/opening-speech-for-second-reading-of-the-civil-law-(amendment)-bill-and-medical-registration-(amendment-bill)-by-mr-edwin-tong-second-minister-for-law-on-6-october-2020.

Ministry of Health, Singapore, Health Sciences Authority, Integrated Health Information Systems. Artificial Intelligence in Healthcare Guidelines [Internet]. Ministry of Health; 2021 [cited 2024 Nov 23]. Available from: https://isomer-user-content.by.gov.sg/3/9c0db09d-104c-48af-87c9-17e01695c67c/1-0-artificial-in-healthcare-guidelines-(aihgle)_publishedoct21.pdf.

Ministry of Science, Technology and Innovation. The National Guidelines on AI Governance and Ethics [Internet]. Ministry of Science, Technology and Innovation; 2024 [cited 2024 Nov 17]. Available from: https://www.mosti.gov.my/dasar/#flipbook-df_74045/1/.

Molnar-Gabor F. Artificial intelligence in healthcare: doctors, patients and liabilities. In: Wischmeyer T, Rademacher T, editors. Regulating artificial intelligence. Switzerland: Springer; 2020. p. 337–60.

Montgomery v Lanarkshire Health Board [2015] UKSC 11.

Rathanamalah d/o Shunmugam v Chia Kok Hoong [2018] 4 SLR 159.

Roe v Minister of Health [1954] 2 QB 66.

Rogers v Whitaker (1992) 109 ALR 625.

Smith H, Fotheringham K. Artificial intelligence in clinical decision-making: rethinking liability. Med Law Int. 2020;20(2):131–54. https://doi.org/10.1177/0968533220945766.

The Legal 500. Singapore: artificial intelligence. The Legal 500 [Internet]. n.d. [cited 2024 Nov 17]. Available from: https://www.legal500.com/guides/chapter/singapore-artificial-intelligence/.

Zainul E. KPJ adopts Watson for oncology to help deliver evidence-based cancer treatment options. The Edge Malaysia [Internet]. 2017 Nov 27 [cited 2024 Nov 17]. Available from: https://theedgemalaysia.com/article/kpj-adopts-watson-oncology-help-deliver-evidenceb ased-cancer-treatment-options.

Zulhasnimar bt Hasan Basri & Anor v Dr Kuppu Velumani P & Ors [2017] 5 MLJ 438 (Federal Court).

Chee Ying Kuek is a Senior Lecturer at the Faculty of Law, Universiti Malaya in Malaysia. She holds a Doctor of Philosophy from Universiti Malaya. Her research interests are in Medical Law and Ethics, Family Law, Probate and Administration. Before joining the academia, she practised as an advocate and solicitor in Malaysia.

Chapter 16
Conclusion: Futures to a Healthier Asia—Challenges and Opportunities

Rong Xiang Ng

Abstract As we reflect at the diverse conversations in 'Future of Asia (Healthcare),' this final chapter put together the threads of tradition and innovation that are reshaping healthcare across the continent. It looked at how social factors, new care models, and technology are all playing a role, and a frank discussion of both the hurdles we face and the chances we must take to make things better, without leaving marginalised society and always involve community, policymaker, researchers and everyone. Hence, this chapters' conversation is not just an analysis; it is a call, really, to all of us, including practitioners, policymakers, and everyone to embrace innovation in a smarter or more creative way, to fight for equity, and to work together, leaving no one behind. The future of healthcare in Asia is not pre-determined and fixed. It is shaped by a series of choices or actions that we are making right now. And it is my hope that this book, and this chapter in particular, can spark and inspire that ongoing conversation, that collaborative motivation, towards a future where everyone in Asia has access to the health and well-being they deserve.

Keywords Asia · Healthcare · Futures · Diversity · Equity

16.1 The Crossroad of Healthcare in Asia

As we are here at the final chapter, we find ourselves at the crossroad of tradition and innovation, facing both challenges and emerging opportunities. The healthcare landscape and trajectory are undergoing profound transformation (Mantaleon 2023; Yeganeh 2019), whereby we have ancient healing practices coexist with cutting-edge genomic medicine (Efferth et al., 2019; Sembiyaa et al. 2025; Singh et al. 2024; Hudtohan and Zhang 2025), community-led initiatives reshape public health approaches which often overlooked (Ayala et al. 2021; Bhatia et al. 2019; Henfrey

R. X. Ng (✉)
Faculty of Medicine, Department of Medicine, Universiti Malaya, Kuala Lumpur, Malaysia
e-mail: rxng@um.edu.my

University of Glasgow, Glasgow, UK

R. Xiang Ng et al. (eds.), *Future of Healthcare in Asia*,
https://doi.org/10.1007/978-3-032-03183-9_16

et al. 2023; Thomas et al. 2024; Vannakit et al. 2020), digital technologies and artificial intelligence (AI) bridging all geographic areas and social divides to deliver care in unprecedented ways and yet (Adhikari et al. 2025; Cornelius 2022; Wong 2025; Khan 2025; Reddy et al. 2022; Shirah and Zakour 2025; Woser 2023), aiming to leaves no one behind (Anastasi et al. 2020; Donoghue and Khan 2019). From rural communities in South Asia to urban centers in the East, from resource-constrained settings to technology powerhouses, the healthcare journey unfolds differently but with shared aspirations toward better health for all in this continent.

Let's take a moment to look back at the key findings from our journey through these three subsections and see what they tell us about the future of Asian healthcare's pathway. The future is not a single path, but rather, we aim to provide multiple possible perspectives and pathways at critical decision points that healthcare stakeholders across Asia will navigate in the coming years.

16.2 Synthesis of Key Themes

16.2.1 Building Healthy Societies

It became clear from our early chapters that healthcare is not an isolated thing; it's deeply ingrained into the society and culture, a point that is consistently echoed in the references that follow (Dutta and Basu, 2008; Lee et al. 2024; Mansyur et al. 2009; May 2013). Health inequities emerged repeatedly across chapters, reminding us that expanded coverage alone cannot remedy disparities rooted in economic, gender, geographic, and other social determinants (Durairaj 2007; Haribondhu et al. 2024; Jayasinghe et al. 2025). The Vietnam case study demonstrated how policy reforms aimed at vulnerable farmers must navigate complex implementation challenges.

Cultural context is equally important in shaping effective healthcare approaches, especially in Asia with rich cultural background (Cipta et al. 2024; Dutta and Basu 2008; Mansyur et al. 2009; Mclaughlin and Braun 1998). The integration of traditional and complementary medicines with contemporary care offers promising possibilities for more holistic care pathways. However, this is not as simple as it is, as several authors noted, significant challenges remain in establishing appropriate regulatory frameworks that ensure safety while honoring diverse healing traditions (Balkrishna et al. 2024; Ijaz and Boon 2018; Park et al. 2022; Rooney et al. 2023; World Health Organization 2017). This tension between standardisation and cultural specificity represents a recurring theme throughout healthcare transformation efforts (Chongsuvivatwong et al. 2011; Ijaz and Boon 2018).

Perhaps the most powerful was the documentation of community-led initiatives, particularly in the HIV response. Chapter 3 highlighted how affected communities, when meaningfully engaged, transform from passive recipients of care to active agents of health system change. This shift in power dynamics from top-down to collaborative approaches offers valuable lessons beyond HIV for addressing stigmatised conditions and reaching marginalised populations as seen in others too (Adebisi et al. 2025; Ayala et al. 2021; Henfrey et al. 2023; Smith, 2014; Yang et al. 2020).

Let us look at how Southeast Asian nations handled pandemic preparedness. The comparative analysis of pandemic preparedness across Southeast Asian nations provided new insights into the intertwining effects between political stability, public health infrastructure, and emergency response capabilities. Pandemic preparedness is about social resilience, not just technical capacity like healthcare infrastructure or resources, medical supplies and surveillance system (Lal et al. 2022; Wernli et al. 2021). But it is far beyond it and necessitates significant social participation and cooperation with public health authorities (Marston et al. 2020). It is a lesson with profound implications as the region prepares for future health threats in an increasingly interconnected world when facing the possibility of future pandemic for an unknown infectious disease, especially when the mode of transmission is unknown, which can be terrifying.

16.2.2 Advancing Health and Care

The second section explored innovations in care delivery; this time, the attention is on mental health, complex conditions in marginalised population, and emerging ethical challenges. The study on mental health services in Indonesia gave us a stark look at the reality: even with better access, stigma and lack of awareness are still huge roadblocks. This mirrors challenges observed across many Asian contexts, where mental health receives growing recognition yet remains underprioritised in both resource allocation and public discourse (Mahomed 2025; Thornicroft et al. 2019).

Emerging scientific understanding of the gut-brain axis offers intriguing new directions for mental health treatment, highlighting how advances in basic science can transform clinical care possibilities. Similarly, the exploration of self-management capacities among people with multimorbidity who use drugs demonstrated the need for integrated care models that address interconnected health challenges rather than treating conditions in isolation.

Novel prevention strategies also featured in this book, whereby exergaming representing just one example of how technology can be repurposed to address physical inactivity in an increasingly urbanized environment. Such innovations reflect creative responses to changing lifestyle patterns and disease burdens, where non-communicable diseases increasingly dominate the health landscape (Habib and Saha 2010).

The prospect of personalised medicine trends revealed both promise and peril (Ackerman et al. 2016; Karaboue and Lacasella 2025; Rose 2013). While genomic approaches offer many possibilities for tailoring treatments to individual characteristics, these advances also raise questions about equity, resource allocation, and we need to strike appropriate balance between use of genomic as personalised medicine and public health measures.

Finally, the discussion on clinical ethics committees underscored the growing importance of structured ethical deliberation in healthcare settings facing rapid technological and social change. As healthcare capabilities expand, so is the complexity of ethical dilemmas faced by practitioners, patients, and systems. The ethics component may be overlooked, and we need to beware of it, especially in the era of healthcare transformation.

16.2.3 Leveraging HealthTech

The final section explores how technology is reshaping healthcare delivery across diverse Asian contexts. Telemedicine applications in prison healthcare and sexual health services demonstrated technology's potential to overcome both physical barriers and social stigma (Abdulai 2022; Edge et al. 2021; Minichiello et al. 2013). Yet these chapters also revealed persistent challenges related to digital infrastructure, privacy concerns, and the need for technology to complement rather than replace human connection in healthcare delivery (Adeghe et al. 2024; Mallipeddi et al. 2025).

The exploration of Hospital-at-Home models showed how technology enables care models that transcend traditional boundaries between institutional and community-based care. Such shifts hold promise for Asia's diverse healthcare contexts, where resource constraints often necessitate creative care delivery solutions.

When we delved into the legal side of AI in healthcare, it gave us a real wake-up call. While everyone is enjoying the hype around technology, it often glosses over some serious ethical and legal questions. It highlighted the complex questions of liability, safety, and governance that accompany technological innovation that we must not take it lightly (Adhikari et al. 2025; Alfiani and Santiago 2024; Chikhaoui et al. 2022; Gerke et al. 2020; McGreevey et al. 2020; Schönberger 2019; Tun et al. 2025). The varying regulatory approaches emerging across Southeast Asia jurisdictions reflect different cultural attitudes toward risk, innovation, and the proper role of technology in society whereby these differences will shape how AI and other advanced technologies are integrated into healthcare systems.

What became clear, again and again, as we looked at these technologies, be it telemedicine, digital health or AI, is that it is not just about the technology itself. We will need the right policies, the systems in place, and people willing to embrace it (Amjad et al. 2023; Adeghe et al. 2024; Goktas and Grzybowski 2025; Mallipeddi et al. 2025). It requires careful integration into existing systems while taking into consideration of various issues.

16.3 Opportunities to Do Better

Strengthening regional collaboration can address transboundary health challenges that no single country can solve alone (Amaya and De Lombaerde 2021; Kumaresan and Huikuri 2015; Sundram 2025). This is seen via enhanced collaboration on disease surveillance networks and data sharing improves pandemic preparedness and response capabilities across borders (Inbanathan et al. 2024; Olaboye et al. 2024; Rahman-Shepherd et al. 2025). The regional infectious disease network established after previous outbreaks demonstrated its value during recent health emergencies (COVID-19), enabling rapid information exchange and coordinated responses.

Besides, regulatory harmonisation across jurisdictions can accelerate innovation while maintaining safety standards (Lim 2018; World Health Organization 2019). Looking ahead, health workforce mobility offers an opportunity for cooperation and collaboration, if agreements are carefully crafted to both address regional distribution needs and safeguard the rights of the workers involved (Carzaniga et al. 2019; Kroezen et al. 2016; Te et al. 2018; World Health Organization 2019). Moreover, collaboration also can extend to the form of research collaborations, involving multiple sites at multiple countries. It can then pool resources to address shared health priorities and generate context-appropriate evidence, through research or investigator working groups or clinical trials. One of the examples is the regional cancer registry involving various cancers. These initiatives showed how collaborative data collection created previously impossible insights into disease patterns and treatment outcomes across diverse Asian populations (Liu et al. 2019; Tan et al. 2025; Wang et al. 2025). As we progress with health challenges increasingly transcend across national boundaries, these cooperative approaches become not just beneficial but essential for effective health protection across the region and for future development.

16.4 Where Does Asia Stand in the World

While this volume has focused primarily on the multifaceted future of healthcare within Asia, it is insightful to briefly consider how these trends compare with those anticipated in other major regions of the world. Examining the healthcare of the United States, Europe, and Africa through a future-oriented lens can provide a broader context for understanding Asia's unique position and the global forces at play.

In the United States, the future of healthcare is likely to be characterised by a continued drive for technological innovation, particularly in areas like artificial intelligence, personalised medicine, and advanced diagnostics (Blease et al. 2020; Frueh 2013; Konar et al. 2022). Asia, with its own unique position, can learn from the US's emphasis on research and development as a potential model for Asian nations looking to become leaders in medical innovation(Conte et al. 2017; Frueh 2013).

Across Europe, the future of healthcare is heavily influenced by established universal healthcare systems and aging populations (Abbing 2016; Szűcs 2024).

The integration of digital health tools within national frameworks, a strong focus on reducing health inequalities and promoting social equity, and the complexities of cross-border healthcare provision are key trends (Majcherek et al. 2024; Quaglio et al. 2018). Asia, with the aspiration to have universal health coverage, can draw valuable lessons from Europe's experiences in balancing comprehensive care with financial sustainability and addressing the long-term care needs of its own aging populations.

The future of healthcare in Africa has a unique set of priorities. It concentrates on building fundamental infrastructure, combating infectious diseases, and leveraging the rapid penetration of mobile technology to extend reach in resource-constrained settings (Al-Worafi 2024; Manyazewal et al. 2023; Megbowon and David 2023). The emphasis on primary healthcare and community-based initiatives in many African nations offers potential insights for Asia's own efforts to strengthen its primary care systems and improve access in rural and underserved areas (Aantjes et al. 2014; Langlois et al. 2020; Sacks et al. 2020).

By considering these global perspectives, the future of healthcare in Asia is not unfolding and progressing in isolation but is part of a larger global evolution, with potential for cross-learning, collaboration, and the sharing of innovative solutions to improve health and well-being worldwide (Braithwaite et al. 2018; Chongsuvivatwong et al. 2011; Husain and Bloom 2020).

16.5 Conclusion: Toward Collaborative Futures

As we conclude our exploration of healthcare futures across Asia, we are back to the original truth that emerged throughout our journey: the future is not predetermined but shaped by countless decisions made today (Tonn 2003; Wack and Business 1985; Wright 2025). The healthcare systems that will serve future generations do not happen naturally, but they are built through current choices about resource allocation, technology adoption, workforce development, and countless other decisions, large and small (Schiavone and Ferretti 2021; Thacharodi et al. 2024; Webb 2025; Willie 2025). Healthcare's futures in Asia will be determined not by any single factor such as technological capabilities, economic resources, and cultures in Asia (Chongsuvivatwong et al. 2011; Schiavone and Ferretti 2021; Webb 2025). Rather, it is molded by complex interplay and intertwining of these factors. Not to forget, it is also influenced by the choices made by stakeholders, policymakers, practitioners and importantly, the communities themselves (Adhikari et al. 2025; Gobat et al. 2025; Mir et al. 2024; Sulistiadi et al. 2024). So, let us use a combination of traditional wisdom, technological innovation, and social organisation to head towards the unknown and mysterious future of healthcare, for a betterment.

We hope that this book has provided not just analysis but inspiration, with peeps of possible futures that might guide action today. The path toward healthier Asian societies is never straightforward. It will require addressing inequities, navigating demographic transitions and balancing competing priorities with limited resources.

By embracing innovation responsibly, prioritising equity and fostering collaborative action across boundaries, the diverse nations of Asia can move toward futures where health and wellbeing are within reach for all. Hence, it is my hope that this edited book represents not an endpoint but a contribution to that ongoing conversation and collaborative journey, a journey that ultimately belongs to the communities, practitioners, policymakers, and citizens working daily to create a healthier tomorrow.

Highlights

- Equity: Core of future healthcare
- Collaboration: Key to regional progress
- Technology: Tool for human-centered care
- Community voices: Drive health solutions
- Asia: Leading global health innovation.

References

Aantjes C, Quinlan T, Bunders J. Integration of community home based care programmes within national primary health care revitalisation strategies in Ethiopia, Malawi, South-Africa and Zambia: a comparative assessment. Glob Health. 2014;10(1). https://doi.org/10.1186/s12992-014-0085-5.

Abbing HR. Health, healthcare and ageing populations in Europe, a human rights challenge for European health systems. Eur J Health Law. 2016;23(5):435–52. https://doi.org/10.1163/15718093-12341427.

Abdulai A-F. Development and application of destigmatising design principles in sexual health-related technologies [dissertation]. Vancouver: University of British Columbia;2022. https://doi.org/10.14288/1.0415384.

Ackerman JP, Bartos DC, Kapplinger JD, Tester DJ, Delisle BP, Ackerman MJ. The promise and peril of precision medicine: phenotyping still matters most. Mayo Clin Proc. 2016;91(11):1606–16. https://doi.org/10.1016/j.mayocp.2016.08.008.

Adebisi YA, Jimoh ND, Bamisaiye A, Ogunkola IO. Rethinking global health research: Decolonization and equity in epidemiology. In: Handbook of Concepts in Health, Health Behavior and Environmental Health. Springer Nature Singapore;2025. p. 1–25. https://doi.org/10.1007/978-981-97-0821-5_127-1.

Adeghe EP, Okolo CA, Ojeyinka OT. A review of emerging trends in telemedicine: Healthcare delivery transformations. Int J Life Sci Res Arch. 2024;6(1):137–47. https://doi.org/10.53771/ijlsra.2024.6.1.0040.

Adhikari S, Ahmed I, Bajracharya D, Khanal B, Solomon C, Jayaratne K, et al. Transforming healthcare through just, equitable and quality-driven artificial intelligence solutions in South Asia. npj Digit Med. 2025;8(1). https://doi.org/10.1038/s41746-025-01534-0.

Alfiani FRN, Santiago F. A comparative analysis of artificial intelligence regulatory law in Asia, Europe, and America. SHS Web Conf. 2024;204:07006. https://doi.org/10.1051/shsconf/202420407006.

Al-Worafi YM. Healthcare facilities in developing countries: infrastructure. In: Handbook of medical and health sciences in developing countries. Springer International Publishing;2024. p. 1–21. https://doi.org/10.1007/978-3-030-74786-2_209-1
Amaya AB, De Lombaerde P. Regional cooperation is essential to combating health emergencies in the global South. Glob Health. 2021;17(1). https://doi.org/10.1186/s12992-021-00659-7.
Amjad A, Kordel P, Fernandes G. A review on innovation in healthcare sector (telehealth) through artificial intelligence. Sustainability (Switzerland). 2023;15(8). https://doi.org/10.3390/su15086655.
Anastasi E, Asiamah B, Lal G. Leaving no one behind: Is the achievement of the sustainable development goals possible without securing the dignity, rights, and well-being of those who are "invisible"? Int J Gynecol Obstet. 2020;148(S1):3–5. https://doi.org/10.1002/ijgo.13031.
Balkrishna A, Sharma N, Srivastava D, Kukreti A, Srivastava S, Arya V. Exploring the safety, efficacy, and bioactivity of herbal medicines: bridging traditional wisdom and modern science in healthcare. Future Integr Med. 2024;3(1):35–49. https://doi.org/10.14218/fim.2023.00086.
Bhatia T, Enoch J, Khan M, Mathewson S, Heymann D, Hayes R, Dar O. Setting targets for HIV/AIDS—what lessons can be learned from other disease control programmes? PLoS Med. 2019;16(2). https://doi.org/10.1371/journal.pmed.1002735.
Blease C, Kharko A, Locher C, DesRoches CM, Mandl KD. US primary care in 2029: a Delphi survey on the impact of machine learning. PLoS ONE. 2020;15(10). https://doi.org/10.1371/journal.pone.0239947.
Braithwaite J, Mannion R, Matsuyama Y, Shekelle PG, Whittaker S, Al-Adawi S, et al. The future of health systems to 2030: a roadmap for global progress and sustainability. Int J Qual Health Care. 2018;30(10):823–31. https://doi.org/10.1093/intqhc/mzy242.
Carzaniga AG, Dhillon IS, Magdeleine J, Xu L. International health worker mobility and trade in services. WTO Work Pap. 2019;(ERSD-2019–13). https://doi.org/10.30875/9ba2dfed-en.
Chikhaoui E, Alajmi A, Larabi-Marie-sainte S. Artificial intelligence applications in healthcare sector: ethical and legal challenges. Emerg Sci J. 2022;6(4):717–38. https://doi.org/10.28991/ESJ-2022-06-04-05.
Chongsuvivatwong V, Phua KH, Yap MT, Pocock NS, Hashim JH, Chhem R, et al. Health and health-care systems in Southeast Asia: diversity and transitions. Lancet. 2011;377(9763):429–37. https://doi.org/10.1016/S0140.
Cipta DA, Andoko D, Theja A, Utama AVE, Hendrik H, William DG, et al. Culturally sensitive patient-centered healthcare: a focus on health behavior modification in low and middle-income nations—insights from Indonesia. Front Med. 2024;11. https://doi.org/10.3389/fmed.2024.1353037.
Conte ML, Liu J, Schnell S, Omary MB. Globalization and changing trends of biomedical research output. JCI Insight. 2017;2(12). https://doi.org/10.1172/jci.insight.95206.
Cornelius K. Contextualizing transformation of healthcare sector in Asia-Pacific in the post-COVID-19 era. UNESCAP Work Pap Ser. 2022. Available from: http://www.unescap.org/kp. Accessed 13 March 2025.
Donoghue D, Khan A. Front matter. In: Achieving the SDGs and 'leaving no one behind': maximising synergies and mitigating trade-offs. ODI;2019. p. 1–2. Available from: http://www.jstor.org/stable/resrep50861.1. Accessed 13 Apr 2025.
Durairaj V. Enhancing equity in access to health care in the Asia-Pacific region: remediable inequities. Report prepared for the UN Regional Thematic Working Group on Health. Geneva: World Health Organization; 2007 May. Available from: https://asiapacific.unfpa.org/en/publications/enhancing-equity-access-health-care-asia-pacific-region-remediable-inequities. Accessed 8 April 2025.
Dutta MJ, Basu A. Meanings of health: interrogating structure and culture. Health Commun. 2008;23(6):560–72. https://doi.org/10.1080/10410230802465266.
Edge C, Black G, King E, George J, Patel S, Hayward A. Improving care quality with prison telemedicine: the effects of context and multiplicity on successful implementation and use. J Telemed Telecare. 2021;27(6):325–42. https://doi.org/10.1177/1357633X19869131.

Efferth T, Xu AL, Lee DYW. Combining the wisdoms of traditional medicine with cutting-edge science and technology at the forefront of medical sciences. Phytomedicine. 2019;64: 153078. https://doi.org/10.1016/j.phymed.2019.153078.

Frueh FW. Regulation, reimbursement, and the long road of implementation of personalized medicine—a perspective from the United States. Value Health. 2013;16(Suppl 1):S1–3. https://doi.org/10.1016/j.jval.2013.06.009.

Gerke S, Minssen T, Cohen G. Ethical and legal challenges of artificial intelligence-driven healthcare. In: Artificial Intelligence in Healthcare. Elsevier; 2020. p. 295–336. https://doi.org/10.1016/B978-0-12-818438-7.00012-5.

Gobat N, Slack C, Hannah S, Salzwedel J, Bladon G, Burgos JG, et al. Better engagement, better evidence: working in partnership with patients, the public, and communities in clinical trials with involvement and good participatory practice. Lancet Glob Health. 2025;13(4):e716–31. https://doi.org/10.1016/S2214-109X(24)00521-7.

Goktas P, Grzybowski A. Shaping the future of healthcare: Ethical clinical challenges and pathways to trustworthy AI. J Clin Med. 2025;14(5). https://doi.org/10.3390/jcm14051605.

Habib SH, Saha S. Burden of non-communicable disease: global overview. Diabetes Metab Syndr Clin Res Rev. 2010;4(1):41–7. https://doi.org/10.1016/j.dsx.2008.04.005.

Haribondhu S. United Nations (UN), Econ Soc Comm Asia Pac (ESCAP). Availability, affordability, and accessibility of universal health-care services in Asia and the Pacific: a policy analysis. 2024. Available from: https://unescap.org/kp/2024/availability-affordability-and-accessibility-universal-health-. Accessed 13 Feb 2025.

Henfrey T, Feola G, Penha-Lopes G, Sekulova F, Esteves AM. Rethinking the sustainable development goals: learning with and from community-led initiatives. Sustain Dev. 2023;31(1):211–22. https://doi.org/10.1002/sd.2384.

Hudtohan ET, Zhang D. Traditional medicine and healthcare in the 21st century. Int J Manag Educ Hum Dev. 2022;2(02):362–5. Available from: https://ijmehd.com/index.php/ijmehd/article/view/1. Accessed 17 Feb 2025.

Husain L, Bloom G. Understanding China's growing involvement in global health and managing processes of change. Glob Health. 2020;16(1). https://doi.org/10.1186/s12992-020-00569-0.

Ijaz N, Boon H. Statutory regulation of traditional medicine practitioners and practices: the need for distinct policy making guidelines. J Altern Complement Med. 2018;24(4):307–13. https://doi.org/10.1089/acm.2017.0346.

Inbanathan FY, Wijesinghe PR, Naidoo D, Buddha N, Salvador EC, Le KK, et al. Assessment of public health laboratory preparedness and response in WHO South-East Asia region during the COVID-19 pandemic: lessons learned and future directions. Lancet Reg Health Southeast Asia. 2024;31: 100496. https://doi.org/10.1016/j.lansea.2024.100496.

Jayasinghe S, Byrne NM, Hills AP. The culture of healthy living: the international perspective. Prog Cardiovasc Dis. 2025. https://doi.org/10.1016/j.pcad.2025.02.001.

Karaboue M, Lacasella GV. Medicalisation in the post-genomic era: transformations in identity, autonomy, and ethics. Clin Ter. 2025;176(1):109–13. https://doi.org/10.7417/CT.2025.5173.

Khan S. Mapping the digital divide: bridging inequalities in access to information and technology. Policy J Soc Sci Rev. 2023;2. Available from: https://ipindexing.com/article/44863. Accessed 1 Apr 2025.

Konar NM, Karaismailoglu S, Karaismailoglu E. Status and trends of personalized medicine research from 2000 to 2020: a bibliometric analysis. Curr Med Res Opin. 2022;38(5):837–46. https://doi.org/10.1080/03007995.2022.2052515.

Kroezen M, Buchan J, Dussault G, Glinos I, Wismar M. Policy brief no. 20: how can structured cooperation between countries address health workforce challenges related to highly specialized health care? Improving access to services through voluntary cooperation in the EU. 2016. Available from: http://www.euro.who.int/pubrequest. Accessed 2 Apr 2025.

Kumaresan J, Huikuri S. Strengthening regional cooperation, coordination, and response to health concerns in the ASEAN region: status, challenges, and ways forward. 2015. Available from: http://www.eria.org/ERIA-DP-2015-60.pdf. Accessed 6 Apr 2025.

Lal A, Abdalla SM, Chattu VK, Erondu NA, Lee TL, Singh S, et al. Pandemic preparedness and response: exploring the role of universal health coverage within the global health security architecture. Lancet Glob Health. 2022;10(11):e1675–83. https://doi.org/10.1016/S2214-109X(22)00341-2.

Langlois EV, McKenzie A, Schneider H, Mecaskey JW. Measures to strengthen primary health-care systems in low- and middle-income countries. Bull World Health Organ. 2020;98(11):781–91. https://doi.org/10.2471/BLT.20.252742.

Lee SSJ, Walker A, Callier SL, Fletcher FE, Galarneau C, Garrison N, et al. Racial equity, diversity and inclusion in bioethics: recommendations from the association of bioethics program directors presidential task force. Am J Bioeth. 2024. https://doi.org/10.1080/15265161.2024.2371116.

Lim JCW. Strengthening health products regulatory systems to enhance access to quality health products in the Asia-Pacific. Ther Innov Regul Sci. 2018;52(6):751–4. https://doi.org/10.1177/2168479018769285.

Liu Y, Uemura H, Ye D, Lee JY, Chiong E, Pu YS, et al. Prostate cancer in Asia: design of a patient registry to inform real-world treatments, outcomes, and quality of life. Prostate Int. 2019;7(3):108–13. https://doi.org/10.1016/j.prnil.2018.12.001.

Mahomed F. Addressing the problem of severe underinvestment in mental health and well-being from a human rights perspective. Health Hum Rights. 2020;22(1):35–49. Available from: http://www.ncbi.nlm.nih.gov/pubmed/32669787. Accessed 5 Apr 2025.

Majcherek D, Hegerty SW, Kowalski AM, Lewandowska MS, Dikova D. Opportunities for healthcare digitalization in Europe: comparative analysis of inequalities in access to medical services. Health Policy. 2024;139: 104950. https://doi.org/10.1016/j.healthpol.2023.104950.

Mallipeddi SR, Goda DR, Yerram SR, Varghese A, Ande JRPK. Telemedicine and beyond: navigating the frontier of medical technology. Technol Manag Rev. 2017;2:37–50. Available from: https://upright.pub/index.php/tmr/article/view/118. Accessed 13 Apr 2025.

Mansyur CL, Amick BC, Franzini L, Roberts RE. Culture and the social context of health inequalities. Int J Health Serv. 2009;39(1):85–106. https://doi.org/10.2190/HS.39.1.d.

Mantaleon D. The evolving landscape of healthcare: challenges, innovations, and a vision for the future. Health Sci J. 2023;17(8):1049. https://doi.org/10.36648/1791

Manyazewal T, Ali MK, Kebede T, Magee MJ, Getinet T, Patel SA, et al. Mapping digital health ecosystems in Africa in the context of endemic infectious and non-communicable diseases. NPJ Digit Med. 2023;6(1):839. https://doi.org/10.1038/s41746-023-00839-2.

Marston C, Renedo A, Miles S. Community participation is crucial in a pandemic. Lancet. 2020;395(19):1676–8. https://doi.org/10.1001/jama.2020.6548.

May C. Agency and implementation: understanding the embedding of healthcare innovations in practice. Soc Sci Med. 2013;78(1):26–33. https://doi.org/10.1016/j.socscimed.2012.11.021.

McGreevey JD, Hanson CW, Koppel R. Clinical, legal, and ethical aspects of artificial intelligence-assisted conversational agents in health care. JAMA. 2020;324(6):552–3. https://doi.org/10.1001/jama.2020.2724.

McLaughlin LA, Braun KL. Asian and Pacific Islander cultural values: considerations for health care decision making. Health Soc Work. 1998;23(2):116–26. https://doi.org/10.1093/hsw/23.2.116.

Megbowon ET, David OO. Information and communication technology development and health gap nexus in Africa. Front Public Health. 2023;11:1145564. https://doi.org/10.3389/fpubh.2023.1145564.

Minichiello V, Rahman S, Dune T, Scott J, Dowsett G. E-health: potential benefits and challenges in providing and accessing sexual health services. BMC Public Health. 2013;13(1):790. https://doi.org/10.1186/1471-2458-13-790.

Mir G, Durrani N, Julian R, Kimei Y, Mashreky S, Doan TTD. Social inclusion and sustainable development: findings from seven African and Asian contexts. Sustainability. 2024;16(11):4859. https://doi.org/10.3390/su16114859.

Olaboye JA, Maha CC, Kolawole TO, Abdul S. Big data for epidemic preparedness in Southeast Asia: An integrative study. Int J Sci Res Updates. 2024;7(2):66–73. https://doi.org/10.53430/ijsru.2024.7.2.0041.

Park JE, Yi J, Kwon O. Twenty years of traditional and complementary medicine regulation and its impact in Malaysia: achievements and policy lessons. BMC Health Serv Res. 2022;22(1):7497. https://doi.org/10.1186/s12913-022-07497-2.

Quaglio G, Figueras J, Mantoan D, Dawood A, Karapiperis T, Costongs C, Bernal-Delgado E. Perspectives: an overview of future EU health systems. An insight into governance, primary care, data collection and citizens' participation. J Public Health. 2018;40(4):891–8. https://doi.org/10.1093/pubmed/fdy054.

Rahman-Shepherd A, Aghogho Evaborhene N, Berman A, Amaya AB, Boro E, Dar O, et al. Establishing the value of regional cooperation and a critical role for regional organisations in managing future health emergencies. Lancet Glob Health. 2025;13(3):e585–92. https://doi.org/10.1016/S2214-109X(24)00500-X.

Reddy H, Joshi S, Joshi A, Wagh V. A critical review of global digital divide and the role of technology in healthcare. Cureus. 2022. https://doi.org/10.7759/cureus.29739.

Rooney EJ, Wilson RL, Johnson A. Integration of traditional therapies for first nations people within western healthcare: an integrative review. Contemp Nurse. 2023;59(4–5):294–310. https://doi.org/10.1080/10376178.2023.2276718.

Rose N. Personalized medicine: promises, problems and perils of a new paradigm for healthcare. Procedia Soc Behav Sci. 2013;77:341–52. https://doi.org/10.1016/j.sbspro.2013.03.092.

Sacks E, Schleiff M, Were M, Chowdhury AM, Perry HB. Communities, universal health coverage and primary health care. Bull World Health Organ. 2020;98(11):773–80. https://doi.org/10.2471/BLT.20.252445.

Schiavone F, Ferretti M. The futures of healthcare. Futures. 2021;134: 102849. https://doi.org/10.1016/j.futures.2021.102849.

Schönberger D. Artificial intelligence in healthcare: a critical analysis of the legal and ethical implications. Int J Law Inf Technol. 2019;27(2):171–203. https://doi.org/10.1093/ijlit/eaz004.

Sembiyaa A, Dharshini J, Rashmika B, Manoj M, Velayuthaprabhu S, Robert AA, et al. Integrated molecular strategies for improving traditional Chinese medicinal plants. In: Biotechnology, multiple omics, and precision breeding in medicinal plants. CRC Press;2025. p. 231–47. https://doi.org/10.1201/9781003475491-17.

Shirah BH, Zakour MO. Digital transformations: a detailed examination of emerging solutions in the Asia-Pacific region. In: Digital healthcare, digital transformation and citizen empowerment in Asia-Pacific and Europe for a healthier society. Elsevier;2025. p. 505–37. https://doi.org/10.1016/B978-0-443-30168-1.00001-3.

Singh K, Gupta JK, Jain D, Kumar S, Singh T, Saha S. Exploring the ancient wisdom and modern relevance of Chinese medicine: a comprehensive review. Pharmacol Res Mod Chin Med. 2024;11: 100448. https://doi.org/10.1016/j.prmcm.2024.100448.

Smith JH. Transformation from below?: the role of civil society organizations in the global governance of the response to HIV/AIDS [University of Bradford]. 2014. Available from: http://hdl.handle.net/10454/7512. Accessed 13 Apr 2025.

Sulistiadi W, Wasir R, Thalib W, Ayuningtyas D, Bawazier N, Buskens E. Building health systems resilience: understanding the social, economic, and cultural impacts of climate change from stakeholders' perspectives in Indonesia. Arch Public Health. 2024;82(1):1403. https://doi.org/10.1186/s13690-024-01403-4.

Sundram P. Network governance in ASEAN: Fostering regional cooperation and integration. Front Polit Sci. 2025;7:1434595. https://doi.org/10.3389/fpos.2025.1434595.

Szűcs P. Demographic shifts and economic challenges in the European Union. Eur Tükör. 2024;27(1):87–106. https://doi.org/10.32559/et.2024.1.5.

Tan MJT, Lichlyter DA, Maravilla NMAT, Schrock WJ, Ting FIL, Choa-Go JM, et al. The data scientist as a mainstay of the tumor board: global implications and opportunities for the global south. Front Digit Health. 2025;7:1535018. https://doi.org/10.3389/fdgth.2025.1535018.

Te V, Griffiths R, Law K, Hill PS, Annear PL. The impact of ASEAN economic integration on health worker mobility: a scoping review of the literature. Health Policy Plan. 2018;33(8):957–65. https://doi.org/10.1093/heapol/czy071.

Thacharodi A, Singh P, Meenatchi R, Tawfeeq Ahmed ZH, Kumar RRS, Neha V, et al. Revolutionizing healthcare and medicine: The impact of modern technologies for a healthier future—a comprehensive review. Health Care Sci. 2024. https://doi.org/10.1002/hcs2.115.

Thomas RM, Spieldenner A, Castellanos E, Chang J, Kim J. Community contexts and knowledge: accountability, leadership and meaningful involvement of communities in programme Science. J Int AIDS Soc. 2024;2024(S2):26281. https://doi.org/10.1002/jia2.26281.

Thornicroft G, Ahuja S, Barber S, Chisholm D, et al. Integrated care for people with long-term mental and physical health conditions in low-income and middle-income countries. Lancet Psychiatry. 2019;6(2):174–86. https://doi.org/10.1016/S2215-0366(18)30298-0.

Tonn BE. The future of futures decision making. Futures. 2003;35(6):673–88. https://doi.org/10.1016/S0016-3287(02)00106-4.

Tun HM, Naing L, Malik OA, Rahman HA. Navigating ASEAN region artificial intelligence (AI) governance readiness in healthcare. Health Policy Technol. 2025;14(2): 100981. https://doi.org/10.1016/j.hlpt.2025.100981.

Vannakit R, Andreeva V, Mills S, Cassell MM, Jones MA, Murphy E, Ishikawa N, Boyd MA, Phanuphak N. Fast-tracking the end of HIV in the Asia Pacific region: domestic funding of key population-led and civil society organisations. Lancet HIV. 2020;7(5):e366–72. https://doi.org/10.1016/S2352-3018(20)30077-1.

Wack P, Business H. Scenarios: uncharted waters ahead. Harv Bus Rev. 1985;63:72–89. Available from: https://hbr.org/1985/09/scenarios-uncharted-waters-ahead. Accessed 12 Apr 2025.

Wang S, Zhou H, Liu Y, Li Y, Nie L. Trends and projections of non-Hodgkin lymphoma burden (1990–2040): a global burden of disease 2021 analysis. BMC Public Health. 2025;25(1):22376. https://doi.org/10.1186/s12889-025-22376-1.

Webb J. Healthcare resource allocation, machine learning, and distributive justice. Am Philos Q. 2025;62(1):33–52. https://doi.org/10.5406/21521123.62.1.03.

Wernli D, Clausin M, Antulov-Fantulin N, Berezowski J, Biller-Andorno N, Blanchet K, Böttcher L, et al. Governance in the age of complexity: Building resilience to COVID-19 and future pandemics. Geneva Science-Policy Interface Policy Brief. 2021. Available from: https://www.preventionweb.net/publication/governance-age-complexity-building-resilience-covid-19-and-future-pandemics. Accessed 13 Apr 2025.

Willie MM. Strategies for enhancing training and development in healthcare management. Adv Hum Resour Manag Res. 2025;3(1):44–59. https://doi.org/10.60079/ahrmr.v3i1.408.

Wong JJ. Closing the digital divide as a public health mechanism: exploring technology infrastructure regulations and trends in the Western Pacific and South-East Asia. Calif West Int Law J. 2024;54(2). Available from: https://scholarlycommons.law.cwsl.edu/cwilj/vol54/iss2/8. Accessed 7 Apr 2025.

World Health Organization. Regional Office for the Western Pacific. Meeting on the contribution of traditional medicine in strengthening primary health care, Manila, Philippines, 14–15 September 2017: Meeting report. 2017. Available from: https://iris.who.int/handle/10665/260010. Accessed 2 Apr 2025.

World Health Organization. Western Pacific regional action agenda on regulatory strengthening, convergence and cooperation for medicines and the health workforce. 2019. Available from: https://www.who.int/publications/i/item/9789290618942. Accessed 13 Apr 2025.

Woser MT, UN ESCAP. Leaving no one behind: access to social protection in Asia and the Pacific. 2023. Available from: https://hdl.handle.net/20.500.12870/6496. Accessed 3 Apr 2025.

Wright A. Back to the future? A Caution J Manag Stud. 2025. https://doi.org/10.1111/joms.13226.

Yang F, Janamnuaysook R, Boyd MA, Phanuphak N, Tucker JD. Key populations and power: people-centred social innovation in Asian HIV services. Lancet HIV. 2020;7(1):e69-74. https://doi.org/10.1016/S2352-3018(19)30347-9.

Yeganeh H. An analysis of emerging trends and transformations in global healthcare. Int J Health Gov. 2019;24(2):169–80. https://doi.org/10.1108/IJHG-02-2019-0012.

Rong Xiang Ng is an Internal Medicine, Infectious Diseases physician and medical lecturer at Universiti Malaya and holds an affiliate lecturer position at the University of Glasgow. He is deeply engaged in understanding and shaping the future of healthcare in Asia, as reflected in this concluding volume. His clinical and research work focuses on optimising clinical pathways and exploring innovative solutions, including digital tools, to address infectious disease challenges across diverse populations. Dr. Ng's contributions to hospital treatment guidelines and research on AI in HIV care and infectious diseases, multidrug-resistant organisms, COVID-19, and sexual health directly reflect the key themes of technological innovation and equitable access explored throughout this book. He also serves as a Working Group Member for the National Center for Global Health and Medicine (NCGM)—ARO Alliance for Southeast and East Asia (ARISE), representing Universiti Malaya Medical Centre (UMMC). As a Fellow of the Asia Pacific AIDS and Co-infections Conference (APACC) and an affiliate of the Centre of Excellence for Research in Infectious Diseases and AIDS (CERIA, Universiti Malaya), his active involvement in research and clinical trials bridges the crucial gap between current clinical practice and the forward-looking perspectives presented in this concluding exploration of the future of healthcare in Asia.

Index

R. Xiang Ng et al. (eds.), *Future of Healthcare in Asia*,
https://doi.org/10.1007/978-3-032-03183-9

MIX
Papier aus verantwortungsvollen Quellen
Paper from responsible sources
FSC® C105338
FSC
www.fsc.org

If you have any concerns about our products,
you can contact us on
ProductSafety@springernature.com

In case Publisher is established outside the EU,
the EU authorized representative is:
Springer Nature Customer Service Center GmbH
Europaplatz 3, 69115 Heidelberg, Germany

Printed by Libri Plureos GmbH
in Hamburg, Germany